URBAN GROWTH AND DEVELOPMENT IN ASIA

Urban Growth and Development in Asia

Volume I: Making the Cities

Edited by
GRAHAM P. CHAPMAN
Lancaster University, United Kingdom
ASHOK K. DUTT
Department of Geography, University of Akron, Ohio, USA
ROBERT W. BRADNOCK
Department of Geography, School of Oriental and African Studies, London, United Kingdom

Routledge
Taylor & Francis Group

LONDON AND NEW YORK

First published 1999 by Ashgate Publishing

Reissued 2018 by Routledge
2 Park Square, Milton Park, Abingdon, Oxon, OX14 4RN
52 Vanderbilt Avenue, New York, NY 10017

Routledge is an imprint of the Taylor & Francis Group, an informa business

A Library of Congress record exists under LC control number: 99073168

ISBN 13: 978-1-138-37013-5 (hbk)
ISBN 13: 978-0-429-42819-7 (ebk)

Contents

PART I

INTRODUCTION

PART II

THE URBAN BASE: CHINA

THE URBAN BASE: THE RUSSIAN FAR EAST

THE URBAN BASE: INDIA

THE URBAN BASE: SAUDI ARABIA

PART III

THE URBAN LAND MARKET

List of Figures

List of Tables

List of Maps

List of Contributors

Amitabh	*Centre for the Study of Regional Development, Jawarhalal Nehru University, New Delhi, India*
Bahammam, Ali	*Architecture & Housing Sciences, King Saud University, Saudi Arabia*
Banerjee-Guha, Swapna	*Department of Geography, Bombay University, India*
Chan, Roger C.K.	*Centre of Urban Planning & Environmental Management, The University of Hong Kong, Hong Kong*
Chapman, Graham P.	*Department of Geography, Lancaster University, UK*
Cheng, Yun	*Centre of Urban Planning & Environmental Management, The University of Hong Kong, Hong Kong*
Chu, David K.Y.	*Geography Department, The Chinese University of Hong Kong, Hong Kong*
Costa, Frank J.	*Department of Geography, University of Akron, Ohio, USA*
Costello, Vincent F.	*Faculty of the Built Environment, University of the West of England, Bristol, UK*
Ha, Seong-Kyu	*Chung-Ang University, Dept of Regional Development, Korea*
Hagishima, Satoshi	*Department of Architecture, Kyushu University, Japan*
Han, Sun Sheng	*School of Building and Estate Management, National University of Singapore*
Harris, Nigel	*Development Planning Unit, University College, London, UK*
Hitaka, Keiichiro	*Kitakyushu Urban Association, Kitakyushu City, Japan*
Hori, Tomoyoshi	*Kitakyushu Urban Association, Kitakyushu City, Japan*
Ikaruga, Shinji	*Department of Architecture, Kyushu University, Japan*
Jacquemin, Alain	*Centrum voor Poloticologie, Vrye Universiteit Brussel, Brussels, Belgium*

Karan, P.P.
Department of Geography, University of Kentucky, Lexington, USA

Kumar, Ashok
Department of Physical Planning, School of Planning and Architecture, New Delhi, India

Kuttaiah, Kalpana
Centre for Public Administration and Public Policy, Kent State University, Ohio, USA

Li, Chun-ju
Institute of Building and Planning, National Taiwan University, Taiwan

Mather, Cotton
New Mexico Geographical Society, USA

Mubarak, Faisal A.
Department of Urban Planning, King Saud University, Saudi Arabia

Noble, Allen G.
Department of Geography, University of Akron, Ohio, USA

Pathak, Pushpa
National Institute of Urban Affairs, New Delhi, India

Rose, Felicity C.
Department of Land Economy, University of Cambridge, UK

Sommers, Brian J.
Central Connecticut State University, Department of Geography, Connecticut, USA

Sommers, Gail Gordon
Centre for Public Administration and Public Policy, Kent State University, Ohio, USA

Spencer, Andrew H.
Consultant, Luton, Bedfordshire, UK; formerly at the Transport Studies Group, University of Westminster, UK

Srinivas, Sampath
Development Planning Unit, University College, London, UK

Timothy, Dallen J.
Central Connecticut State University, Department of Geography, Connecticut, USA

Wang, Hung-kai
Institute of Building and Planning, National Taiwan University, Taiwan

Winarso, Haryo
Development Planning Unit, University College, London, UK

Wong, Shue Tuck
Department of Geography, Simon Fraser University, British Columbia, Canada

Zhang, L.
Department of Geography, University of Washington, DC, USA

Zhao, Simon X.B.
Department of Geography, Hong Kong Baptist University, Hong Kong

Acknowledgements

These two volumes are the proceedings from the 5th Asian Urbanisation Conference, held at the School of Oriental and African Studies in London in August 1997. This was the first conference held in Europe, and London proved a happy venue to attract scholars from North America and Europe as well as from Asia. The level of engagement and the intensity of debate made the conference most worthwhile, and the chapters included here will cumulatively provide the reader with a stimulating portrayal of the processes and outcomes of one of the greatest shifts of population (not just absolutely but proportionately as well) ever to have occurred in human history.

The conference would not have happened and the volumes would not have been produced but for the support of many institutions and people. We thank the School of Oriental and African Studies (SOAS) for hosting the event, Fakhra Ahmed and Alison Henley of SOAS for being girl Fridays and running excellent field trips, the University of Akron for logistic and material support, Dr Charles Clift of the Department for International Development (UK) and the Developing Areas Research Group of the Institute of British Geographers, for subsidising the air fares of some delegates, the India Development Group (UK) for similar support, and the University of Lancaster - really in the guise of Siobhan Waring - for the conference secretarial and subsequent editorial support. We thank Chris Beacock of Lancaster Geography Department for keeping his equanimity as the flood of illustrations and maps inundated his office - with the excellent results evident in these volumes. We are also indebted greatly to him for the final page setting.

Graham Chapman, Department of Geography, Lancaster University, UK
Ashok Dutt, Department of Geography, University of Akron, Ohio, USA
Bob Bradnock, Department of Geography, School of Oriental and African Studies, UK

PART I
INTRODUCTION

Editors' Preface

Why have a conference, let alone a series of conferences, on Asian Urbanisation? The developed OECD countries and most of the Former Soviet Union are highly urbanised, and Latin America is more 50% urbanised. The less urbanised areas of the world include sub-Saharan Africa – with a small part of the total world population – and most of Asia – with urbanisation (apart from in Japan and the Tiger Economies), averaging less than 30%. This Asia includes more than half the world's population. In the last decade or so urbanisation has begun to take off, and the shift of population to the cities represents one of the greatest population movements the planet has ever seen. Projections are by definition untestable at the time they are made, so there is no point in quibbling about the detail too much. But if we accept, as most current projections do, that by 2030 more than 50% of Asia's population will be urban, then over the next three decades more than 500 million people in Asia will have moved - looking for jobs, housing, food and water. They will be both part of a problem and most of the solution – building around them the cities they will live in. Whether this is for better or worse will partly depend on how local and national government and the city dwellers interact with each other, and partly on much broader issues such as environmental stress and the evolution of the global economy and the relationship of those with urbanisation. Simple considerations like these are enough to explain why we should hold such conference(s).

Towns and cities are located in space but also define spaces of their own making, at different scales. Historically and logically they start by being not-rural – so we may see them against their rural background. Immediately there are many complex questions about the trade between town and country, the terms of that trade – in favour or against agriculture or artisanal goods, or industrial goods – and about the cultures and value systems of the two – do towns acquire the cultural habits of their regions , do they become the centres for cultural change? When most of the population is urbanised, so most exchange then takes place between urban places. How do these networks of places operate within a nation – which grow most rapidly, which lag behind and fail to prosper, how and why do they specialise? What can local and national government do to steer these developments? How do clusters of places evolve – as megalopolis, as conurbation, as metropolis with attendant satellites? How do the national networks interconnect internationally – mostly through dominant 'world' cities, or in more diffuse patterns? How do these complex networks of cities produce wealth, abstract wealth, redistribute wealth, at all scales, from local to global?

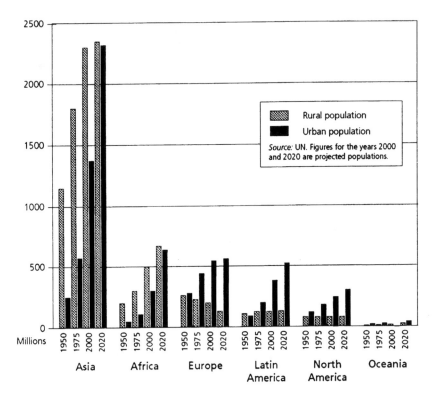

Figure P.1   World urban and rural population projections
Source: UN Projections, www.fao.org/NEWS/FACTFILE.

There are thus so many questions about the functions, size, location and prosperity of cities, but none of this yet touches on the experience of moving to them or living within them. In some sense our lives are the sum of our daily experiences. What are the daily experiences to be found within these cities? The answers are of course conditioned by many physical, social and cultural dimensions. Physically the environments of cities vary considerably. Coastal Bombay which suffers few winter temperature inversions will never have the same appalling air as inland New Delhi does in winter. The social and cultural dimensions are signified by the bodies we have, the labels and credentials we display, saying whether we are young or old, male or female, literate or illiterate, skilled or not, with inherited social and economic capital or not. They are also conditioned by the average wealth and the quality of government of these places - which may provide us with legitimate tenure, or no squatters rights, and which will affect the level of provision of basic services like water, power, sanitation and garbage collection, or the maintenance of environmental standards like air quality - which in turn affect the health and productive capacity of the population. Towns and cities are also places of cultural investment – they have been the major centres of artistic patronage since the earliest towns – and our experience of them as visual landscapes is a major part of our visual daily life.

The chapters in these two volumes touch on all of these issues. They provide the reader with a stimulating portrayal of the processes and outcomes of one of the greatest shifts of population (proportionately – not just absolutely) ever to have occurred in human history. The whole is something of an impressionist painting – by the end of reading these two volumes any reader must have some sense of the whole panorama and drama. The range of countries covered and the range of approaches adopted is very wide, so that it is possible to see very different experiences of urbanisation at a multitude of scales, yet all contributing to the broader picture. Each chapter is like one of the flecks of paint in an impressionist painter's picture, contributing to the whole.

However, as with the painting, so much more of 'reality' is left out than is included, that it is necessary to stress that the 'whole picture' created is still a limited and partial picture of all that could be written or said. Some Asian countries are not represented, some topics not dealt with (for example air pollution only gets passing reference.) This is not only inevitable, but actually a stimulus to further research, and a call for the contributors to the 6th conference to be held in Chennai (Madras) in January 2000 to help fill in the gaps. This then will be followed by the 7th Conference in Athens, Georgia, U.S.A. in 2002, and the 8th Conference in Japan in 2004.

In organising the books we have grouped chapters by topics, and within each topic we have started with the more general historical and spatial surveys, and then narrowed down to the more specific case studies. Where possible we have also grouped papers by the geographical area of interest.

1 Over-view: The Future of Urbanisation in Asia

Nigel Harris

1.1 Introduction

It is surprising how preoccupied the world is with potential disasters, with fears. It often seems as if things have never been so bad: the end of the world is nigh or the end of civilisation 'as we know it' seems always to be with us. A moment's reflection ought to dispel this pervasive foreboding. Leave aside the horrors of the Great Plague in Europe of the fourteenth century, the twentieth century before the present is a catalogue of horrors far greater that we now face - imagine being born in 1900 in Europe with the prospect of fighting in two devastating world wars and also being possibly unemployed for much of the time in between, before and after – and to imagine something worse, in addition being poor, German and Jewish. The parents and grandparents of modern Europe had to have heroic stoicism.

But the moment's reflection that I ask for might suggest that this regression has been outweighed by progress: we have taken two steps forward for each one backward. Since 1820, world population has increased five times; income per capita for the world by eight times; world income by forty times; and world trade by 540 times.[1] That process of improvement was most marked in the two great surges of growth in the world economy, 1840 to 1870, and 1950 to 1974, but improvement continued at a slower pace after that - especially up to 1914 in the first phase. Today, growth patterns are much more varied - the spectacular growth of east Asia and south-east Asia came after 1974 and lasted until 1998. Perhaps it is just this heterogeneity of the economic growth picture which allows simultaneously pessimism and great optimism.

The period between these surges of growth was marked by an immense increase in national autarky. In many respects, this autarkic pattern continued into the period of the second phase of growth, completing nearly a century of a world economy dominated by States (between roughly 1870 and about 1970). The world economy came to appear as composed of a set of national fiefs, not global markets. At times, it seemed, the State politics, not profit-seeking, motivated and activated the system. In important respects, civil servants su-

---

1 *The figures are taken from an address by Paul Streeten at the International Development Centre, Oxford, on the occasion of a celebration to mark his 80th birthday, 26th June 1997.*

6

perseded the role of capitalists, expropriated capital or directed it to ends set by the State. War, the preparation for war and the recovery from war, dominated much of the period, and war is essentially a State-dominated process.

If States, not markets, seemed to direct the system, it was understandable that States were, on the other side of the political divide, the primary focus for messianic hopes of liberation. Thus, the State was Janus-faced, the embodiment of oppression and of liberation. By the 1950s and the anti-colonial revolutions, everyone had become 'socialist', by which they meant that they embraced a belief that there were no economic or social problems which could not be solved by means of State planning. The State came to epitomise the embodiment of collective freedom. It was a position summarised in the heroic proclamation of Kwame Nkrumah (1961) to the leaders of the new 'Third World' of the 1950s:

> Aim for the attainment of the Political Kingdom; that is to say, the complete independence and self-determination of your territories. When you have achieved the Political Kingdom, all else will follow.

Only now, with the benefit of hindsight, can we see that this 'socialism' was little more in practice if not intention than a preparation for global capitalism, a phase of State-incubation of the capitalist chicks, now when grown large enough, flown the coop to join the world aviary. Above all, the globalisation of the last decades has seen mobile international capital championed as the new liberator, the new stimulus to growth and wealth, and an altar on which national autarky must necessarily be sacrificed. It is a moot point whether this paradigm will survive the economic crises of 1998 in such a stark form, but it is less controversial that growth will return.

The heart of this process of world economic growth, the heart of world civilisation, is the city, and the process of urbanisation, warts and all, measures the efficacy of that growth and of the old modernist aspiration of progress. Cities are instruments of economic transformation, great concentrations of human and physical capital. They are the cause and the context of cultural and moral transformations, of the self-perpetuating processes of change, of a self-transforming intelligence. The indices of social improvement are measured at their clearest here.

On the other side, because they are so important, they measure the greatest severity in terms of social catastrophe, the site of the worst horrors that we inflict on ourselves - Berlin in 1945, Beirut and Phnom Penh in 1990, Mogadishu in 1993, Sarajevo in 1995. The city is a concentration of power and therefore attracts some of the most destructive forces in the contest for power.

1.2 Asia

For much of the history of the world, Asia - with much of the world's population - has been the heart of world civilisation, marked by the three great cultural complexes of east Asia (China, Korea and Japan), south Asia (the Indian sub-continent), and west Asia (from Persia to the Middle East). Only relatively recently, from the fourteenth century, did the rise of Europe displace Asia, followed in the nineteenth century by the emergence of North America. As late as 1820, China's estimated share of world gross product was 30 per cent (by 1950, it was 7 per cent [World Bank, 1997]). That displacement led, as everyone knows, to an extraordinary decline in Asia, the destruction of its great empires at the hands of the Europeans and later Americans (and in the twentieth century, by the Japanese, a remarkable countermovement), colonialism and demoralisation. In its heyday, Asia naturally had the world's largest cities. The symbol of the rise of Europe and North America was the growth of their great cities.

What is extraordinary about the rise of Japan after World War II, followed by the four Little Tigers, then south-east Asia, and now China, is that that long drawn out decline has been reversed. There are even speculations that suggest the return of Asian dominance - and in particular, the emergence of China and its 'economic area' as, in a foreseeable period, the largest concentration of economic activity in the world. In the early 1990s, the World Bank identified the new economic cluster of east and south-east Asia as constituting a third pole of growth in the world system with the capacity to offset perceived economic decline in the other two (Europe and North America) (World Bank, 1994 and 1995). Certainly, the crash of 1998 has reinforced the importance of resecuring growth in these emerging markets for the health of the world economy.

However, Asia is also the home to the largest concentration of poor people in the world, in that great band across central India (from Madhya Pradesh to West Bengal) to Bangladesh. The heterogeneity of the growth process in the fastest growing countries is notorious -alongside the remarkable growth of Guangdong or Fujian and the even more extraordinary decline in the proportion of China's population which is poor (down by 200 million since 1978), is the desperate poverty of parts of other provinces (for example, Guizhou) - 70 million live below the Chinese government's official poverty line (22 per cent of the population by the World Bank's criterion). In India, Maharashtra or Gujarat are already nearly at the level of middle income countries – but Orissa and Bihar are still very low income states.

The overall growth underpins the spectacular change in demographic measures - shown in Table 1.1.

Table 1.1 Infant mortality and life expectancy in Asia, 1970-1995

	1970 Infant Mortality	1970 Life Expectancy	1995 Infant Mortality	1995 Life Expectancy
East Asia/Pacific	77	59	40	68
South Asia	138	49	75	61
Middle East/North Africa	135	53	54	66

* per 1,000 live births

Source: World Bank **Development Report** 1997, Tables 1 and 6: 215 and 225.

1. 3 Urbanisation

Asia is still very rural - indeed, more than three fifths of the world's country folk live here. But we are already in the run in to an Asia which is predominantly urban. If the projections are to be believed, all regions of Asia except the south will have a majority of its people living in cities and towns by the year 2025 (and the proportion will be 48 per cent even in south Asia). Of course, this is partly a statistical artefact - if China and India employed the definitions common in Europe or Latin America, possibly 50 to 60 per cent of their respective populations would already be urban.

Asia is marked, however, by the great size of the population aggregates. In 1950, there were some 226 million urban dwellers in the region; by 1990, just under 1,000 million; and projected for 2025, some 2700 million, or an increase of 12 times in 75 years. Indeed, the 1950 urban population of Asia was barely 8 per cent of what it is projected to be in 2025. Quantity must indeed be a transformation of quality with figures like these, a radical change in whole ways of life and civilisations.

Furthermore, increasingly the urban population will live in large cities, and cities of an increasing average size (the economic optimum of city population size seems to increase inexorably). One indirect way of illustrating this is to note the Asian participation in the 30 largest cities in the world and the average size of those cities. In 1950, six of the 30 largest cities were in Asia (with an average population size of 5 million); in 1990, there were 16 (with an average size of nearly 10 million); and projected for 2010, there will be 19 (average size: 15.5 million).

Again, the figures are statistical artefacts, a function of arbitrary boundaries and changing definitions. As is well known, the growth of city populations will almost certainly not mean increasing population densities, but, as elsewhere in the world, declining densities as boundaries are extended. Thus the growing population is increasingly spread out, often with an absolute decline in the population of inner city areas. Mumbai's Fort North (Mumbai used to be known as Bombay), a central city ward, began to lose population absolutely as early as the decade up to 1971. In the 1980s, while Jakarta's annual population growth was 3.1 per cent, the three outer districts had rates of growth of 11.7, 20.9 and 198.8 per cent. As a result, many cities have been seeking to extend their boundaries, or certain administrative powers, to capture more of the spread - the Bangkok Metropolitan Region now extends up to 100 kilometres from the urban core.

The spread of big city population encompasses smaller towns and cities in the surrounding districts, producing much larger semi-urbanised and economically interconnected regions. Perhaps one of the most striking prototypes here is the Pearl River delta, with Guangzhou (and six and a half million people) at one end, Hong Kong (with eight and half million) at the other, and between a scatter of rapidly growing towns and cities, all constituting an interconnected economic region. Within the region is possibly the fastest growing city in the world -Shenzhen, with 30,000 village and fisherfolk in the late 1970s, but now a place of skyscrapers and three million people. Indeed, it might be seen as no more than a suburb of Hong Kong (so that the real Hong Kong population is more like eleven to twelve million).

Along with the growth of Asia's big cities, excluding for the moment many of those in what we used to call 'the Centrally Planned Economies', has gone a remarkable transformation of their economies, most vividly illustrated in the universal process of de-industrialisation. Again, Hong Kong provides the most striking illustration - between a half and two thirds of the employment of Hong Kong's manufacturing companies is now outside the boundaries of the city, much of it in China. If you travel north from Delhi on the Meerut road, it is now lined with factories on greenfield sites in rural areas. The big cities themselves have gone beyond mere industry, which they sub-contract to the regions around.

Among the beneficiaries of the big city loss of manufacturing capacity are smaller cities and towns. Half the world's cities of one million or less population are in Asia, and one third of India's urban population lives in cities with 100,000 people or fewer. The rapidity of transformation of smaller urban settlements with industrialisation can be seen in the extraordinary growth of China's Town and Village Enterprises (which by 1990 had overtaken in value of output the State-owned sector, employing over 100 million workers and overwhelmingly the largest proportion of the big city manufacturing workforce). In the Indian textile and garment industries, the decline of the registered mill sector in historically two major centres of the industry, Ahmedabad and Mumbai (Mumbai's mill employment as recently as 1970 was over 220,000; now it under 60,000), has been paralleled by the rise of the unregistered power loom industry outside the city, in the one case in Surat, in the other Bhiwandi. It is not only a shift in location, but also scale, output, organisation, the class of owners and the ethnic origin and language of the workforce.

Other smaller centres have experienced extraordinary growth under the impact of industrialisation. Tiruppur in Tamil Nadu in India is a well-known example. The sheer speed of change frequently means we are always out-of-date. On the periphery of Dhaka a world-class garment-export industry has been built since 1980, now employing some 700,000 workers, the majority of them women. Or take Bangalore, successively a leading centre of electronic hardware and now of software. The high speed of change transforms the patterns of land use in the city - the decline of the Mumbai textile industry[2] has left parts of the inner city area, Parel and Dadar, littered with semi-derelict mills. The rapid growth of Mumbai's financial sector has had effects of the opposite kind, producing a building boom and some of the highest urban land prices in the world in the southern tip of the city (and the relocation of back-office functions to a new business area in the north of the city).

The available data is often so aggregated (or relates to provinces rather than cities) that we can rarely see in quantitative terms the underlying patterns of specialisation operating here. The temptation is thus to see cities as simply competitive whereas more disaggregated data might show them to be complementary, making the inputs to each others output. We are driven to depend on anecdote or just what can be observed to see the niche markets concerned. Take another small city, this time in Pakistan, Sialkot, which is the major centre for the manufacture of the world's soccer balls - and Scottish bagpipes! But the statistics, even at their most disaggregated, only mention 'leather goods'.

1.4 The Servicing City

The decline in big city manufacturing has been matched by the growth of new service sectors in the city, producing ultimately perhaps a pure servicing city, based now upon logistics, on the management of flows of people,

---

2 *From its beginnings in the 1860s, the Mumbai mills recruited their workers from the villages of a poor arid coastal area, Ratnagiri, in the south of the present Maharashtra State. The Bhiwandi industry now depends upon Telugu-speaking workers from Andhra Pradesh to the east of Maharashtra. The Ahmedabad mills depended upon Gujarati workers; the Surat power loom industry is manned by Oriya labour from Orissa in eastern India.*

goods, information and finance. Such a role depends heavily on the quality of infrastructure. As a result, cities are fiercely competing to establish pre-eminence in key elements of infrastructure. In east Asia, the contest to capture a hub role in air transport - and thus the downstream activities which are made possible by air traffic - has led to a boom in - and heavy borrowing for - the construction of new airports (as in the building of Kansai, but also in the construction of large airports in Seoul and Hong Kong; a planned new airport in Shanghai; of course, the Pearl River delta is littered with airport projects). In container facilities for sea trade, Singapore and Hong Kong have emerged with surprising speed as larger than any others in the world, and are continuing to upgrade their ports capacity to defend this position against rivals (which include Kobe, Kiaohsiung and, in aspiration, Shanghai). In telecommunications, something similar is occurring as most large cities seem to aspire to become 'international financial centres'. Some of these ambitions are embedded in major capital projects, symbols of dominance in the new servicing economy, as in Pudong or the now-postponed projects of Kuala Lumpur. The more comical aspect of this competition occurs in the contest to build the highest tower in the world, much like inter-war New York, or, even earlier, fouteenth and fifteenth century Tuscany (San Gimignano).

Asian cities - with the possible exception of Singapore - have yet to embark on the economically conscious development of other services for export as has occurred in cities in Europe and North America - medical, health and educational services (for foreigners, rather than the inhabitants), sports, culture and entertainment. But computer services - from software programming to data loading and processing - are growing in, for example, Manila, Bangkok, Shenzhen and Mumbai.

Services require the movement of people - the consumer to the service supplier (as with tourism), or the service supplier to the consumer (as with, say, maintenance engineers). But the international mobility of labour is everywhere impeded formally by immigration controls, thus stimulating the emergence of what seems to be a growing 'informal sector', illegal migration - Indian software programmers who slip into San Francisco on a tourist visa for a week's work in Silicon Valley. The de facto growth in mobility strengthens the city role in air traffic. The world pioneers here are possibly the Filipinos, supplying a major part of the world's domestic servants and sailors (others come, in particular, from Sri Lanka and parts of India, Pakistan and Bangladesh). From the Bahrein Duty Free Zone to the Hamburg-Newcastle sea ferry, Filipinos provide an instant labour force.

The issue is going to grow as the advanced world, closely followed by the Newly Industrialising Countries, South East Asia and China, ages. With a contracting labour force and increasing dependency, high rates of economic growth are required, among other things, to cover the growing burden of supporting the aged. Yet the labour force is declining (both because of ageing, but also as a result of shorter working hours, increasing periods in education, increasing holidays, earlier retirement). In manufacturing, capacity can be relocated to where the labour force is available as is now well underway for Japan and the four 'Little Tigers'. But services for the population have to be provided where the population is: as a distinguished migration expert once observed, the dustbins of Munich cannot be emptied in Istanbul. It is possible to substitute equipment for some of these tasks, but for many tasks undertaken by unskilled labour, the costs of doing so outweigh the likely returns, so the jobs - maids, cleaners, hospital and transport workers, porters, security guards etc., without whom a modern economy cannot function - remain undone.

Already a number of Asian economies are dependent on immigrant labour - for example, in Malaysian tin mining and rubber tapping, in Taiwanese and Korean manufacture and construction. The long struggle of the Singapore government to sustain high growth without increasing the proportion - and in some periods, striving to reduce the proportion - of immigrants in the labour force, illustrates the difficulties.

Ageing makes the problem much worse since it shifts consumer demand towards services, particularly labour-intensive 'caring' services (hospitals and clinics, convalescence and retirement homes, home health care, social services etc) at just the time when the active labour force, and particularly the young labour force, is shrinking. The better off aged can retire to countries where the labour force is still expanding as is to some extent happening in Europe and North America, with aged migration to Spain, Portugal and North Africa for the one, and Mexico and Central America for the other. But this will affect only a relatively wealthy minority, assuming the destination countries will accept these immigrants. For the rest, immigrant labour becomes the difference between a protected old age and misery.

If the service economy and ageing forces the relaxation of immigration controls, the last great bastion of national isolation from global economic integration will begin to be eroded. The role of the State - and its shadow, the 'nation' - will need increasing amendment. But on the other hand, increasing worker migration offers the hope of increasing remittances and thus the much more rapid improvement of some of the poor areas of the world, as we have seen in Jalandhar and Kerala in India, in Sylhet in Bangladesh, in the Thai North East, etc.

Thus, even without the expected rapid growth in tourism (demand in tourism destinations is shifting from the developed to the developing countries), the new air transport facilities will be more than fully utilised. The cities' role in providing the location for airports and, even more, the junction point between transport modes, will be strengthened.

1.5 Methodological Qualifications

There are two brief qualifications that need to be made here. First, that the change in economic geography and activity which is occurring rapidly in parts of Asia undermines some of our traditional concepts - for example, familiar assumptions about the urban/rural distinction are threatened by the emergence of semi-urbanised regions, with rural manufacturing and a declining proportion of the rural labour force engaged in cultivation (thus, for example, the 'man-land ratio' as a measure of potential welfare ceases to have much meaning where land is no longer a key constituent of rural income generation, and this in turn undermines the idea that there is much sense in the concept of 'population pressure'). The distinction between 'manufacturing' and 'services', used here, has always been a fudge (services are no more than a residual after the enumeration of sectors which produce a tangible output) and is increasingly so in a modern economy where manufacturing companies have spun off internal service functions to independent enterprises and where a growing service sector is crucial for the growth of manufacturing productivity.

The second observation is that global integration has undermined the effectiveness of parts of public policy, and this has gone with what seems to be a significant growth in statistically unrecorded activity. There are various inevitably imprecise estimates, but possibly between 20 and 60 per cent of a city's economic activity may be unrecorded, and this may include - as it does in Italy - important export industries. Thus, judgements on the level of economic activity, of employment and of poverty, based upon official data may be very doubtful, and indeed, a boom can occur when all the figures point to slump (as did indeed occur in Italy in 1979). But this remark should not be taken to mean that the crash of 1998 is just a figment of the official figures.

Within the statistically unrecorded activity are straightforwardly criminal activities. The value of the world narcotics trade alone is put at some $500 billion annually, and the total for all illegal activity together must be very much larger. Much of this is concentrated in cities. In a small country, the illegal can loom very large - for example, the macro economic significance of the Myanmar's Shan States' heroin trade which supplies perhaps 60 per cent of consumption in the United States. The implications of this hidden economy can be seen when black money seeks to become white - in new towns, property development and tourism projects along the main routes for narcotics movement in Myanmar. The border crossing point between Myanmar and China's Yunnan province is experiencing rapid urbanisation with new housing, hotels and a free trade zone, and Mandalay is experiencing a building boom, both perhaps partially or wholly a by-product of drug revenues.

In the nexus in Mumbai between corrupt policemen, politicians, gangsters, land speculators and construction companies, cities can be physically and socially refashioned from unseen sources. The black economy is credited with financing the withholding of grain from the market, precipitating famine; with causing anti-Muslim riots in Mumbai (to drive Muslims out of particular areas to permit development projects) and financing Bollywood (the murder of a string of Mumbai film producers is attributed to their failure to repay loans of black money). The effectiveness of macro economic policy is undermined, and even more, social order and transparency - the drug vote in parliaments, the decisions of Ministers, all may be tainted by the very large sums of money at stake. Violence on the streets undermines the viability of the new cosmopolitan servicing city as fatally as the disease implications of polluted water supplies and defective sewerage systems. Thus the old agenda of the quality and ethics of life in cities returns as a precondition for economic success.

1.6 Social Change

The operation of criminal networks, so important in many cities, may be interwoven with the sheer social turmoil that follows rapid urbanisation, particularly where governments are inert or corrupt. Then the notorious problems of sheer physical survival in Asia's large cities - in housing, water supply, sewerage and the environment - are intermixed with a cultural or moral crisis. This is particularly true among sections of the intelligentsia and where, as at present, secular political alternatives (nationalism, socialism and communism) have either collapsed or been severely weakened. While the idea of an ideological vacuum is a doubtful metaphor, the rise of militant and revolutionary religious politics - from the Taliban (the hard-line Muslim government of Afghanistan) to the BJP (the Hindu nationalist party of India) - seems to be meeting a new need or, at least, arming newly politicised layers of the population. And these creeds often have special appeal to the intelligentsia, particularly those that are in the spearhead of modernisation - the stronghold of the Muslim Brotherhood in Cairo University is not among theologians, but in the Faculties of Medicine and of Engineering (Wickham, 1997: 120). The university population has grown immensely in the past twenty years, and there is much unemployment among graduates. The Brotherhood provides one of the few means to employ those with practical skills in its extensive network of social and medical services to the poor. In many ways, 'fundamentalism' often encompasses much of the old programme of social democracy (the Sunni Brotherhood in Egypt, the Shi'ite Hiz'bollah in Lebanon). Perhaps, like the growth of Puritanism in seventeenth century England to the point of Civil War, the rise of the Narodniks in Tsarist Russia, the Naxalites in India or guerrilla warfare in the Latin America of the 1960s an underlying theme in the rise of fundamentalism is the immense growth of the educated unemployed of rural origin but migrating to the cities.

Social and moral turmoil can then connect with political fractures. Especially this is possible because the rise of Asia will displace the existing dominant powers and this is always a dangerous process - witness the effects of Germany threatening the entrenched position of Britain and France in the first half of this century. However, there is some glimmer of hope in that global economic integration is helping to disarm the State, even if there is still not much sign of this in Asia (and there are continuing nasty little territorial disputes throughout the region). Hopefully this will not lead to the spectacular brutalities of the twentieth century world wars. But we may have to grit our teeth in the rough ride as we cross over from the familiar predominantly rural Asia to a new Asia of towns and cities.

References

Nkrumah, Kwame (1961): Hands off Africa, *Accra.*

Wickham, Carrie Rosefsky (1997): 'Islamic mobilisation and political change: the Islamist trend in Egypt's professional associations', in Joel Beinen and Joe Stark (Eds.), Political Islam, Essays from the Middle Eastern Report, *MERIP/IB Tauris.*

World Bank (1994): Global economic prospects and the developing countries, *A World Bank Book, Washington DC.*

World Bank (1995): Global economic prospects and the developing countries, *A World Bank Book, Washington DC.*

World Bank (1997): China 2020: Development Challenges in the New Century, *World Bank, Washington DC.*

PART II
THE URBAN BASE: CHINA

2 Reconsidering the Current Interpretation on China's Urbanisation under Mao's Period: A Review on Western Literature*

L. Zhang and Simon X.B. Zhao

Editors' note:
This chapter in part hypothesises why China had industrialisation without urbanisation. This is the opposite of what is sometimes claimed for India: urbanisation without industrialisation. Theorists have debated the merits of the two social-economies - urban or rural- in order to justify their urban or rural bias. This paper illustrates Mao's rural bias, but, nonetheless, in order to achieve industrial progress China had to strengthen its urban centers. Industrial progress and rural bias do not go well together. India also had a theorist, Mohandas Gandhi, who warned of the evils of modern industrialization, and who favoured village-based handicrafts and other industries.

2.1 Introduction

Whereas in the developed countries the processes of rural-to-urban migration and urbanisation are historically attributable to industrialisation, in the developing countries the rapidity of rural-to-urban migration and urbanisation has far exceeded industrial employment growth. The Chinese road to urbanisation is unique, duplicating neither the experience of the developed world, nor the situation in many other currently developing countries (Zhao and Zhang 1995; Zhang and Zhao 1998). It is noted in China that "under-urbanisation"[1] has existed for decades, being a state in which the increase in urban population lags behind the growth of industrialisation. For years under the planned economy, China tended to have a smaller part of its population

* *This chapter is a part of the authors' research project "Re-examining the current interpretations and the level of China's urbanisation". The authors would like to acknowledge financial support from Hong Kong Baptist University (Grant Ref: FRG/95-96/II-84) and research assistance by Miss Pui Kwan Ho.*

1 *"Under-urbanisation" could be defined as achieving industrialisation without a parallel growth of urbanisation. In the socialist context, under-urbanisation is achieved through a series of restrictive institutions and policies. For the discussion on the Chinese case, see Maoxing Ran and Brian Berry (1989). Underurbanisation policies assessed: China, 1949-1984. Urban Geography vol.10, no.2, pp.111-120. Also, Yu Depeng (1995), Shengji chengshihua jincheng de dingliang bijiao (A quantitative comparison of urbanisation process among provinces). Renkou Yanjiu 1995 no.1.*

living in urban areas and a smaller proportion of labour transferred from rural to urban sectors, even though industrialisation seemed to increase substantially (Orleans 1982; Whyte 1983; Ran & Berry 1989). In the most recent period when the central planning system has been partially marketised, data from survey and census have indicated a lower magnitude of rural-to-urban migration and a lower level of urbanisation than is common in many free market economies at a comparable level of development (SSB 1988).

The purpose of this chapter is to gain a better understanding, both conceptually and empirically, of why, in the People's Republic of China (PRC), the progress of industrialisation, defined as an increasing share of industrial output in the total national output, does not necessarily lead to increased urbanisation. Past studies sought explanations in the ideological preferences of the policies incorporated in development strategy. We propose that analysis can be supplemented with reference to relationship between party and state interests and the inherent characteristics specifically of the Chinese socialist system, which lead to unavoidable state control. Unlike other current interpretations, this research focuses on the decisive role of the state and its interaction with social and economic forces on rural-urban migration and urbanisation. This research argues for an alternative conception of state primacy, the so called "state bias", as the moving force behind rapid progress of industrialisation and slow urbanisation.

2.2 'Rural-bias' and 'Urban-bias': Two Orthodoxies

A number of theories have been developed to interpret China's "under-urbanisation", two of which have been particularly influential. One, which dominated up to the early 1980s, is the perception of the Chinese pattern of limited urbanisation as a product of "rural bias" based on an anti-city ideology (Salaff 1967; Schwartz 1973; Meisner 1974; Eckstein 1977; Gurley 1970, 1976, 1979; Merrington 1978; Cell 1979, 1980; Buck 1981; Ma 1976, 1977; Murphey 1976, 1980; Nolan and White (1984); Lardy 1978; Kojima 1978; Lewis 1955; Chen 1972; Parish 1987). The other, which challenged the rural school's insufficient treatment of China's development strategy, sees the low level of urbanisation as a consequence of a special kind of "urban bias" which is shaped by the Chinese strategy of industrialisation and the practical considerations of "urban manageability" or "military preparedness" or "containing urban costs" (Kirkby 1985; Cannon 1990). Since the mid-1980s, the "urban bias" theory has prevailed.

The "rural bias" school states that the ideology of "anti-urbanism" or "pro-ruralism" inherited the rural essence of the Chinese Communist Party

(CCP) and integrated a Marxist analysis of antagonistic class contradiction between the rural and the urban. It was noted that the CCP had deep peasant roots and that the Chinese revolution was historically one of rural-based peasant communist rebellions (Lieberthal 1995). It was instinctive for this peasant party to show, after it had seized power, its pro-rural sympathies. From the perspective of rural-urban antagonisms, rural-urban relations embody three forms of exploitation: first, political-administrative-military cities *(cheng)* populated by corrupt bureaucrats and their clientele, battening onto the surplus exacted from the peasantry through taxation; second, commercial cities *(shi)* populated by merchants and userers who exploited the peasantry through trade and financial manipulation; third, the foreign-dominated "treaty port" cities, which were bridgeheads of imperialist exploitation (Harrison 1972; Murphey 1974; Ma 1976; Prybyla 1987). It was the task of the revolution to eliminate these forms of urban dominance and exploitation. In this view the low level of urbanisation was a product of anti-urban ideological preference.

Ideologically, Maoists believed that the level of living standards should be raised only on an egalitarian basis. Mao's famous statement of "simultaneous development of industry and agriculture" and the goal of ultimately eradicating rural-urban differentials was mentioned as the theoretical assertion for the "rural bias" school. An official line of "taking agriculture as the foundation of the economy and industry as the leading factor" was cited as strong evidence of giving development priority to the countryside. Empirical evidence used to support the proposition of anti-urbanism included reducing agricultural taxation, developing pro-agricultural industries and rural industrialisation, improving rural living standards, promoting rural small industries in agricultural machinery, chemical fertilisers, and relocating urban intellectuals and youth to the countryside under the slogan of constructing a new socialist countryside and abolishing the "three great differences": those between the city and the countryside, between manual and mental labour, and between manufacturing and farming (Chen 1972; Bernstein 1977; Ma 1977). At a certain historical period, the notion of anti-urbanism offers a thoughtful explanation for the peculiar pattern of rural-urban migration and urbanisation in China.

Opposed to this notion of anti-urbanism, a group of scholars, notably Kirkby, Cannon, and Chan, have recently produced a different perspective on the complexity of the process of urbanisation under the socialist industrialisation strategy. This school of thought, known as the "urban bias" model, suggests, despite the above rhetoric, that actually there is a more economic rather than ideological interpretation. As they argue, urban bias represented a realistic description of some crucial economic policies concerning sectional interests in China, such as a "scissors gap" in pricing system (underpricing agricultural products and overpricing industrial products) (Oi 1993; Naughton 1996), "urban bias" in industrialisation (Kirkby 1985; Chan 1994), and the provision of consumption and employment privileges for urban citizens (Chan 1994, 1996b). Cannon contends that the Chinese approach to spatial planning was based upon strategic thinking for national defence or military preparedness, as evidenced by the heavy investment in the "third line" regions during the 1960s and early 1970s. Taking into account the characteristics of central planning and the favourable measures to protect the privileges of the urbanites, both Kirkby and Chan propose that the socialist imperative to industrialise tried to maximise industrial output while simultaneously minimising social and economic costs. This meant maintaining urban order/manageability and minimising total costs by controlling urban population sizes, giving shape to China's particular pattern of urbanisation. This new economic school seems to prevail in current research on Chinese rural-urban migration and urbanisation, as progressively revealing the existence of urban-favoured policies.

Analysing the Chinese development strategy, Chan (1988, 1994b) argues that the economic growth pattern of China is to maximise industrial output while minimising consumption and standards of living. This enables the paramount objective of rapid accumulation which is essential for rapid industrial growth. The policy is thus not anti-urban – it is simply anti-consumption. The logical extension of Chan's argument is that to speed up industrialisation, urban costs must be minimised by any possible means, notably by "urban closed" and "urban restricted" policies, such as tight control of rural to urban migration; rustication of urban residents to the countryside; intensive use of urban infrastructure and limited expansion of the service sector. The existence of the rigid household registration (hukou) system and official or quasi-official revelations of wide disparities between rural and urban areas as well as discriminatory policies against agriculture and rural population are perceived as strong empirical evidence to support the "urban bias" view.

Despite the different bases of explanations, there are some commonalities in the two theoretical approaches. In both, the conflicting relationships between industry and agriculture, between town and country, and between workers and peasants assume critical importance. They have agreed, implicitly at least, upon the significance of resource distribution among sectors in shaping China's urbanisation. Both have explored a strict and antagonistic divide, created and maintained by severe institutional and administrative measures, between the city and the countryside, and have suggested policy discrimination acting like a biased die to tilt economic and social advantages (particularly in resource allocation) to one side and against the other.[2] As such, they have inferred that the pattern of rural-urban migration and the process of urbanisation in socialist China is under state control and is an outcome of development strategy.

2.3 The Role of the State in China

But the reality seems more complicated than even a composite of the explanations of these two schools. While both models represent at least a plausible story of some economic policies and the political economy of sectional interests, the story they have described seems not to be consistently relevant to the Chinese reality in different development periods. Although the distinctive rural-urban segregation is less arguable, the development process exhibits a series of dichotomies which challenge the arguments proposed by the two schools.

In the political realm, there seems to be no clear-cut dominant policy preference towards rural-urban relations favouring a specific side. On the one hand, the rural background of the ruling party does not necessarily lead to political and economic sympathies favourable to the rural side. Although officials from agricultural localities often claimed more resources for their jurisdictions and garnered redistribution away from richer, more industrialised regions (Lardy 1978; Paine 1981), the urbanised leadership's characteristics make the Chinese state politically and economically ambiguous toward the rural areas.[3] On the other hand, some urban-

---

2 *For instance, both schools have agreed that patterns of migration and urbanisation in China are strongly related to the resource allocation between industry/city and agriculture/countryside. But one of the controversial aspects of the debate between the two schools is the extent to which agriculture has provided capital funds for industrial accumulation and urban development, and the desirability and ability of policy intervention to accelerate the rate of this accumulation by manipulating the direction and magnitude of the agricultural surplus flow. While the "rural bias" school emphasizes resources into the countryside from the state through the invisible account (for example, large government subsidies to the agricultural sector on the input side), the "urban bias" school advocates that, in the sectoral allocation of investment and resource extraction by the government, the agricultural sector is largely sacrificed. This argument is often supported by the claim that agriculture makes a factor contribution in the form of underpricing agricultural products selling to the state for financing the non-agricultural sector at the expense of agricultural investment (Lardy 1983; Kun 1992).*
3 *It was true that the vast majority of party members in the early post-revolutionary Chinese state were of peasant origin. Many of them took up important positions in the post-revolutionary government and have remained a crucial force until they were too old to be politically active. But this instinctive 'ruralist' character underwent a urbanized transition in the post-seizure of power situation. State personnel become detached from their social origins and state apparatus moved into urban centres and become a distinct social group in city with their own material basis and social ideology. Furthermore, in the 1950s urbanites, (cont.)*

orientated policies do not necessarily reflect urban dominated politics. While industry seems to have taken the lion's share of state investment, the urban populations themselves cannot benefit from the outcome of industrialisation, which is wholly confiscated by the state and allocated to the areas where state interests can be satisfied. It is true that there was a policy to grant the basic guarantee of essential needs to the urban citizens while there was a relative weakness of interest articulation among the rural population. But at the mass level, urban economic and social privileges are more nominal than factual, since such privileges were fully subject to manipulation by the state. While the urban "Élite" enjoys some privileges over rural people, their prerogatives at the same time are limited by the greater pervasiveness of controls than their rural counterparts. Although urban residence status is birth-ascribed along a maternal line, it is not lifetime secured.[4] Given employment and food guarantees went along with prolonged low living standards, thus urban citizens could not be regarded as the beneficiary group of a rapid growth of industrialisation. The first Five-Year Plan (FYP) was a period of pronounced urban-industrial policy bias, but paradoxically the post-revolutionary state voiced ideological invectives to eliminate the forms of urban dominance and exploitation, and assigned the peasants an active role in carrying out the first and most arduous phase of China's industrialisation (Spulber 1963). For a long time, under enforced high rates of accumulation, the "rational low wage policy" for urban workers was accompanied by even lower peasant income. Small investment in urban infrastructure corresponded with a low flow of resources to rural areas. Along with a relatively high level of industrialisation, per capita incomes remained low. Both urban and rural residents are deprived of the rights to free job selection and consumption. As such, both the countryside and cities are losers. Thus it would be inaccurate to characterise Chinese political interests in terms of a simple contrast between the city and the countryside. Rather, there is a different contrast. In the pursuit of the highest levels of capital accumulation, the trade-off was seen as between the

---

notably from industrial workers and intelligentsia, were increasingly recruited into the top party and   government bureaucracy (Lewis 1963; White 1983) and became influential for decision- making. Though short of precise statistics, this pattern of representation has continued over several decades and has probably intensified over recent years with the party's emphasis on pensioning off of 'aged cadres' and increased recruitment of intelligentsia into the party and government (Nolan & White 1984; Lieberthal 1995). In political terms, which group of   leaders holds power at any one time exerts decisive influence over policy to favor certain localities, particularly under a centralized and hierarchical system of power. Different government   agencies and party sections have their own distinctive 'ideologies' and this internal conflict can sometimes be intensified into political struggle (Oksenberg 1982).
4 This can be illustrated by the cases of youth rustication in the most periods of the 1960s and 1970s and the Cadre May 7 School during the Cultural Revolution.

current generation and future ones.

Chinese development policies also demonstrate some practical ambiguities in regard to area- or sector-biases. The real scenario has reflected the co-existence of apparently contradictory phenomena caused by frequent policy shifting, which analysis should not ignored. While there is increasing evidence indicating a heavy bias towards industry/city, the policies for rural development are also too overwhelming to be disregarded. The Chinese leadership has been concerned about "rural bias" on some occasions and "urban bias" on other occasions. The leadership is historically inconsistent rather than chronologically unilinear in its accounts of the country's industrial and urban development. While rapid industrialisation is definitely the goal that China wants to pursue, in practice the Chinese leadership has tried to balance development in both agriculture/countryside and industry/city, known as "walking on two legs" (developing agriculture and industry, small rural industries and large urban industries, heavy industry and light industry, simultaneously). During different historical periods under different circumstances, the Chinese leadership may choose to focus one of the two considerations or simply to seek a compromise of the two (Lin 1994). One of the "two legs" might have been stronger and longer than the other in the specific circumstance.[5] As a result, the picture of net agricultural contribution to urban industry over the PRC's (People's Republic of China) history is far from clear in the literature. There is a sharp contrast in the estimated direction and the magnitude of intersectoral resource flows (Lardy 1983; Karshenas 1995; Sheng 1993). Moreover, neither the countryside nor cities could take many economic advantages from the progress of industrialisation. During the 1960s China located many heavy industries in the remote regions (known as "three-line" regions) at great expense which benefited neither the countryside nor the city. In regarding to the area- or sector-bias, China's development policies are subject to a great deal of speculation.

At the macro/societal level, both schools have problems conceptualising the relationships between the state and sectoral interests, and between the role of the socialist state and migration/urbanisation. To them, the pattern of urbanisation seems to be interpreted as a spatial effect of the contradictions of sectoral interests like the conflicts between agriculture and

---

5 *In the 1950s, investment policy seems to be highly skewed on the industrial side. Four-fifths of the investments were allocated to heavy industry in the first phrase of industrialisation. (See First Five-Year Plan for development of the national economic of the People's Republic of China in 1953-57. Beijing: Foreign Languages Press, 1956, p.16). But faced by a economic crisis caused by a failure of Great Leap Forward, the Chinese leadership cut back the scope of industrial development and diverted a significant share of resources to agriculture in the early of 1960s.*

industry and between town and country, rather than as a spatial outcome of the contradictions of the socialist system - such as the conflicts between advanced property relations a backward economy, and between the state and civilian sectors. Furthermore, China's urbanisation tends to be viewed as a merely "economic" phenomenon concomitant with industrialisation and to be fundamentally shaped by development strategy (defined as investment patterns in which resources are actually allocated), rather than to be a functional interplay of the socialist system and the development policies determined by the state's primacy. According to both schools, slow urbanisation in socialist China is a response of development strategy designated by the state. The state, to the extent that it is analysed independently at all, tends to be described as playing a mainly catalytic or impediment role from outside the process of urbanisation. Industrialisation, whether rural-orientated or urban-based, is taken as an independent variable to determine a dependent variable — the pattern of migration and urbanisation. To their logic, good or bad urbanisation policies are seen as emerging from rational or irrational decision-making about industrialisation strategy, overlooking the institutional requirements inherent with the socialist system.[6] State involvement has been discussed merely as specific forms and degrees of intervention, without penetrating the arenas of state primacy and the innate requirements of the socialist system which determine the policy agenda in the first place. It could be argued that state interest has primacy on any development strategy in the sense that sectoral interests between agriculture and industry, between town and country, and between workers and peasants are all subordinate to the state interest. In terms of practical implementation of development strategy, the state inevitably assumes comprehensive responsibilities for allocating resources and becomes the sole distributor and arbiter of conflicting claims upon those resources in the context of socialist property relations. There are sectors with special claims which directly serve the state interest and may divert resources from other parts of the economy. The requirements of the military and the growth bureaucracy growth are two typical examples. The role of the state, therefore, does not act as an outside intervening variable which could only temporarily but not fundamentally deviate the pattern of urbanisation as development proceeds, but is decisive in shaping the pattern of rural-urban

---

6 *The Chinese socialist system, broadly defined, constitutes a group of political and economic elements that are configured in the way to express and serve the specific goals of the Chinese ruling party (communist party). Traditionally, these elements include the structure of power characterized by a one political party dominance under the organisation principle of "democratic centralism", Marxist ideology, state-ownership domination, centrally planned economy and other institutional mechanisms for bureaucratic control (Kornai 1992; Lieberthal 1995).*

migration and the process of urbanisation in socialist China.

2.4 'State-bias' as the Most Important Bias

Upon closer scrutiny of current interpretations, four points should be singled out in order to build a more logical and informative framework to analyse the under-urbanisation nature of China's case.

First, the political and economic balances between urban and rural settings are not as clear-cut as the models propose. While a plausible case can be made for an "urban bias" interpretation, evidence can also be found to support the idea of "rural bias" in China's urbanisation, depending on which aspects of the "rural bias" or "urban bias" thesis one wishes to stress. There is therefore a need to accommodate the co-existence of apparently contradictory phenomena. Although the Chinese rural-urban dualism is economically and socially evident and is institutionally and administratively enforced, to the extent that there is a dichotomy it is not the case that rural-urban relationships can be represented by a simple zero-sum game, in which one side gains only to the disadvantage of the other.

Second, the decisive role of the state in rural-urban migration and urbanisation should be adequately addressed. The non-zero-sum nature of the contest between sectors and regions may suggest that the state has acted to enforce "the party-state interest" (the inherent characteristics of the Chinese socialist system) which transcends any other interests. There has been a strong tendency for the state to act for its own perceived interests as a whole. In this context the state's administrative interventions in urbanisation have reflected not only national priorities, overriding sectoral or territorial interests, but also the innate requirements of the socialist system the state intends to build.[7] It is unconvincing to attribute restricted migration and under-urbanisation entirely to the area or sectoral conflicts or to urban cost-

---

7 *It can be argued, for example, that control over rural-to-urban migration and urbanisation, particularly labor migration, may be a requirement of the socialist system to deal with its innate problems, not a necessity of economizing urbanisation costs. In many cases, the restricted policies for migration and urbanisation are passive reactions to the problems of socialist planning system, rather than well designed responses to the quest for rapid accumulation (Tang 1993). Investment hunger, resource hoarding, and enterprise bureaucratic growth, which are inherent features of the planning system of a resource-constrained economy, forces the formulation of control policies not only for the demand for investment expansion but also for inefficient use of labor and unofficial labor migration. Control over rural-urban migration is not so much due to the necessity of saving resources for accumulation (i.e. the development strategy of putting accumulation at the higher priority than consumption) and economizing on urbanisation costs as due to the problems associated with the operation of central planning system.*

minimisation [8] without carefully analysing the system requirements defined by the state priority.

Third, to fully understand the decisive role of the Chinese state in rural-urban migration and urbanisation, one also needs to pay attention to the crucial issue of resource spending. An underlying question regarding the impact of resource spending on migration and urbanisation is the extent to which the problems of change in economic structure are associated with misallocation of resources between sectors or activities. The orientation of resource spending, particularly the flow of agricultural surplus, is clearly of direct relevance to the interpretation of Chinese urbanisation, given that the debate between the two schools always seems to gravitate toward the issue of the transfer of agricultural surplus. But in the absence of clearly defined property relations, the state inevitably assumes comprehensive responsibilities for allocating resources and becomes a powerful distributor and an arbiter of conflicting claims upon those resources. From the point of view of state primacy, it seems that skewed resource distribution has conflicted not so much between industrial/urban and agricultural/rural expenditures as between the state and civilian expenditures. In key areas of development choice, policies which are indeed vitally necessary to promote economic development and efficiency, social equality and living standards of people have often been stifled or weakened by state interests. It would be an over-simplification, therefore, to regard the patterns of migration and urbanisation as a matter of capital expenditure between agricultural/rural and industrial/urban sectors alone.

Fourth, one needs to reconcile the actuality of Chinese development strategy. This may be perceived from the nature of Chinese industrialisation with reference to several questions. Industrialisation was, and still is, the only method known to the human race for achieving substantial increases in labour productivity and rising living standards (measured in material terms) (Post and Wright 1989). However, industrialisation in the socialist context fuels sharp distributional conflicts between industry and agriculture, between consumption and investment, and between employment and productivity. But why did China always emphasise autarkic rapid growth? Why did the Chinese state pursue over a long period a high accumulation rate and preferential

---

8 *The low level of urbanisation in China has been imputed to a series of distinguishing either "anti-urban" measures advocated by the "rural bias" school or "urban-closed" polices by the "urban bias" school. An important institution for rural-urban separation is the household registration (hukou) system for maintaining population in their birthplaces and curbing rural migration flooding into cities. But the function of hukou system, as I will argue, is not originally designed and is not the sole effective weapon to block migration. Urban residency right is completely subject to the state manipulation and is not fully defined by the hukou system.*

investment in heavy industry? Why did the quantity of production output, rather than indices of efficiency and indicators of living standard, have a top priority? Given the fact that living standards of both peasants and workers grew so slowly over lengthy periods (from the mid 1950s to the late 1970s), where did the outcome of economic growth go and who was the beneficiary of industrialisation? In the literature, where industrialisation is often taken as a synonym with economic development, dynamic urbanisation associated with industrialisation is implicitly assumed. But how did the state-led industrialisation produce an insignificant demand for labour transformation? Searching for possible explanations, one could even begin to question to what extent Chinese industrialisation can be regarded as a 'true' or 'genuine' industrialisation. (This begs a comparative analysis. India too has achieved industrial growth without increases in industrial employment – but in that case it has often been described as urbanisation without industrialisation – the exact opposite of China's apparent situation.)

Therefore, the fundamental reasons behind the peculiar pattern of China's urbanisation in Mao's period - rapid industrialisation without urbanisation - must be explored under an institutional framework of 'State Bias'. That is, under this framework of 'State Bias', both the "urban" and the "rural" were manipulated to serve the paramount objectives and priorities of national development, in which neither the "urban" nor the "rural" were regarded as "goals", but "means" of the national development. This 'State Bias' can be more specifically embedded in the very high rate of capital accumulation in non-civilian or military-led industrialisation, and the huge governmental apparatus which was purely designed for the state control and which consumed massive state fiscal resource. To demonstrate that the industrialisation under such a 'State Bias' would not lead to urbanisation, one must ask 'What was the nature of Mao's industrialisation (military-led?)' and 'Where did its output or economic growth get channelled to (to the state control and bureaucracy?)'. One also needs to ask what was the underlying (or Party's) philosophy behind the 'hukou system', which was misleadingly regarded as the key factor that blocked "rural-urban" migration, and hence, the growth of China's urbanisation during Mao's period.

2.5 Concluding Comment

It is by no means easy to make a scholarly depiction of this 'State Bias', or Mao's military-led industrialisation and the huge state or bureaucratic control which consumed up large portions of the state fiscal budget. To show clearly how this worked both requires major reinterpretation of official figures, and also, ideally, data which we know will never be available in the required form and quantity. However, currently the authors are engaged on research in this topic, thanks to the funding from a Hong Kong Baptist University research grant, to improve as far as possible the basis of our understanding of these matters. The preliminary findings already lead us to believe that the objectives and arguments depicted in this chapter are sound and close to reality.

Following Mao's doctrine that there would be no socialist China without the Chinese Communist Party and that class struggle is a long-term revolution, the Chinese Communist Party must consistently monopolise power and authority on a national basis for political survival. The socialist regime, once established, does not allow other organised Èlites or institutions with recognisable rights and adequate nation-wide authority to check, counter and discipline the party and its agent, the government. Although there are elements which might insert some pressures on the party and government, they are never allowed to launch any serious challenge.

To ensure political survival, the socialist China has in its early stages of development abolished individual or private property rights and created state ownership by force. The property rights, which should serve as a norm to define basic economic relationships amongst people and as a set of basic constraining forces governing the behaviour of people, have been transformed to facilitate the abuse of political power. Under the state ownership, the rights of individual are minimised, at least in practice if not in principle. Although the government is the "people's" government, the individual's relationship to the state has all along been one of dependency and not one based on the rights of individuals. Individuals cannot freely choose opportunities. Over time, there is still little increase in individual autonomy.

The justification in the need to control a society can be found in the lack of property rights. Without that internal constraint, the system of socialist china theoretically requires certain social controls working as external constraints. The socialist regime has to impose severe regulation control with respect to all aspects of economic and social life. Social controls are operated by the restricted rules and regulations, a centralised planning economy, and political mobilisation; and supported by the administrative network and the societal stratification, and accompanied by some political promises. Where these features are not built into the social structure, it is assumed that the socialist regime must generate them. In this regard, the Chinese reality and some practices, such as rural-urban inequalities, household registration, forced migration, control of free migration, labour planning, can be argued not as requirements no just of socialist ideology but of the system.

For example, in the course of state appropriation of ownership of property (i.e. nationalisation of industry and commerce and collectivisation of agriculture), some forms of inequality embedded in the private ownership of property are eliminated. But the societal inequalities in general and the rural-urban inequalities in particular, either in a pecuniary form or in a non-monetary form are far from inevitably disappearing with the demise of private property, because of the social stratification needed by the effective and efficient control. Moreover, the conflict between the need to preserve a difference in income on the one hand and socialist egalitarianism on the other is one of those fundamental contradictions in the system which cannot be resolved at a stroke.

To consolidate its position, the regime must devote some resources to economic growth. After taking over all property from the whole society, the party/state as a sole owner of the means of production is the one who is entitled to administer the operation of the economy. The relations of production are not determined by the demands of the productive forces, but by demands of the party/state administrative structure. State administration of the economy cuts voluntary horizontal relationships among enterprises, and instead imposes a vertical structure similar to a government hierarchy. Furthermore, the feature of the socialist public ownership defines the status of sectoral development, because only the party/state can divert surplus generated by economic development to its special interests. The sectoral relationship favours a specific sector (such as industry or agriculture) at the first glance, but is actually heavily favouring the party/state at second glance. One would expect to see not only spatial and sectoral biases in development policy wherever the party/state involvement in economic affairs has been consistent and extensive, but also wherever the biases that favour the regime's political survival, either domestic or international or both, are changed in different periods.

References

Bernstein, T. (1977) Up to the mountains and down to the villages: the transfer of youth from urban to rural China. *New Haven, CT: Yale University Press.*

Buck, D. (1981) Policies favouring the growth of small urban places in the PRC 1949-1979. In L. Ma & E. Hanton (eds.) Urban development in modern China. *Boulder Co: Westview Press, pp.114-44.*

Cannon, T. (1990) Regions: spatial inequality and regional policy. In T. Cannon & A. Jenkins (eds.) The geography of contemporary China. *London: Routledge, pp.28-60.*

Cell, P. (1979) De-urbanisation in China: the urban-rural contradiction. Bulletin of Concerned Asian Scholars *11 (1), pp.62-72.*

Chan, Kam Wing (1988) Rural-urban migration in China, 1950-82: estimates and analysis. Urban Geography *vol.9, no.1, pp.53-84.*

Chan, Kam Wing (1994) Cities with invisible walls: reinterpreting urbanisation in post-1949 China. *Hong Kong: Oxford University Press.*

Chan, Kam Wing (1996) *Internal migration in China: an introductory review.* Chinese Environment & Development *vol.7, nos. 1&2, pp.3-13.*

Chen, P. (1972) Overurbanisation, rustication of urban-educated youths, and politics of rural transformation: the case of China. Comparative Politics *vol.4, no.3, pp.361-84.*

Eckstein, Alexander (1977) China's Economic Revolution, *Cambridge: Cambridge University Press.*

Gurley, John, (1976) China's Economy and the Maoist Strategy, *New York: Monthly Review Press.*

Gurley, John, (1979) Rural Development in China 1949-75, and the Lessons to be Learned from it', in Maxwell, Neville (ed.) (1979) China's Road to Development, *Oxford: Pregamon Press, pp.5-25.*

Gurley, John, 1970. 'Maoist Economic Development : The New Man in the New China', The Center Magazine, *3(3), pp.25-33.*

Harrison, J. (1972) The long march to power: a history of the Chinese Communist Party 1921-72. *New York: Praeger.*

Karshenas, M. (1995) Industrialisation and agricultural surplus: a comparative study of economic development in Asia. *Oxford University Press.*

Kirkby, Richard J.R. (1985) Urbanisation in China: town and country in a developing economy 1949-2000 AD. *London: Croom Helm.*

Kojima, R. (1987) Urbanisation and urban problems in China. *Tokyo: Institute of Developing Economies.*

Lardy, N. (1978) Economic growth and distribution in China. *Cambridge: Cambridge University Press.*

Lardy, N. (1983) Agriculture in China's modern economic development. *Cambridge: Cambridge University Press.*

Lewis, W.A. (1955) The Theory of Economic Growth. *London : George Allen & Unwin.*

Lieberthal, Kenneth (1995). Governing China: from revolution through reform. *W.W. Norton.*

Lin, G.C.S. (1994) Changing theoretical perspectives on urbanisation in Asian developing countries. Third World Planning Review *vol.16, no.1, pp.1-23.*

Ma, Laurence J.C. (1976) Anti-urbanism in China. Proceedings, Association of American Geographers *vol.8, pp.114-18.*

Ma, Laurence J.C. (1977) Counter-urbanisation and rural development: the strategy of Hsia-Hsiang. Current Scene *vol.15, nos.8-9, pp.1-12.*

Meisner, Maurice (1974) 'Utopian Socialist Themes in Maoism', in John Wilson Lewis (ed.), Peasant Rebellion and Communist Revolution in Asia, *Standford: Standford University Press, pp.207-252.*

Merrington, J. (1978) Town and country in the transition to capitalism. In R. Hilton (ed.) The transition from feudalism to capitalism. *New York: Schocken, pp.170-95.*

Murphey, R. (1974) The treaty port and China's modernisation. In M. Elvin & G. Skinner (eds.) The Chinese city between two worlds. *Stanford University Press.*

Murphey, R. (1976) Chinese urbanisation under Mao. In B. Berry (ed.) Urbanisation and counter-urbanisation. *Beverley Hills: Sage.*

Murphey, R. (1980) The Fading of the Maoist Vision: City and Country in China's Development, *New York: Methuen.*

Naughton, Barry (1996) Growing out of the plan: Chinese economic reform, 1978-1993. *Cambridge University Press.*

Nolan, P. and G. White (1984) Urban bias, rural bias or state bias? Urban-rural relations in post-revolutionary China. Journal of Development Studies 20:52-81

Oi, Jean (1993) Reform and urban bias in China. Journal of Development Studies 29, pp.129-

148.

Orleans, Leo A. *(1982) China's urban population: concepts, conglomerations and concerns. In Joint Economic Committee, Congress of the United States,* China under the four modernisation. *Washington DC.: US Government Printing Office, pp.268-302.*

Paine, S. *(1981) Spatial aspects of Chinese development: issues, outcomes and policies 1949-79.* Journal of Development Studies *vol.17, no.2, pp.133-95.*

Parish, William L. *(1987) 'Urban Policy in Centralized Economies: China' in Tolley, George S. and Thomas, Vinod (eds.),* The Economics of Urbanisation and Urban Policies in Developing Countries, *Washington, DC: The World Bank, pp.73-84.*

Post, Ken and Phil Wright *(1989)* Socialism and underdevelopment. *London & New York: Routledge.*

PRC State Statistical Bureau (SSB) (1988) China Statistical Yearbook 1988. *China Statistical Publishing House.*

Prybyla, Jan S. *(1987)* Market and plan under socialism: the bird in the cage. *Hoover Institute Press.*

Ran, Maoxing & Berry, Brian *(1989) 'Underurbanisation Policies Assessed: China, 1949-1986',* Urban Geography, *10(2), pp.12-26*

Salaff, J. *(1967) The urban communes and anti-city experiments in Communist China.* China Quarterly *no.29, pp.82-109.*

Schwartz, Benjamin, *(1973) 'China's Development Experience, 1949-72', in Oksenberg, Michel (ed.),* China's Developmental Experience, *New York: Praegar Publishers, pp.17-24.*

Sheng, Y. *(1993)* Intersectoral resource flows and China's economic development. *St. Martin's Press.*

Spulber, Nicolas *(1963) Contrasting economic patterns: Chinese and Soviet development strategy.* Soviet Studies *vol.15, no.1, pp.1-14.*

Whyte, Martin King *(1983) Town and country in contemporary China.* Comparative Urban Research *vol.10, no.1, pp.9-20.*

Zhang, L. and Zhao, X.B. *(1998) "Re-examining China's "urban" concept and level of urbanisation",* The China Quarterly, *Vol. 154, pp. 330-381.*

Zhao, X.B. and Zhang, L. *(1995) "Urban performance and the control of urban size in China",* Urban Studies, *Vol. 32:4-5, pp.813-844. Reprinted in R. Paddison, J. Money and B. Lever (eds) (1996)* International Perspectives in Urban Studies 4, Jessica Kingsley Publishers: London.

3　Chinese Urbanisation Policies, 1949-1989

Shue Tuck Wong and Sun Sheng Han

3.1 Introduction

Research on Chinese urbanisation policies has been encouraged by two facts. First, the "Chinese Model" of urban development, which was known for its slow pace of urban population growth and rural outlook of the urban morphology, was indeed a product of such urban policies (Whyte and Parish 1984; Pannell and Ma 1983). The fluctuations in the level of urbanisation resulted from the changes of such urban policies (Ran and Berry 1989; Hsu 1994). The current economic reform program that was responsible for reorienting Chinese development policies inspired the search for a new model of Chinese urbanisation (Kwok, et al. 1990). Second, Chinese urbanisation policies have formed an important part of the strategies of national development. They served to implement the development strategies in the rural and the urban sectors, and provided a regional dimension (Han and Wong 1995). They were blamed for causing fluctuations in Chinese economic growth as well as creating chaos in national development during the Cultural Revolution (Cheng 1990). Faced with the present economic transition from a centrally planned system to a liberalised, market-oriented system, the Chinese government must understand its existing policies and simultaneously formulate new ones (Ginsburg 1993).

This chapter seeks to present an account of the changes and continuities of Chinese urbanisation policies between 1949 and 1989. The policies are categorised into four groups: (1) policies that control the population flow between cities and the countryside; (2) policies that guide the development of cities; (3) policies that guide the development of the countryside; and (4) policies that guide regional allocation of industries. The data used for this research was collected from two main sources. The first source was obtained by personal interviews of eight Chinese urban planners working at central, provincial, and municipal governments. The second source was derived from newly released governmental reports.

3.2 Policies that Control the Population Flow between Cities and the Countryside

One general policy and five sub-policies worked to control the flow of population between rural and urban areas.

3.2.1 The General Policy of Administrative Control over Rural-Urban Migration

The Chinese communist leaders believed that rural-urban migration could be stimulated by China's industrialisation (Yeung and Hu 1992). A large number of peasants was expected to enter cities to help urban industries. In practice, however, the ideological scenario of rural-urban migration was distorted. As the communist military marched into the cities with the slogan of "maintaining the existing agencies, maintaining the positions and level of salary of the existing employees", they soon learned that the urban sector could hardly be capable of absorbing the unemployed in cities. Thus, a further increase of urban population would hinder rather than foster China's economic growth (*People's Daily* April 20, 1953).

The policy of controlling rural-urban migration was formulated in the 1950s, as part of the centralised planning system. It sought to limit free migration by three principal means. First, labour hiring of urban sectors had to follow government plans. In 1950, centralised plans for labour hiring were applied in the state-owned sector (Zhao 1989a). Throughout the 1950s, the government made several attempts to stop unplanned hiring (Guo Wu Yuan April 17, 1953; September 14, 1957). In the 1960s and the 1970s, economic stagnation urged strict control over urban-ward migration. The limited number of opportunities of urban employment were taken up by demobilised soldiers, who had joined the army from the cities in the first place (Zhao 1989a).

Second, every household had to register with a local authority. Household registration was initiated by the central government in the mid-1950s (Zhao 1989a). It prevented free urban-ward migration because immigrants had to demonstrate to urban authorities that they had an admission letter for employment or for schooling (Mallee 1995).

Third, the use of rationing coupons reinforced the administrative power of control over rural-urban migration. Rationing coupons covered a wide range of goods from articles for daily use to expensive manufactured products. By 1968, there were approximately one hundred types of commodities that were distributed by coupons (Zhao 1989a).

The policy to control rural-urban migration continued into the 1980s. Although reform and openness gave enterprises power to hire workers, and reduced the variety of rationing coupons, the central government continued to allocate quotas in labour. The *Household Registration System* continued to be in effect. Besides, the government advanced new policies (i.e., to leave the farm but stay in the home town) to strengthen control over some the side-effects of reform.

3.2.2 The Five Sub-Policies

3.2.2a The control of blind urban-ward migration

The term *blind migration* refers to the unplanned flow of population from the countryside to the cities. It was produced by the forces of either rural-push or urban-pull, or both. The force of rural-push was generated by natural disaster and collectivism. Heavy flooding in central China in the early 1950s pushed many peasants to flee away from their homes, in provinces such as Anhui and Henan, to large cities. Rural collectivism resulted in higher productivity in agriculture and reduced the necessary size of rural labour. The opportunities of industrial employment and better living conditions in the cities generated the force of urban pull. Some peasants could find employment in the construction sector or manufacturing plants because centralised control of labour was weak in the early 1950s (*People's Daily* April 20, 1953). In Guangzhou, between 1949 and 1955 the total population of the city was increased by 534,000 people, of which 70 percent was blind peasant-migrants (*Nan Fang Ri Bao* December 30, 1955). By 1956 the number of blind peasant-migrants in Shanghai reached 50,000. As a result, the city's infrastructure was overburdened and its industrial growth was slowed down (*Jie Fang Ri Bao* December 26, 1956).

The policy of controlling blind migration was first issued by the State Council (Zheng Wu Yuan) in April 1953 (Guo Wu Yuan April 17, 1953). In the countryside, the State Council asked the local governments to convince peasants to stay away from the cities. In the cities, blind peasant-migrants were sent back to the countryside, and the urban sector was prohibited from hiring peasants without government permission. This policy was highlighted again in 1954 because blind migration was still a serious problem in some provinces (Nei Wu Bu and Lao Dong Bu March 12, 1954).

In 1955, Mao Zedong suggested that rural surplus labour should find employment in the countryside through the diversified management of agriculture, rather than entering the cities (Mao 1955). Since the beginning of 1956, the central government experimented with the use of household registration to control blind migration. Until the *Household Registration System* was in full use in 1958, the Party and the State Council continued their efforts to stop blind migration (Guo Wu Yuan December 30, 1956a; September 14, 1957).[1]

3.2.2b The relaxation of the centralised control over labour hiring

This policy was a product of the Great Leap Forward in the period 1958-60. Before 1956, regional governments were in charge of formulating plans for the size of the labour force. In 1956 the central government took over by designing quotas and sources of labour force for each sector (Dang Dai Zhong Guo Bian Ji Bu 1985). Enterprises were allowed to take charge of the production process. However, the size of the labour force of enterprises followed quotas that were set by the central government (Zhong Gong Zhong Yang September 20, 1957).

During the Great Leap Forward, the central government asked local administrations to solve their problems at the prefectures, counties, communes and enterprises levels, and at the same time to decentralise the power of control over finance, the growth of industries, commerce and labour force (Zhong Gong Zhong Yang January 20, 1961b). As a result of this power decentralisation, the size of the non-agricultural labour force increased rapidly (Table 3.1). In the state-owned sector, particularly, the size of labour force reached 50 million in 1960, doubling the size of labour force in 1957.

In 1961 the central government took back the decentralised powers that they gave to the local administrations during the GLF and thus discarded the policy of relaxation of control over labour hiring. Decentralisation of urban population then became the focus in the next policy period.

3.2.2c Hui Xiang: the movement of returning to the villages

The policy of *Hui Xiang* was to reduce the size of urban population. It was a principal part of the strategy of readjustment in the period 1961-63, and was largely a reaction toward the rapid increase of non-agricultural labour force during the Great Leap Forward (1958-60). The over-sized non-

---

1 *In December 1957, just one month before the endorsement of the Household Registration System, the Party and the State Council issued a six-point plan to stop blind migration. These were: (1) reinforcing the ideological education of peasants in the countryside; (2) reinforcing the blockages along principal transportation routes; (3) convincing blind peasant-migrants in cities to return to their home villages, and prohibiting them from begging; (4) prohibiting unplanned hiring in all urban sectors; (5) convincing those immigrants who had not been settled in the local rural areas to return to their home villages; and (6) sending blind peasant-migrants home by a single operation (Zhong Gong Zhong Yang and Guo Wu Yuan, December 18, 1957). In 1958, the State Council issued further instructions to complement its six-points of December 1957, particularly the sixth statement.*

agricultural labour force was accused of being the causes of (1) labour shortage for agriculture; and (2) overburdening government finance and grain supply. The reduction of urban population had thus focused on the size of non-agricultural labour force.

Table 3.1 Changing size of labour force, 1949-1963 (in 10,000)

Year	State-Owned Work Units	Collective Work Units	Individual Businesses	Rural Labour	Total
1952	1580	23	883	18243	20729
1953	1826	30	898	18610	21364
1954	1881	121	742	19088	21832
1955	1908	254	640	19526	22328
1956	2423	554	16	20025	23018
1957	2451	650	104	20566	23771
1958	4532	662	106	21300	26600
1959	4561	714	114	20784	26173
1960	5044	925	150	19761	25880
1961	4171	1000	165	20254	25590
1962	3309	1012	216	21373	25910
1963	3293	1079	231	22037	26640

Source: Guo Jia Tong Ji Juu 1991b, p. 95.

In June 1961, the Central Committee of the Chinese Communist Party called for the reduction in the size of the urban population by 20 million in three years. Newly registered non-agricultural population, i.e., those who joined the non-agricultural labour force after January 1958, from the country-side, had to return to their villages. Workers who had been employed since January 1958 but who were urban residents were not required to be removed from the non-agricultural labour force (Zhong Gong Zhong Yang June 16, 1961a). According to Table 3.1, the non-agricultural labour force was reduced mainly in the state-owned sector. In 1961 alone, state-owned working units reduced their workforce by 8.73 million. By the end of 1963, another 8.78 million were moved from the state-owned sector to the countryside.

Hand in hand with the reduction of non-agricultural labour force were the new definitions of towns and cities. The population threshold for towns was to be raised to 3,000, of which 70% should be non-agricultural. The mini-mum population threshold of cities remained 100,000. These helped to re-duce the official number of cities from 208 in 1961 to 167 in 1964, while the number of towns was reduced from 4429 to 3148 (Guo Jia Tong Ji Juu 1990a).

The forced reduction of the non-agricultural labour force and the changes of urban definition worked together to reduce the non-agricultural population from 133.31 million in 1960 to 115.84 million in 1963. A total of 21.47 million non-agricultural population changed their registration to agricultural status. The goal of a reduction of the non-agricultural population by 20 million was thus apparently achieved. So in 1963 the policy to reduce urban population ended (Dang Dai Zhong Guo De Ji Hua Gong Zuo Ban Gong Shi 1987).

3.2.2d Shang Shan Xia Xiang: the movement of going up to the mountains and coming down to the villages

The policy of *Shang Shan Xia Xiang* (going up to the mountains and coming down to the villages) grew from the period of readjustment. In 1963, the Chinese Communist Party asked the government to prepare for about one million urban youths to participate in agricultural production in the countryside each year, for the next fifteen years in order to control the growth of urban population (Dang Dai Zhong Guo De Ji Hua Gong Zuo Ban Gong Shi 1986). The lack of employment opportunities might be the main reason for Shang Shan Xia Xiang. The Labour Bureaux of cities was faced with enormous difficulties in employing urban youths in the urban sector. Employment opportunities, particularly in the industrial sector, were not growing fast enough. The beginning of the Cultural Revolution further intensified the pressure of unemployment in cities because economic growth was replaced by "class struggle".). Under these chaotic economic conditions there was a consequential decline of industrial production in 1967 and 1968 (Table 3.2). Finally Mao Zedong spelled out his famous instruction in 1968:

It is very necessary for those educated youths to go to the countryside to be re-educated by the peasants... (*People's Daily* December 22, 1968)

This statement provided an ideological rationale that reinforced the policy of *Shang Shan Xia Xiang*. It brought about the large-scale resettlement of urban youths in rural China. By 1975, there were 12 million urban youths resettled in the countryside in the period 1966-75 (*People's Daily* December 22, 1975).

Urban youths in the countryside lacked skills in agricultural production and faced hard living conditions. Peasants were busy trying to survive themselves, showing no attention nor interests in re-educating urban youths, Thus, constant efforts were made by the resettled urban youths to return to the cities. In 1966 and 1967, many urban youths returned to the cities carrying the slogan, "establishing revolutionary ties" (People's Daily July 9, 1967). After the death of Mao Zedong, several rallies were organised in Northeastern, Northwestern and Southern regions to protest against the policy of Shang Shan Xia Xiang. These rallies forced the post-Mao government officially to discard the policy in 1979 (Zhao 1989b).

Table 3.2 Changes of industrial productivity, 1958-1970

Year	Increase from Previous Year (%)	Increase from 1952 (%)
1958	-25.2	64.3
1959	4.3	71.3
1960	34.5	130.2
1961	-34.3	51.3
1962	12.0	69.6
1963	32.5	124.6
1964	25.1	180.8
1965	20.3	237.9
1966	16.1	292.3
1967	-20.0	213.7
1968	-11.7	177.1
1969	25.6	247.9
1970	19.5	315.7

Source: State Statistical Bureau 1991b, p. 63.

3.2.2e Urban Policy 1980: Controlling the growth of large cities and encouraging the growth of small cities (1978-1989)

The Deng leadership sought renewed urban policies to facilitate China's economic growth in the post-Mao era. The control over rural-urban migration continued to be one of the emphases. In 1978, *Urban Policy 1980* (viz., controlling the growth of large cities, developing medium-sized cities rationally and encouraging the growth of small cities) was drafted. Smaller cities were given priorities to increase in their size of population and in their numbers, while large cities were to be controlled against further growth. The rationale behind this policy was both ideological and pragmatic. According to Chinese communist ideology, smaller cities would lead to better rural-urban integration, and the growth of large cities had to be contained to save investment.

Urban Policy 1980 gave the large cities smaller quotas for increases in labour while encouraging the growth of small cities. This was done by three policy measures: (1) lowering the population threshold of official cities and towns; (2) calling for "leaving the farm but staying in the hometown" and (3) allowing selected peasants to settle in the cities.

In 1984 the Ministry of Civil Affairs renewed the criteria for settlements to be qualified for official town status, defining the minimum non-agricultural population for a town to be of 2,000. In 1986 the Ministry lowered the threshold population for a city from 100,00 to 60,000. This was in sharp contrast to the 1963 standard, which was set at the size of 100,000. These changes meant that more and smaller places would have appropriate urban administrative functions.

The slogan, "leaving the farm but staying in hometown" called for the development of rural enterprises. The latter was expected to help peasants to increase income, and at the same time, to provide employment for surplus labour. In 1980, the output value of rural industries made up only 8.4 percent of the national gross output of industries and agriculture. By 1986, this percentage increased to 14.7 percent. To Chinese planners, the idea of encouraging peasants "to leave the farm but to stay in the hometown" became a part of the Chinese style of urbanisation (Fei 1985).

In 1984, The State Council asked the local governments to support the peasants entering the market towns (*Jizhen*) to do business by airing the following statement:

The public Security Bureau shall grant permanent household registration to those peasants who applied to establish industrial workshops or to establish commercial outlets in market towns. They should get Self-Grain Support Registration and be counted as non-agricultural population. They should join *Street Residential Commissions*, and have the privileges and obligations of residents of towns (Guo Wu Yuan, 1984b).

Thus peasants were given the green light to establish workshops and/ or commercial outlets in cities (*Cheng*) (Zhong Gong Zhong Yang and Guo Wu Yuan 1985) and a population group of urban residents who were outside the labour plans of the government began to emerge.

3.3 Policies that Guide the Development of Cities

The policies that guided the development of Chinese cities are grouped under two headings. The first was formulated for economic reasons, i.e., to promote China's economic growth. The second was formulated for ideological reasons, i.e., to organise a socialist economy.

3.3.1 Policies for Economic Goals

Policies in this group include: (1) the restoration and increase in urban industrial output, 1949-52; (2) the establishment of industrial bases, 1953-57; (3) the emphasis on steel output as the key link, 1958-60; (4) the reduction in the scale of industrial construction, 1961-63; (5) the model of Daqing, 1964-77 and (6) using cities as engines of economic growth, 1978-89.

3.3.1a The restoration and increase in urban industrial output, 1949-52

In the period in the immediate aftermath of communist victory, there were as yet no Five-year Central Plans. The policy in this period was ad hoc. It promoted urban industries in order to combat unemployment, to produce industrial goods for the countryside , and to support the military in their consolidation of victory (Liu 1949).

Several military orders were issued by the Chinese Communist Party to its army commanders. The latter were asked to protect industries and to make efforts to restore production (Liu 1949). A slogan, "to transfer consumer cities to producers" was aired with two intentions. First, it was to put urbanites back to work in order to stabilise society and to increase industrial products for the military and the countryside. Second, it was to change the economic structure of some cities from commercial centres to industrial centres (Liu 1949). A large proportion of investment was used to construct housing for urban workers (Table 3.3), to build sewage systems, to provide water supply and public transit systems. Large cities benefited greatly from these efforts. The projects of Long Xu Go in Beijing and Cao Yang residential district in Shanghai converted slums into apartments and housed thousands of urban workers. The improved living conditions stimulated the initiatives of urban workers to support the communist development (Dang Dai Zhong Guo 1990).

3.3.1b The establishment of industrial bases, 1953-57

This policy was one of the three tasks during the First Five Year Plan (1953-57). Implementation of the policy emphasised the construction of 156 projects that were aided by the Soviet Union, and 694 projects that were designed by the Chinese. These projects were located in seventeen of the twenty-nine provinces. The provinces of Shaanxi, Liaoning, Heilongjiang, Shanxi, Henan and Jilin accounted 70 percent of the 156 projects. Large and medium-sized cities such as Shenyang, Jilin, Fushun, Haerbin, Anshan, Qiqihaer, Beijing, Taiyuan, Shijiazhuang, Baotou, Xian, Lanzhou and Chengdu were the principal sites used to accommodate the new industries.

The Ministry of Construction ranked the cities for investment priorities in order to facilitate the implementation of the policy. A large number of non-coastal cities, such as Taiyuan, Datong, Baotou, Luoyang, Xian, Chengdu,

Table 3.3 Proportion of Housing Investment, 1950-1978 (in percentage, total 100 percent)

Year	Productive Investment	Non-Productive Investment	Housing
1950	65.0	24.0	11.0
1951	65.0	24.0	11.0
1952	66.9	22.8	10.3
1953	58.6	28.9	12.5
1958	88.0	9.0	3.0
1960	86.4	9.5	4.1
1965	84.7	9.8	5.5
1970	93.5	3.9	2.6
1978	82.6	9.6	7.8

Note: The productive investments include all investments for the economic sectors such as agriculture, industry, construction, transportation, commerce, and communication. The non-productive investments are used to satisfy people' material and cultural needs. They include investments on infrastructure, health care, social welfare, education, media, research, and on the constructions of financial and insurance sectors, government agencies, party branches, and other social groups. Source: The figures are compiled and calculated by the author using raw data from Guo Jia Tong Ji Juu 1982, p. 309.

Wuhan, and Lanzhou received a large proportion of the development projects. Improving their poor infrastructure was given priority by the central government. Other cities were given limited resources according to their industrial potential (Han and Wong 1995). Major industrial and urban centres were thus laid out in China.

3.3.1c The emphasis on steel output as the key link, 1958-60

During the Great Leap Forward (1958-60) steel production was perceived as the key link to accomplish developmental goals. The state owned commercial sectors were called upon to buy anything that could help to promote the output of steel, while banks gave unlimited loans to the iron and steel industries. In 1959 small refinery stoves mushroomed in both cities and countryside. In 1960, two-thirds of the 2000 counties and cities which had deposits of coal and/or iron ores, established refinery factories. These factories were known as either Xiao Tu Qun or Xiao Yang Qun. Xiao Tu Qun was a group of industries that used indigenous methods. Xiao Yang Qun was a group of industries that used foreign methods. The central government requested that every county and city that had deposits of coal or iron ore to set up at least one base of Xiao Tu Qun or Xiao Yang Qun. Xiao Yang Qun was preferred to Xiao Tu Qun because it used advanced technology and was more productive (Dang Dai Zhong Guo 1987). By 1960, the total number of workers and staff in twenty-one provinces was 18.2 million, of which 6.9 million was in Xiao Yang Qun, 3.18 million was in Xiao Tu Qun. This 18.2 million made up 55.2 percent of the national urban work force.

3.3.1d The reduction in the scale of industrial construction, 1961-63
The industrial leap forward caused shortages of investment, material, equipment and labour (Guo Jia Ji Wei and Guo Jia Jian Wei August 19, 1960). In response to these bottlenecks, a new policy was adopted which restricted the number and scale of new projects. It had four points. First, all proposed projects had to be below an investment ceiling. Second, all projects which exceeded this ceiling, but which had been started should be postponed, except those that could provide instant help to the increase of raw materials and agricultural products. Third, large scale constructions should reduce their scales into medium or small projects. Fourth, non-productive construction such as public buildings, cultural and welfare facilities should be stopped. Many industries that had been set up by rural and urban communes were discarded. The majority of factories operated by the county governments were closed down.

Table 3.4 shows the changes in the number of enterprises from 1957 to 1965. In 1959, the total number of enterprises reached its peak—there were 318,000 enterprises including 99,000 state owned and 219,000 collectives. From 1959 to 1965, the number of enterprises was reduced by half. By 1965, there were only 46,000 state owned enterprises and 112,000 collectives.

Table 3.4 Changes of the number of enterprises, 1957-1965 (1,000)

Year	State Owned	Collectives	Total
1957	58	112	170
1958	119	144	263
1959	99	219	318
1960	96	158	254
1961	71	146	217
1962	53	144	197
1963	47	123	170
1964	45	116	161
1965	46	112	158

Source: State Statistical Bureau 1982. p. 203.

3.3.1e The model of Daqing, 1964-77
This was the principal policy for the growth of urban industries in the

period 1964-77. Daqing was an oil field that was developed in 1964. The workers of Daqing used indigenous methods to overcome difficulties due to the lack of investment, equipment, and hard living conditions. The output of crude oil in Daqing roughly met China's domestic demand. During the era of international conflicts between China and the Soviet Bloc, and between China and the U.S., Chinese leaders viewed Daqing as the model of Chinese industrialisation, i.e., it increased output with little or no investment.

The Daqing model also satisfied the communist ideology of rural-urban integration. Family members of Daqing workers were organised to cultivate the farms in the oil field. Muddy houses in rows represented the principle of plain living. During the twenty years from 1960 to 1980, there were only 1,642,000 square meters of apartment buildings constructed in Daqing (Table 3.5). In 1964, Mao called out that "industries learn from Daqing". This slogan was displayed widely on the gates of Chinese factories in the 1960s and in the 1970s. Daqing's spirit and physical form, or the agropolitan appearance, became a model for other cities.

Table 3.5 Housing construction in Daqing, before and after 1980

Year	Building Area (sq. m)*	Housing Area per capita** (sq. m)
1980	1,642,000	4.01
1985	4,842,000	8.91

Source: The construction figures are calculated by the author. * Data based on The Planning Office for the Northeastern Region, The State Council, 1987. *The Economy of the Northeastern China*, Vol.1, pp. 377-90. Beijing: Zhong Guo Ji Hua. ** These figures are calculated on the basis of non-agricultural population (not including counties) of 1980 and 1985, respectively. The population data is from Guo Jia Tong Ji Juu, 1990a. *China: the Forty Years of Urban Development*, pp. 110-1.

A direct consequence of the Daqing Model is the reduction of investment on urban infrastructure. Table 3.6 shows the proportions of investment used for urban infrastructure. In 1952, 3.76 percent of the investment for capital construction was used for urban infrastructure, while in the mid-1970s, it accounted for only 1.09 percent of the capital construction investment. Up to the mid-1970s, the proportion of investment for urban infrastructure kept declining .

3.3.1f Using cities as engines of economic growth, 1978-89
Chinese leaders sought to stimulate national development by promoting the growth of cities in the 1980s. Cities were expected to function as economic organisers, as multi-functional centres, and as catalysts to open to the world.

The Chinese government believed that using cities as economic organisers would help to solve the problems of poor co-ordination between rural and urban sectors, and between "strips" and "areas".[2] This policy required the use of cities as growth centres to promote the development of their rural hinterland. Cities were expected to organise production and circulation for their local regions. The policy called for the formulation of economic zones/ regions that were of various size/scale and types (Zhao 1982).

Table 3.6 Proportion of investment for urban infrastructure, 1952-1985

Period	% of GDP	% of Investment of Fixed Assets	% of Capital Construction
1952	0.24	3.76	3.76
1st FYP	0.27	2.33	2.42
2nd FYP	0.40	1.96	2.13
1963-65	0.22	1.81	2.14
3rd FYP	0.14	1.09	1.35
4th FYP	0.15	0.85	1.09
5th FYP	0.29	1.61	2.19
6th FYP	0.61	3.40	5.31

Note: FYP: Five-Year Plan. The 1st FYP was in the period 1953-57; the 2nd FYP was in the period 1958-62; the 3rd FYP was in the period 1966-70; the 4th FYP was in the period 1971-75; the 5th FYP was in the period 1976-80; the 5th FYP was in the period 1981-85; the 6th FYP was in the period 1986-1990.
Source: Guo, G.Q et al. 1988. p. 265.

In order to implement this policy, city governments were asked to administer most of the industries and enterprises in their areas of jurisdiction. Urban enterprises that were administered by the ministries of the central government were expected to use local services in casting, thermal treatment, mechanical repairs and electroplating for production, and to use local commercial outlets, schools and housing for livelihood. In selected provinces, experiments were made to discard prefectures and to give cities the responsibility in supervising their tributary counties. Further, city governments were encouraged to stimulate the growth of their free markets, which were vehicles to organise the circulation of commodities. From 1979 to 1990, the number of free markets in cities increased from 2,226 to 13,106. The latter attracted peasants and generated intensive interactions between cities and their surrounding settlements.

---

2 *Here "strips" refer to the central government ministries and the hierarchies of administration under these ministries. "Areas" refer to jurisdictions of local governments.*

The economic structure of cities was altered by the growth of light industries and the growth in the tertiary sector. Light industries grew from 43.1 to 48.9 percent of the total industrial output value between 1978 and 1989. It was anticipated that the growth of light industries would activate the economy by providing a large variety of goods, benefit urban residents through providing more employment, and help China to communicate with the world market by increasing export.

Cities were used as catalysts of development to attract foreign investment. After four Special Economic Zones (viz., Shenzhen, Xiamen, Zhuhai and Shantou) were set up in 1980, fourteen coastal cities were given special status as Open Coastal Cities in 1984. In many large- and medium-sized cities development zones were also established and granted special privileges. Foreign investors were granted tax exemption, government subsidies, and made their own plans for labour hiring outside of government planning quotas. The host government had little or no intention of controlling how foreign investors conducted their businesses. Foreign investment was seen by the Chinese government as not only a possible solution to its shortage of capital, but also as a vehicle to import advanced technology to China.

3.3.2 Policies for Ideological Goals

3.3.2a Democratic reform, 1949-52

The policy of democratic reform called for the elimination of the existing organisations and the establishment of new administrations according to communist principles. Democratic reform uprooted the feudal labour contractors (*Ba Tou*) and established worker's commissions. The San Fan (three anti- movement) aimed at crushing graft, waste and bureaucracy. Urban governments were called to uproot counter-revolutionary, feudal organisations completely. New systems of administration were set-up accordingly (F.C. Li May 2, 1952).

3.3.2b Reshaping the ownership structure of industries and commerce, 1953-57

This policy was one of the three tasks of the First Five Year Plan (1953-57). The reconstruction of capitalist industries used national capitalism as a way to control the state supply of materials and the marketing of outputs. National capitalism had three forms: the beginning, the intermediate and the advanced. In the beginning form, private industries were forced to sell their finished goods to the state. The latter then distributed these goods to the market. In the intermediate form, private industries signed contracts with the state to produce goods. The state supplied the raw materials and bought the outputs in return. In the advanced form, co-management was practised between the private and state sectors. Co-management enhanced state control over production and labour-management relations (*People's Daily* November 11, 1953). By the end of the transition, around 1956, private industrialists were forced to give up their factories.

In the commercial sector, the transition began with the owners of wholesale businesses. In 1950, the government started to Tong Go Bao Xiao (i.e., monopoly and monopsony by the state) on some important agricultural products such as grain, cooking oil and cotton. *Tong Go Bao Xiao* forced private businesses in the wholesale sector out of the economy. Private retailers were converted to agents of the state, to operate as commercial outlets. A large number of stores of the same type of goods were merged (e.g., those selling medical equipment and printing machines were grouped into one category, and those selling stationery, clocks and eyeglasses were grouped into another category). As a consequence, the number of commercial outlets kept on shrinking in the period of ownership transition (Table 3.7).

Table 3.7 Total number of commercial outlets in selected cities, 1952-1957

Cities	1952	1957
Shenyang	28119	20865
Jinan	35031	16000
Shijiazhuang	11093	4627
Anyang	3732	1636
Linfen	598	334

Note: Commercial outlets in suburban counties were excluded.
Source: Guo Jia Tong Ji Juu, 1990a, pp. 292-3.

Handicraft workshops made up the majority of industries in the small cities. They were important suppliers of tools of production and articles for daily use for the countryside (*People's Daily* April 14, 1954). The reconstruction of the handicraft industry was to organise the individual producers into semi-collectives and then into collectives. The former was organised on the basis of private ownership, i.e., members of a semi-collective owned their tools individually but worked together so that the production and marketing could be planned. In collectives, members gave up their private ownership of tools to set up collective ownership. By the end of 1956, the total number of the members of collectives reached 5.3 million, which made up 92 percent of the total number of handicraft workers. Collectives in the handicraft industry reached 100,000 in number.

3.3.2c Yi Da Er Gong: Rushing toward pure state ownership, 1958-60

This policy was a product of the Great Leap Forward, when the Chinese communist leaders called for a rush toward communism. Communes were established that contained the population of several collectives and included industries, commerce, agriculture, schools and military. Thus, they were big in size and dominated by the state (*Yi Da Er Gong* literally means: first, big and second, state-ownership). Communes not only were in control of production and livelihood of peasants, but also represented the lowest level of the government. For Mao, communes exemplified the proper form of the socialist China and thus he pushed the establishment of communes nationwide (Zhao 1989a).

Yi Da Er Gong applied to cities in two aspects. First, individual producers and industrial collectives were pushed toward state-ownership. The Chinese Communist Party called to increase the intensity of control and to restructure individual ownerships. Individual producers had to join collectives. At the same time, industrial and commercial collectives were converted into state enterprises, i.e., *Zhuan Chang Guo Du. Zhuan Chang* literally meant the change of industries (in ownership). *Guo Du* referred to the transformation of the Chinese society (toward communism). Second, urban communes were used as organisational forms of *Yi Da Er Gong*. It was expected by the end of 1960 that urban communes would replace all cities except Beijing, Shanghai, Tianjin, Wuhan and Guangzhou (Zhao 1989a).

3.3.2d Reinforcing the responsibility system in urban industries, 1961-65

Yi Da Er Gong undermined the initiatives of the workers through egalitarian distribution. Ignorance over regulations in industrial production caused an increase in accidents and lowered the quality of output. In 1961, the government requested the revolutionaries (i.e., red guards and mass organisations) to establish new systems before smashing the old ones (*Xian Li Ho Pe*). Workers were requested to follow regulations, such as those related to safety, quality checking and material administration, among others (*People's Daily*. February 22, 1961). At the same time, the responsibility system was stressed. Wages, bonus, and promotions were determined on the basis of personal performance, and the performance of the production line or team (*People's Daily* December 17, 1961). As Table 3.8 shows, the social productivity and the industrial productivity of labour recovered from their low points of 1961.

3.3.2e Ji Zuo: Pure state ownership and deregulation, 1966-77

The term Ji Zuo referred to thoughts which hastened the advent of communism. During the ten years of the Great Cultural Revolution, Chinese leaders took short cuts to reach sooner the long term goals of communism. Private producers, market mechanisms were viewed as capitalistic in nature and were contained or discarded. As a result, egalitarianism and the "iron rice bowl" (Tie Fan Wan) became common in the economy. Egalitarian distribution among the enterprises undermined the development potential of the better enterprises, and egalitarian distribution among workers within the enterprises undermined their incentives.

Using the same productivity indicators, it is clear that Ji Zuo policy resulted in slow growth of the economy . In six of the ten years industrial productivity actually declined.

Table 3.8 Changes of productivity in China, 1960-1976

Year	Changes of Social Productivity over Previous Year	Changes of Industrial Productivity over Previous Year
1960	-0.1	34.5
1961	-28.9	-34.3
1962	-6.6	12.0
1963	8.5	32.5
1964	12.6	25.1
1965	12.8	20.3
1966	12.8	16.1
1967	-10.5	-20.0
1968	-9.7	-11.7
1969	14.9	25.6
1970	18.7	19.5
1971	3.4	-2.9
1972	0.8	-4.3
1973	6.8	1.6
1974	-0.9	-5.8
1975	6.1	6.8
1976	-4.5	-12.2

3.3.2f Reforming the structure of ownership and the state enterprise relations, 1978-89

The strategy of reform and openness signalled the green light for the growth of collectives and of private ownership, and relaxed the strict control of enterprises by the state.

There were two reasons for the revival of private ownership and collectives. First, most articles of daily life were produced by private producers and collectives[3] Recognising their importance, Chinese leaders perceived that private and collective operations were complementary to the state sector, and should be allowed to grow (*People's Daily* July 19, 1979; August 4th, 1979; January 9, 1983). Second, the growth of private businesses helped the state to employ the urban youths who came back to cities after the policy of *Shang Shan Xia Xiang* was discarded.

The intention to increase the *Zi Zhu Quan* (self-determining power) of urban enterprises was announced in July 1979, when the State Council endorsed five new regulatory documents.[4] These documents allowed enterprises to share revenue that exceeded the revenue quota, to hire workers according to indices that were made by state labour planning, to decide the institutional structure of the enterprise and to select the middle and upper middle level administrative staff, and to sell or rent the spare fixed assets (Zhao 1989b). In July 1980, the State Council decided to enlarge the *Zi Zhu Quan* of enterprises in production planning, pricing and marketing, and to give enterprises the power to determine administrative structure and personnel arrangements (Guo Wu Yuan 1980). It was frequently asserted that enlargement of the *Zi Zhu Quan* of enterprises was the central focus of economic reform (Zhong Gong Zhong Yang, 1984).

3.4 Rural Reform

In the countryside, the book of *Hu kou* (household registration) was not given to individual households. A village had one household registration book that contained all villagers (Guo Jia Tong Ji Juu 1988). Peasant travelling outside to visit relatives or to do business had to carry introductory letters that carried the stamp of the villager's commission.[5] The policy of rural reform completely changed this by returning the basic accounting unit to house-

---

3 *For example, see* People's Daily *July 17, 1978. It is reported that Wuhu had reduced more than 400 types of commodities as a result of the lack of collectives.*
4 *These documents were: (1) Some Regulations to Enlarge the Management Zi Zhu Quan of State-owned Industries and Enterprises; (2) Some Regulations on Li Run Liu Cheng of State-owned Enterprises; (3) Temporary Regulations on Collecting Tax on Fixed Assets of Industries and Enterprises; (4) Regulations to Raise the Zhe Jiu Luu of Fixed Assets of State-owned Industries and Enterprises and to Improve the Use of Zhe Jiu Fee; (5) Temporary Regulations to Introduce Mortgage of Floating Capital in State-owned Industries and Enterprises.*
5 *This was similar to the collective registration book for urban workers who were assigned to cities where none of their family members lived.*

hold size. The idea behind this was to discard egalitarianism that had hindered the increase of rural productivity. The responsibility system was the major means to carry out this change. By 1981, 90 percent of the production teams introduced the responsibility system (Zhong Gong Zhong Yang 1982). In 1983, the government was called on to dismantle the rural communes (Zhong Gong Zhong Yang 1983). Experiments to dismantle rural communes were made in 69 counties (Zheng and others 1987). By mid-1985, all the communes were dismissed. Instead, town governments were set up in the place of the previous communes. The 56,000 communes were replaced by 92,000 town and township governments (Zheng and others 1987).

Rural reform uncovered the problem of surplus labour that was previously disguised by the commune system in the agricultural sector. The destruction of the communes and the encouragement to diversify agricultural production increased the output value of the agricultural sector and made farmers less dependent on the land. Many farmers shifted to non-agricultural employment, such as selling farm products in cities. In 1985, the central government carried out a policy which legitimised the migration of some peasant businessmen to cities. Many city governments responded promptly to this policy by changing the registration of these peasants. With the stopping of control over rural-urban migration at the commune level came the rise of a large "floating population" in the cities. The "floating population" that was made up of peasants accounted for 20 percent of the population of large cities, such as Beijing, Shanghai, Tianjin, and Guangzhou (Table 3.9). Those suburban districts where cheap housing was available became overwhelmingly occupied by peasant migrants. Peasant migrants lived in these areas and

Table 3.9 Floating population in Beijing, Shanghai, Tianjin and Guangzhou (1,000)

Year	Beijing	Shanghai	Tianjin	Guangzhou
1979				234
1980				306
1984	300	1,020		500
1985	600			
1986	900	1,830		800
1987	1,150			1,145
1988	1,310	2,091	1,129	1,170

Note: The floating population refers to those population that lived in a city without registration of residence in that city.

Source: The data is collected by interviewing key informants.

used them as bases for their informal employment in cities (i.e., garbage collecting and shoe repairing).

3.5 Policies that Guide the Regional Location of Industries

This group of policies includes two types: those emphasising inland regions and those emphasising the coastal region. The inland regions include the western region and the central region of the Sixth Five Year Plan (1980-85), while the coastal region is also known as the eastern region.

3.5.1 Policies Emphasising Inland Regions

This policy had been applied for almost thirty years before 1978. Its purpose was to balance the distribution of industries and to reduce the unevenness of development among Chinese regions. The Chinese government believed that the uneven distribution of production forces was against Marxist principles and was the cause of social tensions between Chinese and other ethnic groups. It was thought that the uneven distribution of industries was inefficient in a planned economy.

In order to implement this policy, new industries were mainly allocated to inland regions. During the 1950s, only five of the 156 projects that were aided by the Soviet Union were allocated in the East Region. The reason was that the East Region was the forefront exposed to the Western world. In

Table 3.10 Allocation of investment for capital construction, 1953-1990

Period	Coastal Region	Inland Region	Western Region
1st FYP (1953-57)	44.1	34.3	21.6
2nd FYP(1958-62)	40.8	35.9	23.3
1963-65	37.4	35.1	27.5
3rd FYP(1966-70)	29.4	31.6	38.0
4th FYP(1971-75)	39.4	33.3	27.3
5th FYP(1976-80)	45.7	32.7	21.6
6th FYP(1981-85)	50.5	31.2	18.3*
7th FYP(1986-90)	58.1	25.6	15.9**

Source: * The figures are calculated by the author, based on raw data put out by The Statistical Bureau 1987, pp. 50-1.
** Guo Jia Ji Wei Tou Zi Yan Juu Suo, *Zhong Guo Tou Zi Bao Gao 1992* (Beijing: Zhong Guo Ji Hua, 1993), 26.

contrast, 56 projects were allocated to the Northeast Region where China's traditional industries concentrated. This region was close to the former Soviet Union and was regarded as the safe base of socialism. The Northwest and the North regions were also major recipients of industrial projects (Table 3.10). These two regions hoped to establish major inland industrial bases. Baotou in Neimenggu was one city which was the focus of the establishment of iron and steel industries in the North Region in the 1950s. Among the 694 projects that were designed by Chinese, 472 were in the inland, which made up 68 percent, while only 32 percent were along the coast (Lu 1989).

The emphasis of investment on inland regions made the government overlook coastal industries. Mao Zedong in his speech on the *Ten Great Relations,* in 1957, pointed out that investment for the coastal industries had to be increased as the coastal industries were much stronger than new industrial bases in inland regions. However, these comments had little effect on the allocation of investment. As Table 3.11 indicates, investment in the coastal region kept on dropping until the period of the Fourth Five-Year Plan (1971-75). Even in the 1970s, investment in the coastal region did not exceed 50 percent of the total investment, despite the fact that the industrial output of coastal industries was far beyond 50 percent of the total.

Since the mid-1960s, the policy to emphasise inland regions in industrial location was further focused on remote, mountainous regions known as Third Lines. The Third Line division of China was a strategic grouping of the provinces which resulted from two major threats to the Chinese communists, namely, the threat from the USSR and the threats from the U.S.A. Provinces along the coast and along the northern border made up the front line that could be easily hit in warfare. The Third-Lines, including *Da San Xian* (in the central region) and Xiao San Xian (in remote locations of provinces) were safe for building industries.

The construction of Third Lines continued until 1977. Since 1972, more investment shifted back towards the following the re-establishment of diplomatic relations between China and the United States.

3.5.2 Policies Emphasising the Coastal Region

In the 1980s, China applied an uneven growth strategy which emphasised growth in the coastal region. The coastal region was selected to experiment with policies of reform and openness, and to get resources from the central government finance because the industries there were stronger, the labourers were more skilful, and the urban infrastructure was better than in inland regions. It was hoped that growth would spread from the coast to the inland regions, and would change unbalanced to balanced growth with higher economic status (Zhong Gong Zhong Yang 1985a).

New tax and other privileges were introduced in the four Special Economic Zones (viz., Shenzhen, Zhuhai, Xiamen and Shantou), the fourteen Open Coastal Cities (viz., Dalian, Qinhuangdao, Tianjin, Yantai, Qingdao, Lianyungang, Nantong, Shanghai, Ningbo, Wenzhou, Fuzhou, Guangzhou, Zhanjiang, Beihai), the two open peninsulas (viz., Liaodong and Jiaodong), two open river deltas (Yangtze River Delta, Zhujiang River Delta), one triangular area (viz., the triangular area formed by Zhangzhou, Quanzhou and Xiamen) and numerous development zones in the coastal provinces (Map 1). By 1989, almost all the counties/cities that were along the coast were given special privileges to allow tax exemption and to relax government controls.

The investment of capital construction shifted further to the coastal region. During the Sixth Five-Year Plan period, more than 50 percent of the total investment on capital construction was allocated to the coastal region. In the late 1980s, the percentage of investment for capital construction approached 60 percent in the coastal region.

3.6 Summary and Conclusions

Since the Chinese communists took over in 1949, the development of Chinese cities has been guided by a number of deliberately designed policies and strategies. Although the influences of these policies and strategies on Chinese urbanisation are widely discernible in the literature, a systematic understanding of the urbanisation policies is still lacking. The demand for such a knowledge, however, is escalating during the past fifteen years as China opens herself to the world market and determines to achieve rapid economic growth. Scholars, individual businesses and institutions inside and outside China, as well as the Chinese government are all enthusiastic about the experiences, lessons and characteristics of the evolution of cities under Chinese socialism.

This chapter has contributed to the understanding of Chinese urbanisation studies by presenting systematically a detailed account of Chinese urbanisation policies. Based on personal interviews and analyses of major government policy documents, four groups of policies used by the Chinese government to control urbanisation were identified. These four groups of policies included those that were designed to control the population flow between cities and the countryside, to guide the development of cities, to guide the development of the countryside, and to guide regional industrial location. They were all deeply rooted in the Chinese ideology of the rural-urban relations and were used to implement the experimental strategies for national development. They were interlocked, first as a reaction to previous policies in time and a response to the strict control of urban-ward migration, and second, as a mutual reinforcement in facilitating the development of the Chinese economy.

Chinese urbanisation policies worked together as a means of controlling Chinese urban development.[6] They acted like a check-and-balance system. While one set of urban policies attracted migrants to the cities, another set pushed them away from cities. Rural collectivisation assisted the organisation of the rural labour force through collective household registration. Regional development policies directed the flow of urban population among regions and provinces through industrial allocation. Together these push-and-pull forces regulated the in-and-out migration of people that shaped the process and pace of Chinese urban growth and development.

While many of the urban policies were changed, two policy intentions were continued over the study period. The first was to control the growth of large cities and to encourage the growth of small cities. The second was to control the mobility of peasants and to keep them in their hometowns. At this point, the continuity of communist ideology is evident. Rural-urban integration, that is rooted in the idea of spatial and aspatial equities, continues to be the goal of China's development. The means to accomplish this goal are reflected either in the hindrance of the growth of large cities (if not a lip service in policies) or in the promotion of the growth of small cities.

As government control on social and economic affairs are further relaxed, possibilities will probably emerge to make way for a reconstruction of the socio-political system. Ideas of equity and efficiency will continue to generate debate in Chinese national development and urbanisation policies. Ad hoc policies may be used to solve problems and to achieve goals in the short term. However, in the long run, the principles of equity and efficiency must be better co-ordinated and cohesively defined, as effective urbanisation policies will continue to be needed as the growth of China's urban population accelerates.

---

6 *The impact of these policies on Chinese urbanisation is tested by using multiple regression analysis. The results reveal variations in policy influences but generally support the idea that urbanisation policies had a direct bearing on the growth of Chinese cities (Han 1994).*

References

Buck, David D. *1981. Policies Favouring the Growth of Smaller Urban Places in the People's Republic of China, 1949-1979. In* Urban Development in Modern China, *edited by L.J.C. Ma and W. Hanten, pp.114-44. Boulder, Col.: Westview Press.*

Chan, K.W. *1989.* Economic Growth Strategy and Urbanisation Policies in China, 1949-1982, *Research Paper No. 175. Toronto: Centre for Urban and Community Studies, University of Toronto.*

_____. *1992. Post-1949 Urbanisation Trends and Policies: an overview. In* Urbanizing China, *edited by G.E. Guldin, pp. 41-63. New York: Greenwood Press.*

_____. *1994.* Cities With Invisible Walls. *Hong Kong: Oxford University Press.*

Cheng, X. *1990. Problems of Urbanisation under China's Traditional Economic System. In* Chinese Urban Reform: What Model Now? *edited by R.Y.W. Kwok, et al. pp. 65-77 London: M.E. Sharpe, Inc.*

Dang Dai Zhong Guo Bian Ji Bu *(Editorial Board, Contemporary China). 1985.* Dang Dai Zhong Guo De Jing Ji Guan Li *(Economic Administration in Contemporary China). Beijing: Zhong Guo She Hui Ke Xue.*

_____. *1987* Dang Dai Zhong Guo De Ji Ben Jian She *(Shang) (Capital Construction in Contemporary China). Beijing: Zhong Guo Shehui Kexue.*

_____. *1990.* Dang Dai Zhong Guo De Cheng Shi Jian She *(Urban Construction in Contemporary China). Beijing: Zhong Guo She Hui Ke Xue.*

Dang Dai Zhong Guo De Ji Hua Gong Zuo Ban Gong Shi *(Office of Planning, the Editorial Group of Contemporary China). 1987* Zhong Hua Ren Min Gong He Guo Guo Min Jin Ji He She Hui Fa Zhan Ji Hua Da Shi Ji Yao *(Major Events in the National Economy and Social Development, People's Republic of China) (Abbreviated as Da Shi Ji Yao). Beijing: Hongqi.*

Farina, M.B. *1980. Urbanisation, De-urbanisation and Class Struggle in China, 1949-79. In* International Journal of Urban and Regional Research *4: 487-501.*

Fei, X.T. *1985.* Xiao Cheng Zhen Si Ji *(Four Essays on Small Cities and Towns). Beijing: Xinhua.*

Ginsburg, N. *1993. Review Essay on China's Coastal Cities: Catalysts for Modernisation.* Annals of the Association of American Geographers *83: 558-60.*

Guo Jia Ji Wei *(Planning Commission of the State Council), (1961). 1987. Guan Yu Chong Xin Tiao Zheng Ji Ben Jian She De Tong Zhi (Notice about Readjustment of Capital Construction). In* Da Shi Ji Yao, Dang Dai Zhong Guo, *p.171. Beijing: Hongqi.*

_____, and Guo Jia Jian Wei *(Construction Commission of the State Council), (1960). 1984. Guan Yu Suo Duan Ji Ben Jian She Zhan Xian Bao Zheng Sheng Chan De Cuo Shi (Measurements to Reduce the Scale of Capital Construction in order to Facilitate Production)." In* Da Shi Ji Yao, Dang Dai Zhong Guo, *pp.155-4. Beijing: Hongqi.*

_____, Tou Zi Yan Juu Suo *(Research Institute of Investment). 1993.* Zhong Guo Tou Zi Bao Gao (Investment Report, China) 1992. *Beijing: Zhong Guo Jihua.*

Guo Jia Lao Dong Zong Juu *(The State Bureau of Labour), Jian She Zong Juu (The State Bureau of Construction), Gong An Juu (Public Security Bureau) and Gong Shang Guan Li Juu (Administrative Bureau of Industries and Commerce), (1981). 1987.. Guan Yu Jie Jue Fa Zhan Cheng Zhen Ji Ti Jing Ji He Ge Ti Jing Ji Suo Xu Chang Di Wen Ti De Tong Zhi (Notice on the Provision of Land and Space for the Growth of Collective and private Businesses in Cities and towns). In* Zhong Guo Jing Ji Guan Li Fa Gui Wen Jian Hui Bian *(shang), Zhong Guo She Ke Yuan Fa Xue Yan Jiu Suo, 382. Beijing: Faluu.*

Guo Jia Tong Ji Juu *(State Statistical Bureau). 1982.* Zhong Guo Tong Ji Nian Jian (China Annual Statistics) 1982, *Beijing: Zhong Guo Tong Ji.*

_____. *1985.* Zhong Guo Tong Ji Nian Jian (China Annual Statistics) 1985, *Beijing: Zhong*

Guo Tong Ji.
_____. 1987b. Zhong Guo Tong Ji Nian Jian (China Annual Statistics) 1987, *Beijing: Zhong Guo Tong Ji.*
_____. 1987c. Zhong Guo Tong Ji Zhai Yao (A Statistical Survey of China) 1987, *Beijing: Zhong Guo Tong Ji.*
_____. 1988e. Zhong Guo Cheng Shi Tong Ji Nian Jian (China Annual Urban Statistics) 1988, *Beijing: Zhong Guo Tong Ji.*
_____. 1990a. China: the Forty Years of Urban Development. *Beijing: Zhong Guo Tong Ji.*
_____. 1991b. Zhong Guo Tong Ji Nian Jian (China Annual Statistics) 1991, *Beijing: Zhong Guo Tong Ji.*
_____. 1992b. Zhong Guo Tong Ji Nian Jian (China Annual Statistics) 1992, *Beijing: Zhong Guo Tong Ji.*
_____. 1993c. Zhong Guo Tong Ji Zhai Yao (A Statistical Survey of China) 1993, *Beijing: Zhong Guo Tong Ji.*
Guo Wu Yuan (The State Council), (1953), 1989. Guan Yu Quan Zhi Nong Min Mang Mu Liu Ru Cheng Shi De Zhi Shi (Instructions on Convincing Peasants not to Migrant to Cities Blindly). *In* Da Shi Ji, edited by D.Q. Zhao, p.409. *Kaifeng: Henan Renmin.*
_____. (1956a), 1989a. Guan Yu Fang Zhi Nong Min Mang Mu Wai Liu De Zhi Shi (Instructions on Preventing Peasant Emigration from Their Hometowns). *In* Da Shi Ji, edited by D.Q. Zhao, pp. 418-9. *Kaifeng: Henan Renmin.*
_____. (1957), 1989a. Guan Yu Fang Zhi Nong Min Mang Mu Liu Ru Cheng Shi De Tong Zhi (Notice on Preventing Blind Peasant-migration toward Cities). *In* Da Shi Ji, edited by D.Q. Zhao, p.940. *Kaifeng: Henan Renmin.*
_____. (1980), 1987. Guan Yu Kuo Da Qi Ye Zi Zhu Quan Shi Dian Gong Zuo Qing Kuang He Jin Hou Yi Jian De Bao Gao (Report on the Experiment of Enlarging the Self-determining Power of Enterprises and on the suggestions for Further Progress). *In* Zhong Guo Jing Ji Guan Li Fa Gui Wen Jian Hui Bian (Shang), edited by Zhong Guo She Ke Yuan Fa Xue Yan Jiu Suo, p.160. *Beijing: Faluu.*
_____. (1984b), 1984. Guan Yu Nong Min Jin Ru Ji Zhen Luo Hu Wen Ti De Tong Zhi (Notice about the Registration of Peasants Entering Market Towns). *In* Fa Gui Hui Bian (1984, 1-12), edited by Guo Wu Yuan Fa Zhi Juu, pp. 88-9. *Beijing: Faluu.*
_____. (1984d), 1983. Guan Yu Jin Yi Bu Kuo Da Guo Ying Qi Ye Zi Zhu Quan De Zan Xing Gui Ding (Temporary Regulations for further Enlarging the Self-determining Power of State-owned Enterprises). *In* Jian Chi Gai Ge, Kai Fang, Gao Huo, edited by Zhong Gong Zhong Yang Shu Ji Chu, pp. 207-10. *Beijing: Renmin.*
Guo, G.Q. et al. 1988. Cheng Shi Hua Yu Cheng Shi Ji Chu She Shi Jian She (Urbanisation and the Construction of Urban Infrastructure). In *Zhong Guo Cheng Shi Hua Dao Lu Chu Tan,* edited by W.J Ye, et al. 254-94. Beijing: Zhong Guo Zhan Wang.
Han, S.S. 1994. Controlled Urbanisation in China, 1949-1989. *Ph.D. dissertation. Department of Geography, Simon Fraser University.*
_____ and S.T. Wong. 1994. The Influence of Chinese Reform and Pre-Reform Policies on Urban Growth in the 1980s. Urban Geography 15: 537-564.
_____ and S.T. Wong, 1995. Revolution, Space and Urbanisation: the ideological foundations of Chinese urban develpment. (mimeo).
Hsu, M.L. 1994. The Expansion of the Chinese Urban System. Urban Geography, 15: 512-534.
Jie Fang Ri Bao. December 26, 1954.
Kwok, R.Y.W., William L. Parish, A.G.O. Yeh with X.Q. Xu, eds. 1990. Chinese Urban Reform: What Model Now?. *London: M.E. Sharpe, Inc.*
Li, F.C. (1952), 1989a. Zai Quan Guo Lao Mo Hui Shang De Jiang Hua (Speach in the Conference of National Models of Labour)." *In* Da Shi Ji, edited by D.Q. Zhao, 28. *Kaifeng: Henan Renmin.*

Liu, Shaoqi 1949. Xin Min Zhu Zhu Yi De Cheng Shi Zheng Ce *(*Urban Policies in Democratic China*). Tianjin: Shu Dian.*

Lu, D.D. 1989. Zhong Guo Gong Ye Bu Juu *(*Industrial Location in China*). Beijing: She Ke Yuan.*

Ma, L.J.C. 1974. Urbanism in China. Proceedings of Regional Science Review 8: 114-18.

Mallee, H. 1995. *Chinese Household Registration System under Reform.* Development and Change. 26: 1-29.

Mao, Z.D. (1955), 1977. Zhong Guo Nong Cun She Hui Zhu Yi Gao Chao De An Yu *(Notes on the Peak of Socialist Movement in Chinese Countryside).* Mao Xuan Vol.5, edited by Zhong Gong Zhong Yang Wen Xian Bian Ji Shi, pp. 253-4. Beijing: Renmin.

_____. (1962). "Untitled speech." Quoted from Yu and Sun, (1982): 64.

Nan Fang Ri Bao. December 30, 1955.

Nei Wu Bu *(Ministry of Internal Affairs) and Lao Dong Bu (Ministry of Labour), (1954), 1989a.* Guan Yu Ji Xu Quan Zhi Nong Min Mang Mu Liu Ru Cheng Shi De Zhi Shi *(Instructions for Continuous Efforts to Convince Peasants not to Migrate to Cities). In* Da Shi Ji, *edited by* D.Q. Zhao, p.410. Kaifeng: Henan Renmin.

Pannell, C. and Ma, L.J.C. eds. 1983. China: The Geography of Development and Modernisation. *London:V.H. Winston & Sons.*

Parish, William L. 1987. *Urban Policy in Centralised Economies: China. In* The Economics of Urbanisation and Urban Policies in Developing Countries, *edited by George S. Tolley and Vinod Thomas, pp. 73-84. Washington, D.C.: The World Bank.*

People's Daily. February 6, 1950.

_____. July 15, 1951.

_____. April 20, 1953.

_____. November 11, 1953.

_____. April 14, 1954.

_____. February 22, 1961.

_____. December 17, 1961.

_____. February 16, 1964.

_____. July 9, 1967.

_____. December 22, 1968.

_____. December 22, 1975.

_____. July 17, 1978.

_____. July 19, 1979.

_____. August 4th, 1979.

_____. January 9, 1983.

Ran, M.X. and Berry, B.J.L. 1989. Under-urbanisation policies Assessed: China 1949-84. Urban Geography, 10: 111-9.

Salaff, J. 1967. The Urban Communes and Anti-city Experiment in Communist China. China Quarterly 29: 82-110.

Salter, C.L. 1974. Chinese Experiments in Urban Space: the quest for an agro-politan China. Habitat 1: 19-35.

The State Council 1987. The Economy of the Northeastern China, Vol. 1. Beijing: Zhong Guo Ji Hua.

Steiner, H.Arthur 1950. Chinese Communist Urban Policy. American Political Science Review 44: 47-63.

Whyte, M.K. and Parish, W.L. 1984. Urban Life in Contemporary China. Chicago: University of Chicago.

Yeung, Y.M. and X.W. Hu, eds. 1992. China's Coastal Cities. Honolulu, Hawaii: University of Hawaii Press.

Zhao, D.Q. 1989a. Zhong Hua Ren Min Gong He Guo Jing Ji Zhuan Ti Da Shi Ji (Important

Economic Events in the People's Republic of China) 1949-66 *(Abbreviated as* Da Shi Ji).
Kaifeng: Henan Renmin.
Zhao, D.Q. 1989b. Zhong Hua Ren Min Gong He Guo Jing Ji Zhuan Ti Da Shi Ji (Important
Economic Events in the People's Republic of China) 1967-84 *(Abbreviated as* Da Shi Ji).
Kaifeng: Henan Renmin.
Zhao, Z.Y. (1982), 1987. *Ji Ji Wen Tuo Di Jia Kua Jing Ji Ti Zhi Gai Ge De Jin Cheng (Speed
up the Process of Economic Reform Vigorously and Safely). In* Jian Chi Gai Ge, Kai Fang,
Gao Huo. *Zhong Gong Zhong Yang Shu Ji Chu, pp.165-64. Beijing: Renmin.*
Zheng, D.R., M.X. Han and X.L. Zheng 1987. Zhong Guo Jing Ji Ti Zhi Gai Ge Ji Shi (An
Empirical Record of Chinese Urban Reform). Beijing: Chun Qiu.
Zhong Gong Zhong Yang *(The Central Committee of the Chinese Communist Party), (1951),
1989a. Guan Yu Nong Ye Sheng Chan Hu Zhu He Zuo De Jue Yi (Cao An) (Decisions to
Promote Mutual Help in Agricultural Production). In* Da Shi Ji, edited by D.Q. Zhao, pp.
261-2. *Kaifeng: Henan Renmin.*
_____. (1957), 1989a. Guan Yu Gai Jin Gong Ye Guan Li Ti Zhi De Gui Ding (Cao An)
(Decisions to Improve Industrial Administration). In* Da Shi Ji, edited by D.Q. Zhao, p.900.
Kaifeng: Henan Renmin.
_____. (1958a), 1958. Guan Yu Zai Nong Cun Jian Li Ren Min Gong She Wen Ti De Jue Yi
(Decision to Establish People's Communes in the Countryside). In* Fa Gui Hui Bian (1958,
7-12), Fa Zhi Juu, pp.1-5. Beijing: Faluu.
_____. (1961a), 1989a. Guan Yu Jian Shao Cheng Zhen Ren Kou He Ya Suo Cheng Zhen
Liang Shi Xiao Liang De Jiu Tiao Ban Fa (Nine Measurements to Reduce Urban
Population and to Reduce the Volume of Sale of Grain). In* Da Shi Ji, edited by D.Q. Zhao,
p.944. *Kaifeng: Henan Renmin.*
_____. (1961b), 1989a. Guan Yu Tiao Zheng Guan Li Ti Zhi De Ruo Gan Gui Ding
(Regulations of Adjusting the Administrative System). In* Da Shi Ji, edited by D.Q. Zhao,
918. *Kaifeng: Henan Renmin.*
_____. (1981a), 1987. Guan Yu Ji Ji Fa Zhan Nong Cun Duo Zhong Jing Ying De Bao Gao
(Report on Active Development of Diversified Economies in the Countryside). In* Jian Chi
Gai Ge, Kai Fang, Gao Huo, Zhong Gong Zhong Yang Shu Ji Chu, pp. 77-81. Beijing:
Renmin.
_____. (1982), 1987. Quan Guo Nong Cun Gong Zuo Hui Yi Ji Yao (Notes from National
Conference of Agricultural Development). In* Jian Chi Gai Ge, Kai Fang, Gao Huo, edited
by Zhong Gong Zhong Yang Shu Ji Chu, pp. 129-37 Beijing: Renmin.
_____. (1983), 1987. Guan Yu Dang Qian Nong Cun Jing Ji Zheng Ce De Yi Xie Wen Ti
(Issues Related to the Contemporary Economic Policies in the Countryside). In* Jian Chi
Gai Ge, Kai Fang, Gao Huo, Zhong Gong Zhong Yang Shu Ji Chu, pp. 169-84. Beijing:
Renmin.
_____. (1984), 1987. Zhong Gong Zhong Yang Guan Yu Jing Ji Ti Zhi Gai Ge De Jue Ding
(Decision on Economic Reform)." In* Jian Chi Gai Ge, Kai Fang, Gao Huo, Zhong Gong
Zhong Yang Shu Ji Chu, pp. 229-54. Beijing: Renmin.
_____. (1985a), 1984. Guan Yu Zhi Ding Guo Min Jing Ji He She Hui Fa Zhan Di Qi Ge
Wu Nian Ji Hua De Jian Yi (Suggestions for Formulating the Seventh Five Year Plan). In*
Shi Yi Jie San Zhong Quan Hui Yi Lai Jing Ji Zheng Ce Wen Xian Xuan Bian, Zhong Guo
She Hui Ke Xue Yuan Gong Ye Jing Ji Suo, pp. 366-7. Beijing: Zhong Guo Jing Ji.
_____. (1985b), 1987. Zhong Gong Zhong Yang Guan Yu Jin Yi Bu Huo Yue Nong Cun
Jing Ji De Shi Xiang Zheng Ce (Ten Policies to Activate the Economy in the Countryside).
In* Gao Huo, edited by Zhong Gong Zhong Yang Shu Ji Chu, pp. 269-277. Beijing: Renmin.
_____ and Guo Wu Yuan (The Central Committee of the Chinese Communist Party and the
State Council), (1963), 1990. Guan Yu Tiao Zheng Shi Zhen Jian Zhi, Suo Xiao Cheng Shi
Jiao Qu De Zhi Shi (Instructions to Adjust the Setup of Cities and Towns and to Reduce the*

Size of Suburbs of Cities). In Dang Dai Zhong Guo De Cheng Shi Jian She, *Dang Dai Zhong Guo, pp. 83-4. Beijing: Zhong Guo She Hui Ke Xue.*

_____ *and* _____. *(1985), 1987. Guan Yu Jin Yi Bu Huo Yue Nong Cun Jing Ji De Shi Xiang Zheng Ce (Ten Policies to Further Activate Agricultural Economy). In* Gao Huo, *Zhong Gong Zhong Yang Shu Ji Chu, 269-73. Beijing: Renmin.*

4 Regional Development in the Yangtze and the Pearl River Delta Regions[1]

Roger C.K.Chan

Editors' Note:
The two most economically developed sub-regions of China are the Yangtze and Pearl River Delta regions. This paper brings forth the nature and characteristics of economic development in both delta regions. Though the development of the Yangtze Delta region, based on Shanghai, has come later, it has the best potential for developing as an industrial hub of East Asia. The Pearl River Delta development, which started earlier than Shanghai, has two foci- one in Guangzhou (Canton) and the other in Hong Kong. Both the Yangtze and the Pearl River regions have attracted foreign investments, both have local Chinese entrepreneurs, and both have port locations. Shanghai, though, has the advantage of river transport further into the interior of China. On the other hand, the Pearl River Delta region is integrating with established internationalist Hong Kong, which is still governed by a more liberal political and economic framework under the philosophy of 'one country – two systems'.

4.1 Introduction

Conventional theories of regional development do not properly take into consideration factors that influence contemporary regional development in less industrialised countries, especially those in Asia. Most theories of regional development have their roots in ideas developed before the 1970s, based almost exclusively on the experience of western industrialised states (Rostow, 1956). Up to the 1970s, there was the acceptance of a unified theory of development, represented best by the theory of growth centres. Critics in the early 1980s began to suggest that this approach was no longer tenable (Stohr and Taylor, 1981). More recently the literature has become alive with studies of post-Fordism (Amin, 1994; Harrison, 1992; Ohmae, 1995), flexible industrialisation (Sabel, 1989), industrial districts, industrial clusters and the new regional development (Porter, 1990).

Many attempts to promote regional development through a traditional Fordist model have failed to stimulate either long term economic growth or the environmental well being of an area (Roberts, 1994; Stohr, 1989). Such models have also frequently regarded the resources of a region as a storehouse to be plundered at will. Perloff and Wingo (1964) provided an early

1 *The author wishes to acknowledge the financial assistance from the University of Hong Kong [CRCG 337/088/0006] and the UK-HK Joint Research Scheme 1997-98 for conducting research work leading to this chapter.*

assessment of the relationship between economic growth and the environment in which they argued that because the resource endowment is continually redefined by changes in the production environment. Changes in demand, production technology and economic organisation factors other than those associated with the best environmental solution exert considerable influence upon the pattern of resource development.

An additional characteristic of the traditional growth-dominated is the domination of functional integration over territorial integration. Whilst it might be expected that regional planning and development would give priority to the provision of spatial coherence, in practice many such exercises have failed to provide the specific guidance necessary in order to direct public and private policy and investment towards territorial integration. Many regions have been planned and developed in a manner principally designed to secure the needs of national growth, sectoral requirements or the desires of multinational companies to maximise profit (Wu, 1992). Although this model of regional development as a form of either internal or external colonialism is well-known, the implications for the environment have generally been ignored (Friedmann and Weaver, 1979).

In China, the 'economic reform and opening-up' measures carried out in the last 20 years have attracted much international capital, especially to coastal regions. The success of the Pearl River Delta region is a case in point. The launching of the Pudong New District in Shanghai aims to revitalise the economy in Shanghai as well as the lower Yangtze Delta region. Here and elsewhere in Asia, shifts in the loci of global economic growth and the radical changes in some socialist countries have spawned a variety of supra-national growth regions. The Singapore-Johor-Riau Growth Triangle between Singapore, Malaysia and Indonesia is one that is supported by the governments involved. Other growth triangles include the Taiwan-Fujian, Hong Kong and the Pearl River Delta region as well as the Tumen Economic Zone (UNDP, 1993).

This chapter aims to highlight the issues to be faced in building an integrated strategic framework for the study of sustainable regional development. It is built around two case studies. Both of them demonstrate the consequences of urbanisation and development undertaken without prior thorough considerations. From a sustainable perspective, they also provide a number of lessons which may help to bridge the chasm of imagination separating the exhortation of 'thinking globally' from the harsh realities presented by the attempts to 'act locally'. The first study of Shanghai and the lower Yangtze delta region reflects the experience of an old urban region amidst the challenge of restructuring in a more sustainable manner. The second case of the Pearl River Delta region exemplifies the tensions evident in regions undergoing rapid industrialisation of an unsustainable nature. These studies point to the necessity for new models of regional development, that incorporate a more integrated type of supra-national and regional development that crosses national boundaries.

4.2 Shanghai and the Yangtze Delta Region

In China, the effort since 1949 has been on modernisation, namely, to transform a semi-colonial and semi-feudalist and war-torn national economy into one of the leading economic powers in the world. China looked to 'Big Brother', the former Soviet Union, for initial inspiration and assistance, but was left at the crossroad of development after 1960 when Sino-Soviet relations turned sour. A self-reliance development strategy was then formulated. In his writing entitled *The Ten Great Relations*, Mao Zedong reflected on the course China should take up on the road to nation building. The emphasis was on industrial development, with a preference on heavy industries (Chan, 1994). However, not much was achieved as the national economic construction was made to play second fiddle to political aims during the Cultural Revolution. Since the introduction of the Open Door policy since in 1978, market forces have formed part of the planned strategy and also played a significant role in reshaping the urban land use and landscape of the big Chinese metropolises. In the course of these changes, the urban configuration of major cities has also evolved and changed to accommodate industrial development plans under national and regional development strategies.

4.2.1 Post-1949 Development in Shanghai

Shanghai was initially a small river-port township along the mouth of the Huangpu river that joins the Yangtze river. Under the Treaty of Nanking in 1842, the city became one of the five treaty ports to be opened up for foreign trade. The first foreign settlement was established in 1848. Thenceforth, the city was transformed from a walled-city to a city with a built-up area of 82.4 sq. km. in 1949.

In the early days of the socialist construction, Shanghai was not among the key cities designated for preferential investment. The geo-political location of the municipality renders her too vulnerable to foreign attack in the event of warfare. Development was focused on the interior and the Northeast region of the country during the First Five Year Plan. In the meantime, there was conscious effort to relocate the working population from the city centre to the suburbs. New workers communities were built with a comprehensive range of facilities. New quarters were built near the factories and industrial districts so as to reduce the cost and time involved in the journey to work. Large scale housing development for Shanghai was triggered-off. More than 20,000 housing units were earmarked in nine different locations (Fung, 1981, p.282). But the new satellite towns were unable to attract voluntary migration out of the old urban areas where services and other amenities concentrated. Investment on housing stock as well as infrastructure, such as road and other means of communications, was evidently lacking. These inadequacies, however, tend to be universal rather than being unique to Shanghai. People choose to stay in the city core in spite of deteriorating standard of living. In the 1980s, about 27% of the city dwellers were regarded as 'households of poor living standard' while some 3 million sq. metres of slum areas were identified. More than 0.8 million households have to take coking coal balls as their fuel source. Industrial development took place at the expense of the environment so that the once prided Suzhou Creek has since degenerated into the dirty nullah of Shanghai.

4.2.2 Development in Shanghai since the 1980's

The 1984 Master Plan is probably the first comprehensive plan for Shanghai since 1960s. Spatially speaking, it aims to renew and reinforce the urban centre with parallel development of the north (Baoshan-Wusong area) - south (Jinshanwei-Chaohejing area) axis of Shanghai. The population for central Shanghai is earmarked at 6.5 million. The central city will be divided into 11 zones, all with independent commercial centres. The central city and its outer zones will be separated by green belts comprising agricultural land, parks and botanical gardens and zoos. About 30-40 km. from the central city is a set of satellite towns, including Minhang, Jiading, Songjiang, Wujing and Anting. The estimated population of the satellite towns is 1.5 million, the majority of which is to be resettled from the central area. Urban sprawl is a noted feature of the rapid urban development since 1980s. Four of the ten counties have been upgraded to districts, making a total of 14 districts. Second-level cities are planned in the suburban areas too.

In February 1985, the State Council approved the *General Report on the Strategy to Develop Shanghai's Economy*. In order to realise the objectives stated in the Report representing a long-term effort on the part of the municipality towards a new development strategy, a General Development Plan was ushered in 1986. The Plan consists of three sections: (1) guiding principles for Shanghai's economic development; (2) the policies and tasks for Shanghai's Economic Development Strategies; and (3) measures for reforming the urban economic system. Five measures are suggested for bringing to fruition the following objectives: (1) to promote the "open to the outside world" policy; (2) to remodel the traditional industry; (3) to develop new industries such as hi-tech and bio-tech; (4) to develop the tertiary sector from being 30% of the GDP to 60% by the year 2000; and (5) to renew the old urban districts and to construct new urban areas (*Chen,* 1986).

The Pudong development plan was officially announced in 1990. Four economic and technological development zones are to be set up with a total area of 44 sq. km. Preferential policies to attract foreign investment are also introduced. In 1993, the new district was enlarged to 522 sq. km. and 18.9 sq. km. was actually developed by 1995 with USD 9.14 billion foreign direct investment *(Shanghai Statistical Yearbook* 1996).

The opening up of Pudong has given the impetus for improving the overall transportation in Shanghai. The city's inner ring road (a highway constructed above the existing one along the Zhongshan road) was put into operation in late 1994. In April 1995, the first line of the Metro began to serve the commuters. The ring road, an elevated orbital except the section in Pudong, together with the underground railway system, has opened up a new spectrum in transport geography for Shanghai. By early 1990s, the road and bridges network in Shanghai have been improved significantly as a result of intensive planning and development since 1978. A total of RMB 13.25 billion was spent on urban infrastructural projects in 1993 as compared to RMB 1.75 billion in 1985.

On 11 November 1994, the Municipal Government of Shanghai sponsored the International Conference on the development strategy of Shanghai towards the 21st Century. The objective of developing the metropolis into the international economic and financial centre of China in the next century was discussed and a report under the same title was published (Cai, 1995). The promotion of the urban spatial location and sectoral restructuring was regarded as the means to achieving the target of economic development.

The report highlights the advantages the city stands to gain from the increased manufacturing activities in the Asia-Pacific region and the rapid economic growth in East Asia. China, under the current reform and open era, will compliment and fit into the development trend in the region. At the same time, Shanghai, with the development plan of Pudong, will become the focal point of development of both China and the region as a whole due to its strategic location. To that end, Shanghai will have to catch up with other international metropolitan cities in South-east and East Asia in the coming decade in a number of areas. The strategic development objectives to be attained by the municipality by the year 2010 are:

- A total urban area of 6,300 sq. km. with a multiple-centre, multiple-function megalopolis.
- A GDP of RMB 150,000 per capita is to be achieved by 2010, with an average growth rate of 11.4% between 1995 and 2000, and 9.8% for the first decade in the next century. This will bring to fruition the targeted GDP of RMB 2,000 billion.
- A tertiary-oriented economy with emphasis being laid on finance, trade, information exchange, service sector and attracting multinational corporations to set up regional headquarters in Shanghai.
- A total population of 14 million (with an additional 4-5 million 'floating population') meaning an urbanisation level of more than 80%.
- Restructuring the urban land use with a 5 sq. km. new city business centre which spans across the Bund (in Puxi) and the Lujiazhui area (in Pudong).
- An infrastructure development that includes a modern highway system, port facilities and a new international airport at Pudong. (*Wah Kiu Yat Po*, 23 November 1994, p.24)

With Pudong as the new district of development, Shanghai city in turn forms the 'Dragon's Head' of the Yangtze Delta region. Administratively, the region includes Shanghai city, southern Jiangsu province and northern Zhejiang province. The region is thus represented by an area of 99,610 sq.km. with a population of 73.71 million people, of which 30 million are urban dwellers. It is the most developed economic region in China with high degree of urbanisation. The urban hierarchy includes three super-large cities (Shanghai, Nanjing and Hangzhou), four large cities (Wuxi, Suzhou, Zhangzhou and Ningbo), 17 medium-size cities and 30 small cities. In 1982, the State Council gave approval for the setting up of the Shanghai Economic Zone. The zone covers 10 cities in the Yangtze Delta and 55 counties. It was subsequently enlarged to include parts of Jiangsu, Zhejiang, Anhui and Jiangxi provinces in 1985. Map 4.1 shows the administrative boundary of the region.

Urban development (in the form of sprawl and rural urbanisation) has been occurring fast. There are two development corridors in the Yangtze delta region. The northern one, running from Shanghai to Nanjing, has 34 cities (63% of the region's total) with a GDP of RMB 396.7 billion. The southern corridor, ranging from Shanghai to Hangzhou, has 19 cities (35% of the region's total) and a GDP of RMB 240.4 billion. Map 4.2 presents the extent of urban agglomeration in the region while Table 4.1 shows the socio-economic composition of the key cities in the region.

The pattern of urban development is reflected by key transportation arteries of the Shanghai-Nanjing, Shanghai-Hangzhou and Hangzhou-Ningbo

Map 4.1 The Shanghai Economic Zone

railway systems. The 248 km. highway between Shanghai and Nanjing has shortened the travel time between the two places while also augmenting their economic interaction. Towards these ends, the city authorities in the northern corridor have been working closely together with Shanghai Development along the southern corridor, however, is hampered by the indecision of the local authorities to improve its inter and intra-regional transportation network.

The economic success of the Yangtze delta region largely hinges on

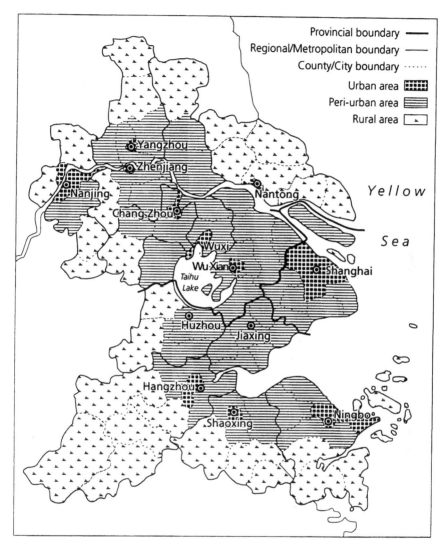

Map 4.2   Urban, peri-urban and rural areas of the Shanghai Economic Zone

the promotion of township enterprises during the early years of 'economic reforms and open policy'. Township enterprises are encouraged in order to contain the influx of surplus rural forces into the big cities. This also complements the 'spill-over' from the process of industrial restructuring going on within Shanghai.

Environmental degradation is another development concern in the Yang-

Table 4.1 Socio-economic profile of selected city regions in the Yangtze delta

City region	Total Population (million)	Total Area (sq.km.)	Population Density (person/ sq. km.)	GDP(RMB billion)	GDP (RMB, per capita)
Shanghai	13.01	6,341	2,052	246.3	18,943
Nanjing	5.21	6,516	801	53.85	11,086
Hangzhou	5.97	16,596	360	76.2	12,797

Source: Report on the Development of the Yangtze Delta Region (*1997*).

tze delta region. Intensive industrial development with inadequate safeguards has adversely affected environmental quality. The demand for land for urban and infrastructural development has put pressure on highly productive farmland in the delta region. The loss of good quality farmland paves the way for real estate and manufacturing development in the urban fringe and rural areas. Water and air pollution seem to be the environmental costs which people are ready and willing to accept in order to bring about economic affluence. The high cost of environmental conservation frequently aggravates the conflicts between local authorities within the region. Take for example the development in Suzhou and Wuxi, carried out at the expense of the water quality in Lake Ta (Ning, 1997). We shall return to discuss the search for a sustainable development strategy after looking at the second case study of the Pearl River Delta region.

4.3 The Pearl River Delta Region

The Pearl River Delta (PRD) region is a traditional agricultural region that has experienced rapid and significant industrialisation during the past decade. It is located in Guangdong province of south China, and comprises the Pearl River Delta Open Economic Area (PRDOEA), with some 28 cities and counties. The institutional profile of the region is rather complex: it includes the provincial capital city Guangzhou (itself a coastal open city), two Special Economic Zones (SEZs) of Shenzhen and Zhuhai, four bounded zones (two in Shenzhen SEZ, one in Guangzhou, one in Zhuhai) and three state-level export and technological zones. The region covers 26% of the total land area of Guangdong province. Guangzhou is the political, administrative and economic centre of Guangdong, and the transport and cultural centre of southern China.

Since 1980 the two SEZs have provided a laboratory for experimenting with different strategies for economic development in China, including the promotion of export-processing industries through the application of preferential treatments such as tax holidays. Table 4.2 elaborates on the Actual Utilisation of Foreign Direct Investment.

Table 4.2   Actual utilisation of foreign direct investment (US$ billion)

Year	Amount	Year	Amount
1979	0.038	1988	1.43
1980	0.114	1989	1.476
1981	0.182	1990	1.654
1982	0.213	1991	1.946
1983	0.269	1992	3.221
1984	0.628	1993	6.415
1985	0.833	1994	8.299
1986	0.943	1995	8.579
1987	0.837	**TOTAL**	**33.077**

Source: Report on the Development of the Pearl River Delta Region *(1997), p.3.*

The region is at the confluence of the three main tributaries of the Pearl River drainage system and is one of the most fertile regions in Guangdong province. Until the 1970s agriculture dominated the economy of the region (producing rice, sugar-cane, fresh fruit and fish) together with handicraft industries including silk and pottery. With the introduction of the 'Open Door' policy and economic reform measures, the special policy status granted to the region has been fully exploited to ensure the maximisation of economic productivity and the attraction of foreign investment. Township enterprises and joint ventures have mushroomed, and bold attempts have been initiated to reduce central planning and deregulate prices (Cheung, 1993).

The PRD region is attractive to medium-to-small investors from Hong Kong wishing to escape labour shortages and high land prices (Hong Kong Federation of Industries, 1992). Hong Kong businesses currently employ more workers (over 3 million people) in the region than they do in Hong Kong (under 650 000 people).

Since the late 1970s the PRD region has been one of the most rapidly developing regions of China and a cornerstone of the economy. With about one-third of the province's total population, the region's GDP reached RMB389 billion in 1995. Total population in the region amounted to 28.12 million. With a total area of 42,700 sq. km., the average population density is 659

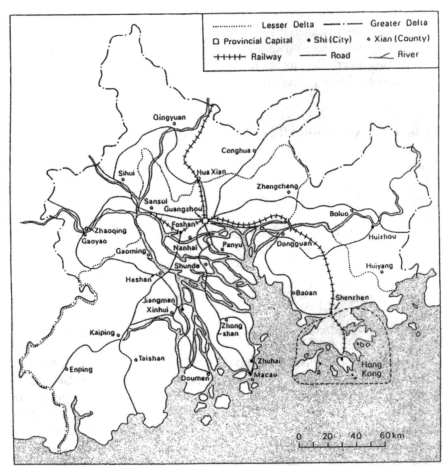

Map 4.3 The Pearl River Economic Zone

persons per sq. km.

During the past decade the economic structure of the region has been transformed faster than in the whole of the preceding 30 years of socialist government. The percentage of GDP composition for the primary, secondary and tertiary sectors in 1990 was 15:46:39, compared with the national percentage distribution of 29:44:27 (Zhao Jianhua, 1992). Rates of economic growth within the region are uneven with traditional economic centres like Guangzhou and Panyu being dwarfed by newly emerged cities such as Shenzhen and Zhongshan. Rapid rural industrial development is also evident with industrial output in the rural area accounting for 48% of the regional total (Zhao Jianhua, 1992). The pattern of development is selective with greater prosperity in the areas between Guangzhou and Shenzhen and Guangzhou and Zhuhai than elsewhere.

4.3.1 Infrastructure Development in the Pearl River Delta

Like its counterpart in the Yangtze delta region, the success of the PRD region has been reinforced by infrastructural development. Local and provincial authorities have spent substantial resources in constructing highways; during the last five-year plan (1991-95) RMB1.5 billion was earmarked for improving the transport network of Guangdong province, including the construction of an underground railway system in Guangzhou. By 1996, there were 280km. highway, over 1,000km. first class motorway and 1,800km of second class motorway in the PRD region. It is a showpiece of highway network in south China. Additionally, light rail networks were at the inception stage in Guangzhou and Shenzhen. High speed trains now runs between Guangzhou and Hong Kong while electrification work is underway along the Beijing-Guangzhou, Beijing-Kowloon lines. Five civilian airports (Guangzhou, Shenzhen, Hong Kong, Zhuhai and Macau) all of them newly constructed or under active planning and construction stage will turn south China into a regional aviation hub.

The container port of Kwai Chung in Hong Kong handled a record 12.5 million TEUs (twenty-foot equivalent units) in 1995, with 40% of the cargo originating from the Pearl River Delta region. The Huangpu port of Guangzhou handled 72.9 million tons of cargo in 1995, marking it one of the largest ports in the region. The rapid growth of trade has stimulated the building of additional river and seaports in the region. The combined capacity of Yantian and Shekou ports is currently under 1 million TEUs, but is projected to increase to over 1.7 million TEUs by 1996, whilst the Gaolan port in Zhuhai is also planned.

In the light of the rapid development, the PRD region is evolving into a metropolis within a short period of time. Table 4.3 shows the key economic features of the major urban corridors:

There has been an accompanying exploitation of land resources, driven by the rapid pace of infrastructural developments, resulting in the massive conversion of farmland to construction sites for housing and factories. Between 1980 and 1990 the total area of arable land dwindled from 15.67 million mu (1 mu = 0.0667 hectares) to 13.69 million mu, a loss of over 12%. In some areas, the loss of land was as high as 25%.

It is common for the ownership of land plots earmarked for development to be transferred several times as a result of speculation, but without

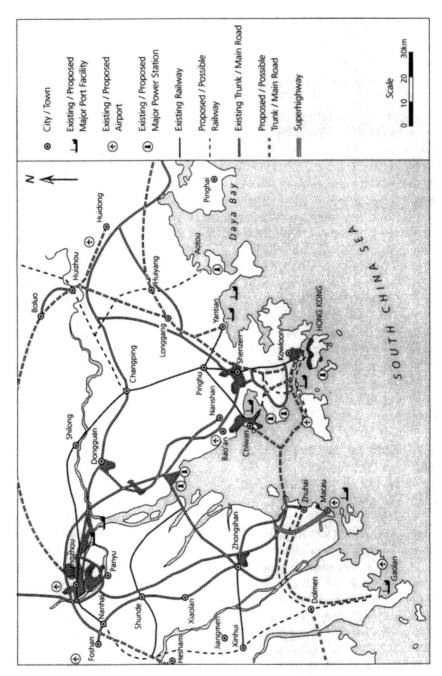

Map 4.4   Transport in the Pearl River Economic Zone

Table 4.3 Urban corridors in Pearl River Delta Region

Urban Corridor	Total Population (million)	Non-agriculture population (million)	Total Area (sq.km.)	Population Density (person/ sq. km.)	GDP (RMB billion)	GDP (RMB. per capita)
Guangzhou-Fushan	9.43	5.22	11,570	815	141.3	14,984
Hong Kong-Shenzhen	8.42	3.11	5,563	1,513	1172.8	139,383
Macau-Zhuhai-Zhongshan	2.25	1.09	3,284	684	84	37,409

Source: Report on the Development of the Pearl River Delta Region *(1997), p.15.*

construction work being carried out. Alarmed by the rapid depletion of farm-land, the Guangdong provincial authority has recently issued a directive to reverse this trend. Idle areas of land have been purchased by local authorities to allow the return of such land to arable use through its redistribution to the peasants.

4.4 Regional Development: Economic Growth before the Environment?

The idea of sustainable development is a novelty for the Chinese economy in its current mode of development (Roberts and Chan, 1997). The country experimented with different socialist development strategies during the years of Communist rule. However, although the Chinese claimed leader-ship in certain areas of scientific proficiency and industrial production, the overall rate of economic performance was disappointing. In contrast, the 'Open Door' policy is development led. Under this policy the economic costs for environmental maintenance are disguised under separate accounts, but it has been estimated that the annual economic loss due to environmental degrada-tion amounts to US$11 billion. In practice only a diminutive RMB10 billion (0.7% of GDP) is set aside for environmental improvement. In their study of the environmental quality of Guangdong, Neller and Lam lament the alarm-ing rate of the degradation of the physical environment. Not only is the quality of air and water deteriorating, but there is also an evident failure to enforce legal requirements. In their view, the various aspects of environmental protection policy are applied randomly and there is little chance that environmental problems will be tackled in a comprehensive manner (Neller and Lam, 1994).

Aggressive measures to provide infrastructure and land are practised to enhance the inward investment potential of the region. In order to stimulate local enterprise, higher authorities do not intervene. This flexible attitude, whilst assisting in the growth of the economy, encouraged environmental damage and imposes heavy environmental costs. These flexible policy measures, together with fiscal incentives, attract outward processing industries from Hong Kong, but when the favourable terms enjoyed by joint ventures expire, they shut down their original production lines and start another company within the PRDOEA, a phenomenon known as 'factory hopping'. It is difficult to convince these low-value-added operations, which operate on very short-term production horizons, to convert to environmentally sound production techniques. Among other considerations, one of the reasons prompting the establishment of overseas production lines in China is the desire to avoid more stringent environmental legislation enforced elsewhere. Local authorities welcome these operations because of their low set-up costs, their high economic productivity and their ability to absorb redundant agricultural labour.

However, serious flooding in Guangdong in recent years has been attributed to the uncaring attitude towards the environment, including the absence of adequate planning to guide rapid urban development. A series of natural hazards demonstrates the lack of institutional arrangements in municipal management and the inability of local authorities to execute stated regulations. Although the environmental vulnerability of the PRC region is partly due to its geomorphologic configuration, it is mainly due to ignorance regarding the adverse effects of the man-land interaction. It is difficult to ensure a balance between rapid development and environmental protection in a region as unique, yet fragile, as the PRD. Assuming that the local authorities retain their flexible stance on policy, further deterioration of the environment will eventually become an obstacle to sustainable development. As noted by Gibbs (1994), the devolution of decision making to local authorities should be supported by the appropriate resources. Concepts such as Strategic Environmental Assessment and Ultimate Environmental Thresholds can assist in policy formulation and construction.

4.5 Agenda for Regional Sustainable Development

Although it is acknowledged that some previous attempts at regional planning and development have encapsulated the concerns inherent in the concept of sustainable development, it is also apparent that much of past and current practice has been concerned with the achievement of other objectives. In the case of the less developed nations Drakakis-Smith has referred to the spurious nature of the relationship between urbanisation and development, often based on the "old and false surrogate GNP per capita" (Drakakis-Smith, 1995, p. 660). In Western economies, Simonis (1993) has identified the continuing obsession with 'tonnage ideology' measures of success.

This suggests that it is essential to clarify the concept of sustainable development as it applies to regional and urban planning, and to identify criteria that may be used to guide its achievement. The most obvious entry point into this debate is through the work of the World Commission on Environment and Development (1987) and its definition of sustainable development as "development that meets the need of the present without compromising the ability of future generations to meet their own needs" (p. 8). Jacobs (1991) and Healey & Shaw (1993) point to the three essential components of this concept:

- the inclusion of environmental concerns in economic policy making;
- intergenerational and intragenerational equity;
- the need to adopt a balanced model of development rather than an approach aimed solely at the promotion of economic growth.

However, many strategies for sustainable development fail to achieve the integration of economic and environmental concerns, and the transformation of cultural attitudes and behaviour, necessary for ecological modernisation. Strategies which consider only the institutional or technocratic dimension of ecological modernisation fail to recognise the need for fundamental changes in the ideology and priorities of planning necessary to avoid the "absurd dilemma" of having "to choose between jobs and ecology" (Lipietz, 1992, p. 53). Whilst sustainable development provides a philosophy, ecological modernisation offers a practical pathway towards avoiding such dilemmas.

Equally important in the design and implementation of planning and development for sustainable development is the role of strategy. Thinking about long-term needs and the evolution of economic, social and environmental conditions is a prerequisite for effective planning, and this is even more so the case if the intended output from the planning process is a plan which meets the requirements of sustainable development and ecological modernisation. For most of the past two centuries the proponents of economic growth, especially in the western world, have regarded the environment as a storehouse of resources and as a free dumping ground for waste. Changing course will take time, and this implies a minimum planning horizon of 20 to 25 years, way beyond the normal target date for many public policy activities (Begg, 1991).

A final element of considerable importance in the development of an approach to sustainable planning is the need to ensure the integration of policy and the desirability of embedding environmental concerns across the full span of sectoral activities. This is a difficult task in a single region or locality, and is even more challenging at a transnational level where, for example, responsibility for a particular function is likely to be vested in different sorts of agencies with varying degrees of power and responsibility.

Finally, it is important to screen development proposals in order to ensure that new activities do not add to the existing stock of environmental problems or negative externalities. The case for introducing strategic environmental assessment (SEA) has been argued for some considerable time (Glasson, 1995), and the adoption of SEA is now acknowledged as a vital step in avoiding further environmental degradation. It also assists in reducing the waste of resources associated with projects that fail to satisfy the criteria used to judge the sustainability of proposed developments.

In 1949 China decisively rejected Western capitalist development models in its reaction to foreign imperialism. It became introverted and withdrawn, and obsessed with social ideological rectitude, that reduced any other considerations of economic efficiency or of ecological sustainability to mere footnotes. In its pursuit of economic wealth, it has now become more pragmatic and more aligned with market forces, but it has not yet integrated environmental philosophy into its regional development plans. To avoid future greater costs, it is advisable to integrate these values now. Thus a model which reflected the problems and opportunities of urban regional development in China would have to acknowledge its socialist past and semi-socialist present, the often weak property relations implied and a weak sense of guardianship of the common resources of the environment, while simultaneously guiding private profit with public accountability, greater equity, and environmental sustainability. This is a very great challenge.

References

Amin, A. ed. 1994: Post Fordism: A Reader. *Oxford, Blackwell.*

Begg, H. 1991: "The challenge of sustainable development", The Planner, *Vol.77, no.22, pp.7-8.*

Cai, Laxian, 1995: Shanghai: Moving into the 21st Century, *Shanghai: Reinmen chubanshe. (in Chinese).*

Chan, Kam Wing 1994: Cities with Invisible Walls. *Oxford, Oxford University Press.*

Chan, R.C.K. 1993: "The Impact of China's Open Policy on the coastal cities and regional disparities" in Tan, K.C., Taubmann, W. and Ye, Shunzan (eds.) Urban Development in China and South-East Asia. *Studiengang Geographie, Fachbereich 8, Universitat Bremen, pp.49-64.*

Chan, R. C. K. 1996: "Urban development strategy in an era of global competition: The case of South China", Habitat International, *vol.20, no.4, pp.509-523.*

Chen, Minzhe 1986: "Development strategy of Shanghai city", City Planning Review, *Vol.10, no. 6, pp.21-24.*

Cheng, J.Y.S. and MacPherson, S. (eds.) 1995: Development in Southern China. *Hong Kong, Longman Press.*

Cheung, P.C.Y. 1993: "Pearl River Delta development" in Cheng, J.Y.S. and M. Brosseau (eds.) China Review 1993. *Hong Kong, The Chinese University Press.*

Drakakis-Smith, D. 1995: "Third world cities: sustainable urban development", Urban Studies, *Vol.32 (4-5), pp.659-673.*

EAAU (East Asia Analytical Unit) 1992: Australia and North-East Asia in the 1990s: Accelerating Change. *Department of Foreign Affairs and Trade, Canberra, Australian Government Publishing Service.*

Frank, A.G. 1978: Dependent Accumulation and Underdevelopment. *New York, Monthly Review Press.*

Friedmann, J. and Weaver, C. 1979: Territory and Function. *London, Edward Arnold.*

Froebel, F. and J. Heinrichs, et al. 1980: The New International Division of Labour. *Cambridge, Cambridge University Press.*

Fung, Kai-iu, 1981: "The spatial development of Shanghai", in Howe, C. (ed.)., Shanghai: Revolution and Development in an Asian Metropolis. *Cambridge, Cambridge University Press.*

Fung, Kai-iu, Yan, Z.M. and Ning, Y.M. 1992: "Shanghai: China's World City" in Yue-man Yeung and Hu, X.W. (eds.), China's Coastal Cities, *Honolulu, University of Hawaii Press.*

Gibbs, D. 1994: "Towards the sustainable city" Town Planning Review, *Vol.65 no.1, pp.99-109.*

Glasson, J. 1995: "Regional planning and the environment, time for a SEA change", Urban Studies, *Vol.32, no.4-5, pp.713-731.*

Harrison, B. 1992: "Industrial districts: Old wine in new bottles?" Regional Studies, 26 (5): 469-483.

Healey, P. and Shaw, T. 1993: "Planners, plans and sustainable development", Regional Studies, *Vol.27, no.8, pp.769-776.*

Hong Kong Federation of Industries 1992: Hong Kong Industrial Investment in the Pearl River Delta. *Hong Kong, Hong Kong Federation of Industries.*

Jacobs, M. 1991: The Green Economy. *London, Pluto Press.*

KRIHS (Korea Research Institute for Human Settlements) 1996: Industrial Cooperation and Regional Development in Northeast Asia. *Seoul, KRIHS.*

Leeming, F. 1993: The Changing Geography of China. *Oxford, Backwell.*

Lipietz, A. 1992: Towards a New Economic Order. *Cambridge, Polity Press.*

Lo, F. C. and Yue-man Yeung eds. 1996: Emerging World Cities in Pacific Asia. *Tokyo, United Nations University Press.*

MacKaye, B. 1928: The New Exploration: A Philosophy of Regional Planning. *New York, Harcour Brace.*

Markusen, A. R. 1987: Regions: The Economics and Politics of Territory. *Totowa, New Jersey, Powman & Littlefield.*

Massy, D. 1984: Spatial Division of Labour: Social Structures and the Geography of Production. *London, Macmillan.*

NAEF (Northeast Asia Economic Forum) 1995: 5th Meeting Northeast Asia Economic Forum, Niigata, Conference Proceedings. Honolulu, Northeast Asia Economic Forum.

Neller, R.J. and Lam, K.C. 1994: "The Environment" in Yueng, Y.M. and Chu, D.K.Y. (eds.) Guangdong: Survey of a Province Undergoing Rapid Change. *Hong Kong, The Chinese University Press.*

Ning, Yuemin 1997: Shanghai: An Emerging World City in Asia-Pacific Region. Unpublished manuscript.

Ohmae, Kenichi. 1995: The End of The Nation State: The Rise of Regional Economies. *New York, The Free Press.*

Organisation for Economic Cooperation and Development 1990: Environment Policies for Cities in the 1990s. *Paris, OECD.*

O'Riordan, T and Turner, R.K. 1983: An Annotated Reader in Environmental Planning and Management, *Oxford, Pergamon.*

Perloff, H. and Wingo, L. 1964: Regional resource endowment and regional economic growth. in J Friedmann and W. Alonso eds. Regional Development and Planning, *Cambridge, Mass, MIT Press.*

Pannell, C. and Veeck, G. 1989: "Zhujiang Delta and Sunan: A Comparative analysis of regional urban systems and their development", Asian Geographer, *vol.8, no.1-2, pp.133-150.*

Porter, M.E. 1990: The Competitive Advantage of Nations. *New York, Free Press.*

Report on the Development of the Pearl River Delta Region *1993. China: National Science Foundation Project: 49331010.*

Report on the Development of the Yangtze Delta Region *1993. China: National Science Foundation Project: 49331010.*

Roberts, P. 1994: "Sustainable regional planning", Regional Studies, *28 (8): 781-783.*

Roberts, P. and Chan, R.C.K. 1997: "A tale of two regions: strategic planning for sustainable development in East and West", International Planning Studies, *Vol.2, no.1, pp.45-62.*

Rostow, W.W. 1956: "The take-off into self-sustained growth", Economic Journal, *66: 25-48.*

Sabel. C.F. 1989: "Flexible specialisation and the re-emergence of regional economies" in P. Hirst and J. Zeitlin eds. Reversing Industrial Decline? Industrial Structure and Policy in Britain and Her Competitors. *Oxford, Berg.*

Simonis, U.E. 1993: "Industrial restructuring: does it have to be jobs vs. trees?" Work in Progress of the United Nations University, *Vol.14, no.2.*

Stohr, W. 1989: Regional policy at the crossroads: An overview in L. Albrechts, F. Moulaert, P. Roberts and E Swyngedouw eds. Regional Policy at the Crossroads: European Perspectives, *London, Jessica Kingsley.*

Stohr, W. and Taylor, D. eds. 1981: Development from Above or Below: The Dialectics of Regional Planning in Developing Countries. *New York, John Wiley and Sons.*

Sung, Y.W., Liu P.W., Wong, Y.C. and Lau, P.K. 1995: The Fifth Dragon: The Emergence of the Pearl River Delta. *Singapore, Addison Wesley.*

UNDP 1993: A Regional Development Strategy for the Tumen River Economic Development Area and Northeast Asia. *New York, United Nations Development Programme.*

Wong, K.K. and Chan, H.S. 1994: "The development of environmental management systems" in Brosseau, M. and Lo, C.K. (eds.) China Review 1994. *Hong Kong, The Chinese University Press.*

Wood, C. 1994: Painting by Numbers. *Lincoln: Royal Society for Nature Conservation.*

World Commission on Environment and Development 1987: Our Common Future. *Oxford: Oxford University Press.*

Wu, C.T. 1992: *Policy aspects of export process zones : Lessons from an international study.* Southeast Asian Journal of Development. *19 (1/2): 44-63.*

Xia Guang 1995: "*Environmental problems and policies in China's industrialisation*", Asia Journal of Environmental Management, *Vol. 3, no.1, pp.7-11.*

Yeung, Y M. and Chu, D. K.Y. 1994: Guangdong: Survey of a Province Undergoing Rapid Chang. *Hong Kong, The Chinese University Press.*

Yeung, Y.M. and Sung, Y.W. 1996: Shanghai: Transformation and Modernisation under China's Open Policy. *Hong Kong, The Chinese University Press.*

Zhao Jianhua 1992: "*Economic development of the Pearl River Delta: a retrospect and prospects, in Economic Centre of Pearl River Delta Economic Development and Management*

5 Consideration of Urban Development Paths and Processes in China since 1978, with Special Reference to Shanghai

Felicity C. Rose

5.1 Introduction

Since 1978, China has embarked on a new economic and urban development path, based around a series of fundamental ideological changes, and extensive economic and spatial policies. A chief part of Reform has been the opening of China to the global economy, an action which has increased regional competitive pressures, as well as the availability of investment finance and commercial expertise. In particular, national and local-level officials have placed explicit emphasis on the attraction of foreign capital to assist in the industrial and spatial development of China. Shanghai, which was re-opened to foreign investment as an 'Open Coastal City' in 1984, and which forms the heart of the Yangtse Delta Economic Region, now has the highest level of foreign capital utilised per head of population.

Since the late 1980s, Shanghai has embarked on a rapid path of modernisation based around foreign investment and foreign-oriented economic activity. GDP/capita reached approximately US$2175 in 1995. As development has progressed, however, Shanghai's urban planners have begun to realise the costs of accepting all forms and amounts of foreign capital, especially in terms of the urban environment and urban transport systems. Since 1993, they have intervened at the city level to moderate the inflow of capital, especially in the real estate sector, and to take a more rational approach to each project, with the aim of altering the direction and pattern of foreign investment, and hence its impacts, rather than its amount. The principal role of urban planning in Shanghai today has come to be recognised as the management of change in the face of multiple demands on space from various economic interests.

5.2 Externally-oriented Development: The Role of FDI

Evidence from recently industrialised and newly industrialising economies in east and south-east Asia indicates a positive role for foreign direct investment (FDI) and trade in the economic and urban development of nations and regions. The mechanisms involved here are outlined below. It is apparently on this basis that policies within the 'Open Door' framework have sought to encourage foreign investment in China. Given three decades of low, and even at times negative, economic growth, and comparative industrial underdevelopment, compounded by a neglect of the core urban-industrial areas in favour of more disparate locations, China's available investment capital was very limited in 1978 and her industrialisation process lagging. This, coupled with overemployment in both agriculture and industry, and low technology and productivity levels, spurred a shift in ideology in favour of the opening of the country to international economic forces, and especially to multinational interests, in an effort to restructure the economy and kick-start its growth.

The new endogenous growth theory (first developed by Romer, 1986) is used here as a point of departure for the analysis of foreign investment capital in economic development. According to this model, increases in one or more factor input or productivity will spur output growth, with stronger economies becoming stronger relative to poorer ones by virtue of increasing returns to physical and human capital. Dynamic economies of scale thus cause an upward spiral of development among those regions or nations which enjoy a head-start, and the local development environment combines with accumulation processes to produce strong growth.

Research on economic development has indicated a major role for capital in the growth process (see World Bank, 1993) , and a sizeable portion of China's recent growth can be attributed to capital investment (Hu and Khan, 1997). Capital invested from overseas will bring technological advantages and assist with access to international export markets. Export growth has in itself been shown to boost productivity growth, as new styles of economic organisation and new technology are introduced(The World Bank, 1993). Increasing returns to capital, or productivity growth, result from the interaction between technological and management inputs and capital accumulation (see Romer, 1993). For China, as for many other developing countries, it is the contribution of foreign companies to domestic technological progress and productivity improvements, as well as pure financial capital per se, which makes this source of investment attractive, although domestic efforts to increase productivity levels, essentially through the reorganisation of production activities and increasing the independent decision-making functions of enterprises, has also contributed to China's rapid post-Reform economic growth (see Hu and Khan, 1997).

However, a particular concern must be the sustainability of the growth path, especially given a constantly fluctuating world economy and a fickle investment community. Attempts should be made to exploit inward investment for explicitly long-term urban-industrial gains. The four Asian dragons have demonstrated the scope for success in achieving sustainable economic growth and restructuring through the utilisation of foreign capital. The effect on inward flows of investment on economic growth should best be considered as short-term and positive; domestic-driven reform, especially in the industrial sector, will however also be required in order to achieve long-term deep-rooted improvements in the economy. The latter is likely to be particularly important in countries which are undergoing a transition from socialist to market-oriented economic practices, as is the case in China. China not only has the challenge of developing sluggish industries through modernisation, technological inputs and improvements in labour skills, but also has to deal with an array of problems associated with the ownership of industry and physical property, the welfare responsibilities of economic units to their labour force and its dependants, and a surplus labour supply in factories and in agriculture. Reliance on foreign investment alone in such circumstances is unlikely to produce the economic turnaround sought; rather, a complex mix of market interests and public policy, especially with regard to economic structure and industrial operations, will be required.

5.3 FDI and Urban Development

Indirectly through the workings of the economy, and directly through the construction activities of factories and real estate companies, FDI and the externally-oriented economy will impact on urban space. At a macro level, the distribution of economic development across a country will influence the nature of its urban-industrial development, and its urbanisation pattern. FDI tends to have a natural affinity to either established urban centres or special zones backed by investment incentives. Differing from indigenous investment, FDI tends to be more concentrated and more linked to infrastructure and basic amenities, and related to corporations' previous distribution networks (Wong, 1994). Public and private investment are interdependent such that the scale and spatial distribution of public investment, especially in infrastructure, will impact affect locational decisions. With inward investment driving growth, distortions in the urban hierarchy and pattern are likely to follow investment choices. The mobility of capital and technology is such that the opening of the economy will produce specific urban and regional landscapes, with differential patterns of development and decline among cities and towns affected and passed over (Wong, 1994). The spatial concentration of FDI has been shown to result in renewed urban polarisation across

East and South-east Asia (Wong, 1994). Thus, without effective redistributive policies by State and local governments, a biased urban structure will persist.

At the city/district level, high profile FDI will start to influence the shape and nature of urban form. An increasing trend for cities to adjust to a common cultural pattern has been suggested in the growing literature on globalisation and world city formation (see, for example, Thrift, 1994; Harvey, 1989; Schachar, 1994; Knox, 1993). Cities become representations of inter-national capitalist power structures. Real estate companies will most actively contribute to the changing face of a city, with architecture and design quality, as well as the type and scale of space, reflecting global commercial interests rather than being responsive to local environments. In the absence of local control, a common result is an increase in office supply in downtown areas, with each building trying to distinguish itself from the rest by a plethora of distinctive features designed to assist marketing rather than the legibility of the city. The introduction of foreign retail and entertainment concerns, and prolific advertising, will alter physical faÁades and the atmosphere of the city; the 'Americanisation' of international cities, especially through the in-fluences of American multinationals and television-influenced lifestyle pat-terns, is particularly notable here.

Thus, in an economy heavily influenced by the activities of overseas capital, such activity can be expected to have far-reaching impacts on urban space and society, in both a physical and intangible sense.

5.4 FDI in China

In 1995, over US$ 39 billion of foreign capital was utilised in China, 94% of which (US$ 37 billion) was in the form of FDI (State Statistical Bureau, 1996). 233,564 enterprises nation-wide had registered foreign capital in 1995. On average, 31.5% of total export value was sourced from foreign-invested enterprises, rising to 49% in Tianjin and 46% in Guangdong. Hong Kong, Japan and the USA have consistently been the dominant source coun-tries for foreign investment. Sectorally, 75% of foreign-invested projects are in industry, with 9% in real estate and related business services; by value the shares are 68% and 20% respectively, reflecting the relative scale of real es-tate projects (State Statistical Bureau, 1996).

The regional distribution of FDI has been highly uneven, reflecting historical discrepancies in overall development levels across China, and government policy. Guangdong Province, with three SEZs, the Pearl River Delta Economic Development Zone and Guangzhou Open Coastal City, was the first area to be opened to foreign investment, and still accounts for 28% of total FDI utilised in China (State Statistical Bureau, 1996). The eastern region[1] in total accounts for 88% of FDI, despite having less than a third of the national urban spatial area. Shanghai has the highest rate of FDI per head of population (US$ 204 in 1995), but ranks only fourth in total FDI (3.8%) (ibid) due to its small relative size. North-east China[2] receives relatively low amounts of FDI, despite its historical position as the nation's industrial heartland; the structure of overseas investment in this area also differs from the east and south-eastern coastal provinces[3], with a lower ratio of FDI to other types of investment (90% in Liaoning and close to 80% for Jilin and Heilongjiang, compared with 96-97% in other south-eastern coastal provinces, and a national average of 94%).

Beijing, Shanghai/the Yangtse River Delta, and Guangdong (Guangzhou, the SEZs and the Pearl River Delta) have become the most popular destinations for foreign businesses wishing to establish representative, sales or management offices, and operational business units in China, although for different reasons. Beijing is a common primary office location, as it is perceived necessary to be close to the centre of political power in what is still a very heavily State-controlled economy; Guangdong, with the benefit of the SEZs, became a focus of overseas attention in the 1980s as it was the only option for early investors, and due to its proximity to Hong Kong where the regional or international headquarters of overseas firms were often located, and which was in any case the dominant source of investment in the early stages of opening. Shanghai has an historical reputation and a new national policy focus, is crucially located at the mouth of the Yangtse, and carries service as well as manufacturing functions, making it a potential headquarters location. Cities close to Shanghai, especially in southern Jiangsu (e.g. Suzhou, Wuxi, Taicang), benefit from slightly lower costs compared with Shanghai and good external communications.

Since the late 1980s, there has been a gradual spread in the areas allowed to accept foreign capital, and local-level incentives have been developed. The overall impact of this policy approach to FDI, plus differences in local investment and market environments, has been a concentration of foreign economic activity in certain sub-regions, where investment incentives have been the greatest. Less proactive policies in the early Reform period kept Shanghai's foreign investment and development in check until the 1990s, when Shanghai's role as China's leading urban, industrial and financial cen-

---

1 *Beijing, Tianjin, Shanghai, Liaoning, Hebei, Shandong, Jiangsu, Zhejiang, Fujian, Guangdong, Guangxi, Hainan.*
2 *Comprising the provinces of Liaoning, Jilin and Heilongjiang.*
3 *The Tianjin and Shanghai metropolitan areas and the provinces of Shandong, Jiangsu, Zhejiang, Fujian and Guangdong.*

tre was given formal ratification. Over time, and as costs in core areas have risen, some firms have begun to decentralise at least their manufacturing functions to suburban or county locations, usually with good communications to the urban core. This has benefited, for example, the development of smaller cities and towns in Guangdong, and cities along the Shanghai-Nanjing corridor in southern Jiangsu. Infrastructure, labour skills and local bureaucracy however remain major obstacles to the redistribution of FDI.

Interest from multinationals has been strong in both the manufacturing and tertiary sectors. China has also been successful in securing foreign funds for infrastructure development from a variety of private, as well as public, sources - e.g. the Hong Kong-invested Shenzhen-Guangzhou superhighway (Gordon Wu) and the German-invested Shanghai Subway (Siemens).

5.5 The Role of Public Policy in the Post-Reform Development Process

The national and local State still maintains an influential role in the Chinese development process, despite the extensive marketisation and opening of the economy since 1978. The dominance of ideology, State control and economic planning have however been declining (Yeh and Wu, 1995), in favour of a more pragmatic policy approach. Three strands of domestic public policy are of particular relevance here: urban industrial reform, land supply, and urban and regional planning. All three have been important in determining the nature of the economic growth process and its impact on urban space, and have influenced the role of FDI in China and its urban outcomes.

5.5.1 Urban Industrial Reform

The opening of the economy discussed above was just one aspect of reform. Arguably more far-reaching were amendments relating to agriculture, labour supply, population movement and private property rights which have complemented the reforms in the industrial sector. Institutional and organisational factors are particularly important, including the privatisation of certain industrial activities, the introduction of bankruptcy and redundancy in the State industrial sector in an attempt to boost efficiency, and changing tax burdens for both industries and cities. With the sale of land use rights beginning in certain localities in the late 1980s, the prospect of individuals and enterprises controlling land has re-emerged, and wider property rights have been defined.

State industry has in the main struggled to remain competitive and lo-

cal governments have initiated restructuring through mergers and bankrupt-
cies, or encouraged joint ventures with overseas investors, all in an effort to
stimulate modernisation and increase efficiency and output. In the Shanghai
textile industry alone, 10 State firms were made bankrupt in 1996, while in
the city as a whole layoffs stood at 5% of employment in 1996 (Shanghai
Star).

5.5.2 Land Supply

At the end of the 1980s, the government began to experiment with the
scale of land use rights and introduced the concept of land value. Throughout
the period from 1949 to the 1980s, land had been allocated administratively
by the State to various users, mostly individual work units (danwei). Land
did not have a price. Once allocated, State influence on the ultimate use of the
and was effectively lost; the State could determine the type of work unit controlling the land, but not the use to which they put the land. Land parcels were
usually divided up by work units for their own diverse needs - productive
activity (factory, hospital, university), plus housing and welfare facilities for
their workers. This led to a very mixed urban land use pattern, coupled with
pollution and congestion, and much inefficiency and wastage of land.

In 1988, the Land Administration Law re-introduced the concept of
private property rights in relation to land. The rights to use a certain piece of
land (but not the land itself) can now be sold in a highly regulated market; the
State retained the ownership of land, but leased the rights to use that land to
non-State interests, including foreigners, for a specific period of time. Although land leasing is still in its infancy and subject to strong government
influence, the emerging land market is assisting in land conversion and
(re)development. Much of the chaotic construction activity in China's large
cities has been the result of work units building commercial premises on land
previously not in use, or under-utilised. Most notable here has been high-rise
residential building, retail uses of all scales and office space, changes that
have been linked to reform in other sectors, particularly the commercialisa-
tion of housing. The possibility of the comprehensive restructuring of urban
land use to reflect land value has thus emerged, while the local State has been
given an alternative source of revenue, to be used in the upgrading of urban
infrastructure and public facilities. Some cities, including Shanghai[4], have
predetermined land prices according to a hierarchy of values that relate to

---

4 *Nine land value classes have been applied to the central area of Shanghai, with the highest
values attributed to Huang Pu District, the commercial core of the city. These assigned values
are not however market-determined.*

spatial location within the whole city. As the land market matures, land value can be expected to have an increasing role in determining the distribution of industrial and other economic activity across urban space (Yeh and Wu, 1995) through its impacts on decision-making in relation to profit at the enterprise level.

5.5.3 The Role of Urban and Regional Planning

Until the mid-1980s, urban planning did not have major independent role in Chinese urban development and was approached as an integrated part of macro-level economic planning. Industrial development was the particular focus of economic plans, and other urban functions, including housing and infrastructure, received limited attention, and usually only in conjunction with industrial and enterprise development. Urban plans took the form of physical blueprints designed to map urban-industrial construction. Over the past decade, urban planning has been reinstated as a discipline and a limited legal framework[5] has been established for plan formulation and development control. A hierarchy of urban plans has been introduced, running through national, regional and city levels. At the municipal level, the following plans are to be produced: urban system plan, strategic outline for the comprehensive city plan, comprehensive city plan, district plan and detailed site plan. However, the urban planning process remains immature and suffers a number of limitations in its attempt to effectively promote and control urban development.

Urban plan coverage and effectiveness also vary between cities, depending on local priorities, expertise and resources. Shanghai has one of the more advanced urban planning systems, but still suffers limitations in its attempts to influence urban development outcomes.

Given China's size and the economic development policies aimed at certain regions, regional planning could be expected to have a major role in macro and local level development. However, there is no formal regional planning system and each spatial administrative unit is able to act in its own interests, with control of its own resources. Due to competition between areas, especially in the race for foreign investment, each urban area is adopting an isolationist approach, concerned simply with its own development, rather than

5 *Planning regulations have generally lagged behind actual urban change and economic development. Three stages of regulatory development were seen through the 1980s: Provisional Regulations for Preparing and Approving Urban Plans and Provisional Standards for Urban Planning (1980), Regulations for Urban Planning (1984), and The City Planning Act (1989). The system implemented by these laws and regulations was neither comprehensive nor particularly modern, focusing on the scientific mode of planning (see Ng and Wu, 1995).*

with how that development may be linked with neighbouring areas, with mutual benefits. The result is a duplication of expenditure, especially on infrastructure development[6], and a lack of regional spatial and economic cohesion.

5.6 The Dynamics of Urban Change in Shanghai

Shanghai has since the nineteenth century been China's largest and most economically advanced city. It was the most important of the Treaty Ports, and arguably the city most influenced by international interests. The legacy of the foreign settlements and concessions, which were dominated by the French, British and American business communities, can still be seen in the structure of the city and in its architecture; in the French Quarter, there have been recent attempts to revive the original character of the district through the attraction of French chain stores, cafes and p,tisseries, and, in a limited number of cases, through the restoration of colonial villas and streetscapes. The re-internationalisation of Shanghai is not, however, restricted to the small number of historical quarters, nor is the urban modernisation process dominated by renewal and retrospective; modern Shanghai is developing in a form not dissimilar to Hong Kong, Singapore or Tokyo, as an Asian international city.

Shanghai suffered from a lack of a 'policy push' throughout the early Reform period, which delayed her urban and economic transformation. This was because Shanghai has always been perceived as the powerhouse of not only the regional, but also the national, economy. Beijing wanted to experiment with externally-oriented economics in areas of less economic and political significance, until, once proven, Shanghai could then follow. Thus Shanghai was first opened to FDI in 1984 when it was designated as an open coastal city, four years later than the opening of Guangdong. But Shanghai received a further boost to its external role and economic development as the centre of the new Yangtse River Economic Development Zone, designated in 1985. However it was not until the designation of Pudong in 1990 as a new economic zone with far-reaching incentives for investment and trade that

---

6 *For example, there has been massive airport development over the past ten years within both the Pearl and Yangtse River deltas, with each major city attempting to build a national or international airport; a more rational allocation of resources, involving co-ordination between all the cities concerned, would have produced one major airport, efficiently linked to urban areas by a high-speed surface transport network; instead Shanghai, Suzhou, Wuxi and Nanjing in east China are all building or planning new airports, while Hong Kong, Macau, Shenzhen, Zhuhai and Guangzhou, in and adjacent to Guangdong Province, all have massive new airports, several of which are operating considerably under capacity.*

Shanghai really began to take off. It was styled the 'Dragon's Head' of development in East China helping the Yangtse River Delta to unlock its potential. The past seven years have seen phenomenal growth in the metropolis, both economically and spatially. The economy grew at over 12% in 1996, and Shanghai has consistently outperformed the national average by two percentage points. Table 5.1 summarises Shanghai's economic and urban development from 1978 to 1995.

Table 5.1 Shanghai: Main urban economic indicators

	1978	1995
Population (10,000 persons)	1,098.28	1,301.37
Urban population (10,000 persons)	615.01	954.66
Urban population as % total population	54.00%	73.51%
GDP (100,000,000 yuan)	272.81	2,462.57
GDP per capita (yuan)	2,483.97	18,922.90
Annual GDP growth rate	18.43%	24.88%
Gross output value of industry (GOVI) (100,000,000 yuan)	514.01	5,349.53
GOVI per capita (yuan)	4,680.14	41,104.91
FDI (actually used, US$ 10,000)		324,994.00
FDI per head of population (US$)		249.73
Exports (US$ 100,000,000)	28.93	115.77
Imports (US$ 100,000,000)	1.33	74.48
Balance of trade	23.60	41.29
Urban unemployment (10,000 persons)	5.03	13.62
Unemployment rate	1.20	2.70
Average urban household income (yuan)	560	6,822
Personal computers per 100 households	-	2.2
Black and white TVs per 100 households	75	29
Colour TVs per 100 households	1	109
Refrigerators per 100 households	-	98
Washing machines per 100 households	-	78
Living space per capita (average) (sq.m.)	4.5	8
Population density (persons/sq.km.)	1776	2052

Source: Shanghai Statistical Yearbook, 1986, 1994.

5.7 Shanghai's External Economic Orientation

Shanghai has attracted multinational and transnational companies who are taking advantage of the access it gives to the wider Chinese market. Import values are particularly high, matching high domestic demand for consumer goods; export values are rising, but not as fast as imports. Shanghai appears to be a stronger regional than global competitor, with 57% of exports going to Asia in 1993, compared with 21% to north and south America and 15% to Europe. FDI availability in Shanghai has been growing continuously since 1978, and especially since 1990. Foreign capital used in Shanghai rose from US$ 108.8 million in 1985 to US$ 3175.0 million in 1993. During the Eighth Five Year Plan period (1991-95), FDI comprised 31% of Shanghai's total investment, a figure expected to rise to a third or a half over the next few years. Foreign investment in the first half of 1996 was up 13.3% on the first half of 1995 (China Daily). FDI is predicted to reach $49.5 billion in 1997, up 15% on 1994. At the end of 1995, 14,800 overseas invested enterprises were registered in Shanghai, with a further 4147 representative offices of foreign invested ventures. Forty-six of the world's top one hundred industrial corporations had operations in Shanghai at the end of 1994. In 1995, the approximate breakdown of investors by nationality was: Hong Kong: 46%, Taiwan 15%, USA 12%, Japan 11% and others 13%. 43.7% of representative offices were those of Hong Kong or Macanese companies, with a further 19.6% from Japan (Shanghai Star). This breakdown shows the dominance of investors from the rest of the Asian region, and the limited extent of global linkages. Tables 5.2 and 5.3 provide examples of overseas firms in Shanghai. 5.4 shows how non-state and non-collective enterprises have increased from zero to more than a quarter of all enterprises in less than 20 years. and 5 give a breakdown of Shanghai's industrial structure by organisational type, location and industrial type.

5.8 Economic Growth and Restructuring

The Shanghainese economy has undergone a period of economic restructuring since the mid-1980s. The contribution of the tertiary sector to Shanghai's GDP rose from 18.6% in 1978, to 24.1% in 1985 and again to 33.9% in 1993. 30% of foreign investors are now in the tertiary sector. By 1995, there were 200,000 tertiary sector enterprises operating in Shanghai (Shanghai Star). However, the secondary sector remains dominant in terms of its contribution to GDP and is also the fastest growing sector (5.6). Shanghai's industrial base is strong and in the long run Shanghai is likely to remain

Table 5.2   Overseas enterprises in Shanghai: Selected international firms

Company name	Sector	HQ country
Arthur-Anderson	Financial	USA
Coca-Cola Company	Manufacturing	USA
Credit Lyonnais	Financial	France
Daimler-Benz	Manufacturing	
DuPont	Manufacturing	USA
General Motors	Manufacturing	USA
Gillette	Manufacturing	USA
Hewlett-Packard	Manufacturing	USA
Hitachi	Manufacturing	Japan
Hutchinson Whampoa	Retail	Hong Kong
IBM	Manufacturing	USA
Intel	Manufacturing	USA
Johnson & Johnson	Manufacturing	USA
Kodak	Manufacturing	USA
McDonalds	Restaurants	USA
McDonnell Douglas	Manufacturing	USA
Mitsubishi	Manufacturing	Japan
Morgan Stanley	Financial	USA
Motorola	Manufacturing	USA
NEC Corporation	Manufacturing	Japan
Ricoh	Manufacturing	Japan
Sharp	Manufacturing	Japan
Siemens	Infrastructure	Germany
Sony	Manufacturing	Japan
Unilever	Manufacturing	Holland
Volkswagen	Manufacturing	Germany
Volvo	Manufacturing	Sweden
Xerox	Manufacturing	USA
Yaohan	Retail	Japan

a "dual-function" city, with emphasis on both the manufacturing and tertiary sectors. Nevertheless, urban planners place heavy emphasis on the development of the spatial framework to support Shanghai's tertiary activity.

Shanghai already houses one of the two Stock Exchanges in China and is making great efforts to attract overseas banks to the city. In 1991, the Chinese government permitted authorised foreign banks to open in the city for the first time since 1949. Shanghai is now one of only four cities in China

Table 5.3   Selected international banks in Shanghai

Bank name	*HQ country*
Bangkok Bank	Thailand
Bank of America	USA
Bank of East Asia	Hong Kong
Bank of Tokyo-Mitsubishi	Japan
Banque Nationale de Paris	France
Chase Manhattan	USA
Citibank	USA
Hong Kong and Shanghai Banking Corporation	UK
Industrial Bank of Japan	Japan
Standard Chartered	UK

Table 5.4   Shanghai: Industrial enterprises by ownership type

Year	Total		State		Collective		Other	
	Enterprises	*%*	*Enterprises*	*%*	*Enterprises*	*%*	*Enterprises*	*%*
1978	7962	100	3372	42	4590	58	0	0
1980	7149	100	3164	44	3914	55	71	1
1985	10656	100	4176	39	6141	58	339	3
1990	13220	100	4517	34	7122	54	1581	12
1995	15877	100	3713	23	7871	50	4293	27

Source: Shanghai Statistical Yearbook, 1996.

Table 5.5   Shanghai: Industrial enterprises by location and industry type

	Enterprises		Employment		Gross output value	
	Total	%	Total	%	Total	%
Total	39,908	100%	414.43	100%	4,543.47	100%
Central state	338	0.85	29.82	3.16	753.77	14.66
City	3,427	8.59	184.08	44.20	2,134.56	44.94
District	3,724	9.33	28.48	4.84	203.66	4.48
County	1,433	3.59	12.68	3.04	10.76	0.24
Neighbourhood	673	1.69	2.54	0.61	12.67	0.28
Town	1,367	3.43	14.11	3.87	171.51	3.77
Country	4,387	10.99	60.43	14.51	519.92	11.43
Village	11,367	28.48	63.79	15.32	514.02	11.35
Light industry	25,610	64.17	223.44	53.66	2,092.89	44.02
Heavy industry	14,298	35.83	192.99	44.34	2,454.57	53.98

Source: Shanghai Statistical Yearbook, 1996.

Table 5.6   Shanghai: Industrial structure by sector

Sector	1978 GDP (100m yuan)	1995 GDP (100m yuan)	1978 % share GDP	1995 %share GDP
Primary	11.00	61.68	4.0	2.5
Secondary	211.05	1,409.85	73.4	53.3
Industry	203.47	1,298.97	74.0	52.7
Construction	3.58	110.88	1.3	4.5
Tertiary	50.76	991.04	18.6	40.2
Banking and insurance	3.02	245.45	2.6	10.0
Real estate	0.27	91.29	0.1	3.7
All sectors	272.81	2,462.57	100.0	100.0

Source: Shanghai Statistical Yearbook, 1996.

where foreign banks are allowed to open sub-branches: the others are Dalian, Tianjin and Guangzhou. Pudong is the only location where foreign banks are allowed to undertake Renminbi business. In 1996 four banks - Citibank (USA), the Hong Kong and Shanghai Bank (UK), the Bank of Tokyo-Mitsubishi (Japan) and the Industrial Bank of Japan – were given permission to transact Renminbi business. As the Renminbi becomes convertible, the financial role of cities such as Shanghai, as the gateways to overseas financial and industrial markets, will be boosted. For the moment, however, Shanghai's international financial role is still limited, both regionally and further afield.

5.9 Urban Structure

In terms of basic urban structure, Shanghai is comprised of an urban core, surrounded by 6 suburban counties. Development has followed a deliberate plan in the post-49 period, and especially in the reform era, producing a central core city surrounded by satellite industrial towns and suburban expansion. The suburban areas are now characterised by a series of industria: estates of various sizes and degrees of separation from the suburban centres. Some of these areas have witnessed a clustering of firms of specific types or nationalities - such as the cluster of Japanese electronics firms in Songjiang, and automobile parts around the Volkswagen plant in Anting. The designation of Pudong as a *"xin qu"* in 1990 altered the development focus of Shanghai, shifting the city eastwards onto land previously little considered (by planners or industrialists) as a potential area for development. This has changed dramatically the layout of the central urban core, making the Huang Pu River no longer a boundary but a new urban focal centre.

Shanghai in the 1940s had a number of different urban styles and there was considerable functional differentiation across the urban space. After Liberation, however, following the ideology of the day, differentiation was lost, and Shanghai by the early 1980s was an intensely developed urban area with no clear spatial differentiation between residential, commercial and industrial districts, and between different social areas, and lacking in public open space. Development and planning in the post-Reform period are once again introducing functional and social differentiation to urban Shanghai. A particular goal, in response to major urban environmental problems, is to segregate residential and industrial space, and to push polluting industries to the periphery of the city, and to suburban counties.

In the urban districts, the principle features of Shanghai's current spatial form are as follows:

- multi-centric urban form with a single core and 3 district centres (Xu Jia Hui, Wu Jiao Chang and Putuo District);
- new development zones and areas with special development designations - e.g. Hong Qiao Economic and Technological Development Zone, Caohejing High-Tech Development Zone;
- clusters of high-rise commercial development but as yet no single distinct CBD - e.g. Hong Qiao, the Shanghai Centre/Nanjing Road West, Huai Hai Road Central;
- increasing spatial and socio-economic differentiation between districts;
- large-scale intra-urban infrastructure development - e.g. four subway lines (one completed, one under construction, two under planning), two major river bridges, and several cross-river tunnels linking Puxi and Pudong, and high-speed ring roads.

Redevelopment is inevitably the principal means for change at the local level. Gaining possession of whole street blocks is often difficult, resulting in a mixed landscape of pencil towers standing adjacent to low-rise dilapidated dwellings and shops. Building densities are rising, with no complimentary injections of open space or road width. The result is physically and functionally congested urban space. Another change as a result of redevelopment of this type is a change in street-level activity. In the old living quarters, external space was always very important; now activities are enclosed within glass curtain walls and imposing entrances emphasise the change in scale and a growth in the impersonality of urban form.

Shanghai today represents the consistent grafting of different architectural and urban design styles onto the city space over time - Chinese imperial and cultural tradition, colonialism, communist revolution and commercial reform are all represented in the physical fabric of the city. In the post-1949 period, the buildings of former foreign rule were kept, although their uses were changed and occupancy levels increased. Symbols of State control - national flags, red stars and propaganda slogans - as well as a limited amount of austere Soviet-style architecture, dominated the city before China opened its doors again to the outside world. Today this urban atmosphere is changing. Immigration is altering the social, ethnic and linguistic mix of the city. Commerce and consumerism are the dominant symbols of modern Shanghai, represented especially by the change in the secondary profile from propaganda to advertisements for commercial goods; while domestic products are increasing their advertising presence, the most prominent are the multinationals - Coca-Cola, Pepsi, Mirinda, 7Up, Visa, McDonalds, Kentucky Fried Chicken. As yet, no "Shanghai style" of architecture has developed. Many buildings are clearly inspired by, if not exact replicas of, buildings in other countries - e.g. a building very similar architecturally to La Grande Arche in Paris has been built in Liu Jia Zui; the new buildings on Huai Hai Road East are reflective of those in downtown Hong Kong. The outskirts of the city and its surrounding counties have undergone large-scale industrialisation over the past two decades, with factory development based around specific-purpose satellite towns and industrial estates/districts. While these influences are stronger in certain areas than others, the overall sense is of a city actively striving to return to its cosmopolitan character, almost as a reaction to the socialist state under which it operates.

5.10 Distribution of Urban-Industrial Activity

Three types of overseas-based organisation are evident in downtown Shanghai:
- the representative offices of companies in all sectors - e.g. Marks and Spencers, which, like many other international companies, is watching the market and building contacts before committing to production and retailing;
- the management offices, sales offices and service centres of firms which have manufacturing functions located in suburban Shanghai or neighbouring provinces - e.g. Walls has a sales office in Shanghai and a factory in Taicang, Jiangsu; Hewlett Packard has a management of-

fice and repair centre in downtown Shanghai and a factory in Pudong;
- the operational offices of service and financial sector firms - e.g.
Citibank has several branches in Puxi and Pudong.

The manufacturing operations of foreign firms tend to locate in specifically designated industrial or trade zones, depending on the nature of their operations - e.g. high technology firms locate in Caohejing, and export processing firms in Waigaoqiao. Joint venture operations will often be located in the existing factory of the local partner.

There has been a strong decentralising trend in industrial activity in Shanghai, backed by local government intervention. Since the 1970s, large numbers of industrial enterprises of all sizes have been moved from the downtown area to the outskirts of the city and to surrounding counties. This served as a strong urbanisation force in itself for the rural areas around Shanghai. Three formal satellite towns have been established around Shanghai, each based on a particular industrial sector. Small factories and workshops still persist in the central city, but large scale industry has now been moved out.

When looking at the location of overseas firms within central Shanghai, a certain pattern can be observed. Firstly, in the case of almost all sectors, there is a strong bias towards office location within Puxi rather than Pudong. The most commonly cited reason for wanting offices in Pixie is proximity to existing client networks and service facilities, as well as workers' residential preferences and journey-to-work patterns. But permission for new firms in the foreign sector to begin operations in Shanghai are tied in part to proposed location, especially due to the government's aim to develop Pudong. Government "carrots and sticks" are important in influencing company location decisions: for example, new business licenses are now only being granted to foreign companies operating out of Pudong.

Within Pixie, the distribution of overseas companies matches to a large extent the distribution of quality office space, as well as the location of international hotels, from which many companies still work. In the case of American companies in both the service and manufacturing sectors, the highest concentration of firms is found in the 200040 postal district - the area around the Shanghai Centre which also includes a number of major hotels, plus Nanjing Road West. Another concentration is in the 200335 postal district, where Hong Ciao ETDZ plus the International Trade Centre are located. A 1996 survey revealed that on average the top 6 office buildings in terms of quality were 93% occupied by overseas companies (Jones Lag Woollen). The Shanghai Centre is 100% foreign occupied, including not only business users but also several consulate offices. Companies in the manufacturing sector are located in similar numbers to service providers in the downtown area, although companies in the financial sector seem to be more concentrated in 200040 and 200002 districts while manufacturing firms are more spread out. This pattern reflects high downtown rents[7] which are prohibitive to most Chinese companies.

The location pattern for Chinese companies is different from that for overseas companies. Private firms are increasing in number and size but their financial resources are such that they tend to seek lower quality, lower priced space in less central locations. In the State sector, location is a factor of history rather than design. Overall, Chinese firms are less clustered in certain locations, although this is likely to change, especially with increasing participation in the tertiary sector and a developing land market. Chinese firms have shown more willingness to locate in Pudong, possibly because of large rental differences with Pixie, or, as in the case of banks, according to government direction.

This pattern of activity is typical of a city undergoing economic transition and where government policy and regulations, and the property market, are not stable. As office supply increases and more overseas firms move into production, the presence of manufacturing companies in the downtown area is likely to lessen, while service sector companies will have greater choice in their location. As Liu Jia Zui is completed, financial companies are likely to make the move across the river. The planned development of Pudong International Airport may have an interesting effect on company location; it would seem likely that international companies requiring frequent overseas travel or using air freight will move their operations to Pudong, while companies locating near Hong Qiao airport, to be Shanghai's domestic airport, will be increasingly Chinese.

5.11 Institutional Influences and Urban Planning

The market is not being allowed to determine development outcomes free from government intervention. The first *Comprehensive Plan for Shanghai* was approved by the State Council in 1984. Revisions to the Plan have not yet been ratified at State level, although they have been made and put into practice. Overall, the structure of Shanghai as we see it today is as laid down in the 1986 plan, with one major addition: development in Pudong, which was not envisioned nor suggested by Shanghai's planners in the 1980s. It was a State Council decision in 1990 which laid Pudong open for development and changed the balance of development in Shanghai between Pudong and

---

7 *Grade A office rents in Shanghai are among the highest in East and South-east Asia, and approximate those in Hong Kong and Tokyo.*

Puxi. The basic development principles and pattern represent a hierarchy of development levels: CBD, Central Cultural District (CDC), district centres, inner ring development, outer ring development and 6 suburban towns. Revisions in the mid-1990s have stressed the following areas: (1) expansion of the 3 industrial satellites to populations of 1 million each to give them a degree of self-sufficiency in tertiary functions; (2) raising downtown densities on developed sites to release land for open space; (3) increasing focus on tertiary functions, especially in the central city; (4) integration of land use and transport planning, and increased general emphasis on infrastructure development; (5) greater emphasis on the preservation of buildings of historical value; (6) greater consideration of environmental issues and increased green space planning, and reduced emphasis on "key issues"[8] alone.

It is up to individual districts to implement the section of the plan relevant to their areas and to carry out detailed local level planning and land leasing. Planning for key areas of the city - the central commercial and business districts - tends to be fairly closely monitored at the city level, but in non-key areas city level control and co-ordination has been relatively lax, resulting in inconsistencies with the plan. As the decentralisation policy continues, so these discrepancies are likely to increase, with no city-level body to ensure the comprehensive implementation of the plan for the urban area as a whole. The Comprehensive Plan provides a very poor guide to detailed local level planning, while those responsible at the district level are often not trained professionals. As districts are subdivided and specific areas handed over to development corporations[9] the lack of cohesion and other problems are intensified. The plan becomes responsive to the market, rather than pro-active in directing development. The principal role of the local-level plan has devolved not into a guide or control for development, but into a tool in the marketing of land. The plan, complemented by two- and three-dimensional CAD images, is used as an indication of future improvements in currently very run-down areas in order to attract investment. Outside designated development zones, there is little attention paid to renewal. This reflects the prioritisation of urban development action in the face of limited resources. The government rationale in this approach is therefore that land for high-

---

8 *Industry, housing and infrastructure.*

9 *It is common for districts to identify key sub-areas for development for specific purposes - e.g. retail centres such as Xu Jia Hui and Wu Jiao Chang in Xu Hui and Yang Pu Districts, or high-technology parks, as in Yang Pu. Public sector development corporations are then created by the district government with responsibility for the detailed planning and development of these sub-areas, including infrastructure provision and the marketing and leasing of land. These development corporations have extensive decision-making powers relating to both land development and the attraction of investment and economic activity, but are frequently controlled by politicians rather than development professionals.*

value commercial and industrial uses must first be leased in order to generate revenue for more general district upgrading.

The fragmented land supply, which is mostly in the hands of individual danwei, is also influencing urban outcomes. Suddenly finding themselves with economically valuable surplus land, often in key areas of the city, factories and universities are constructing apartment and office blocks, restaurants and shops, and letting them out for profit. As there is no transfer of land, the government's scope to intervene in this activity is limited. Alternatively, land holders may enter into joint venture arrangements with overseas developers or industrialists; this is becoming a common practice within Shanghai's universities where new buildings are being built on campuses to house foreign business activity - e.g. thirty small-scale German industrialists are currently operating from the German Business Centre at Tong Ji University. Due to difficulties in gaining possession of land, the scope for comprehensive redevelopment is limited, although compulsory purchase has been used for major infrastructure projects. The costs of rehousing residents, which are borne by the developer, can also be prohibitive. Real estate developers are in the main constrained by the land release programme in their choice of development location, although some powerful overseas developers have been able to acquire sites on request. Non-central industrial sites and the new development zones do not have such land supply constraints, as they are in the main constructed on greenfield sites or former agricultural land.

District competition is further affecting Shanghai's urban structure. Districts are offering various concessions to developers in order to attract them to their areas, and various other incentives are also being offered to consumers, such as 'blue cards,' a form of residence permit. Each district is striving to attract foreign investment and hence to market itself as a commercial centre; in some cases however this is incongruous with baseline district characteristics. Shanghai is still in a transitional period, but as the economy and property markets mature, the less attractive districts are likely to find themselves left with unlettable space and functional mismatches. This is especially true in the old industrial districts to the north of Suzhou Creek, which, despite their basic characteristics, are attempting commercial development. The plan calls for CBD development in the Bund area and in Liu Jia Zui and in the long run this locational pattern is likely to be realised, although the timing is unclear.

5.12 Limitations of Urban Planning in Shanghai

Most large cities have now produced, or are in the process of producing, city and district plans[10], but problems have arisen with their implementation. In large part these problems reflect the immaturity of the planning process and the bureaucratic framework within which it operates. Both forward planning and development control are compromised. The principle limitations to an effective urban planning process in Shanghai are:

5.12.1 Quality of Urban Decision-Makers

Although professional planners are increasing in number[11], and universities and design institutes are being brought in as consultants to projects, planners at the district level especially are often under- or unqualified for the job they are attempting. For example, the renewal of Yang Pu District and the creation of a high tech park in the area is led by a Party official with experience in the education of cadres rather than in urban and economic development. University planning education also focuses on physical planning with a disregard for the economic and social development of the city, perpetuating the lack of vision in planning practice. Another problem is that it is often not planners themselves who are managing projects or making ultimate development control decisions, but rather government officials whose interests may not coincide with those of the planners, especially with regard to non-productive space. In practice, plans are usually conceptualised by cadres, with planners being delegated the responsibility for drawing them up on the basis of instruction from the political sphere, rather than with broad consideration of the public interest (Ng and Wu, 1995).

5.12.2 Quality of the Plans

The plans produced, though usually very detailed, do not offer much scope to guide development patterns, especially at a comprehensive level. They are purely physical in nature and the basic design unit is the individual site. There is little attention paid to urban design and the interaction of different spatial elements in the overall form of the city. The main function of district plans is to specify development conditions for specific sites, and/or the basic layout of special development areas and zones. They therefore es-

---

10 *Plan coverage : By the end of the 1980s, 339 designated cities (96%) and 1695 county towns (85%) had produced comprehensive urban plans. 282 and 1293 plans respectively had been approved (Ng and Wu, 1995). Revision of these plans over the past decade in line with development has however been questionable; even Shanghai's current official plan is the one approved in 1986, despite a number of subsequent non-approved plans.*
11 *China had 15,000 urban planners by the end of the 1980s.*

sentially provide the baseline for urban development, rather than any form of co-ordinated guidance and strategy. Implementation of the plan is reliant on external actors - i.e. the investors government officials want to attract - and with no statutory authority the scope for amendment is wide. Most plans, at the city as well as the local level, are very ambitious and often do not relate sufficiently to current needs and realities. Much effort is wasted in preparing long term plans, with realisation dates in the middle of the next century. Plans ought to be revised every five years: yet the Shanghai plan was approved in 1986 and subsequent amendments have yet to be ratified, despite a gap of over ten years. Complimentary sectoral reforms, as have been seen in industry and housing, compound the problem of creating an up-to-date and useful plan (Ng and Wu, 1995), especially as such reform decisions are taking outside the planning system. Planners have no power to co-ordinate spatial and non-spatial policies (ibid).

5.12.3 Clarity of the Planning Process

It is often very difficult for a developer to determine the exact nature of the planning process in a particular area as the land administration and planning authorities appear to adopt a case by case, developer by developer approach to the approvals process. Negotiation is a part of the system and can result in changes in development conditions, negating the impact of the plan per se on development outcomes.

5.12.4 Motives in Planning and Development

District governments are in the main driven by economic imperatives. Pressing urban problems do not receive the attention they need. The city and some central districts have over the past year or so begun to realise the implications of this bias in favour of economic development and taken action to amend plans and urban decision-making behaviour. In particular, emphasis has recently been placed on green space and low cost residential planning. Again however not every district has made the necessary changes, reflecting the level of economic development already achieved, district revenue balances and professional competence. With the power of the plan limited, urban development comes to be manipulated by the few who have access to political power (Ng and Wu, 1995); thus, in a development process dominated by the interests of industry and commerce, political and economic stakeholders control urban outcomes.

Since 1993, there has been a recognition in Shanghai that the rapid pace of growth and change may be having unwanted effects on the city. Recent revisions of the *Comprehensive Urban Plan for Shanghai* express a need to build up local urban resources, especially low-cost housing and health, recreational and transport infrastructure, and improve the urban environment in general. The scope for implementing these new planning ideas, especially at the district level, may be restricted however by the lack of professionalism and a continuing reliance on the attraction of FDI as a means for economic and urban development.

5.13 Conclusions

Urban and economic development patterns in China since 1978 have been a function of the changing political relationship between the nation-state, international capital and the public interest. A combination of domestic reform in the urban-industrial sector and the external-orientation of the economy has facilitated economic development and produced spatial restructuring at the urban and regional levels. The development process has been dominated by a political goal to attract foreign investment to assist urban-industrial restructuring and urban development in general, a goal supported by favourable sectoral and spatial policies and incentives. This has led to a re-orientation of the development process in favour of the interests of capital, especially overseas capital, and hence a restructuring of urban space and land use according to the needs and characteristics of that capital.

Since opening to the regional and global economies through trade and investment, Shanghai has witnessed phenomenal economic growth and extensive urban restructuring. However, the extent to which Shanghai acts as a growth pole for the domestic economy can be questioned. In the absence of regional planning, the locational proximity and economic linkages in this area are not being fully exploited, and competitive rather than co-ordinated development is observed. This is leading to a duplication of resource expenditure, and may affect the region's long-term ability to compete internationally for investment. In essence, Shanghai's role within the Yangtse River Delta is as a provider of central tertiary functions and a bridge to overseas markets.

This case study of Shanghai prompts a number of policy recommendations. The role of urban planing in guiding and controlling development needs to be strengthened, with closer attention paid to the public interest relative to business and political interests. Some improvements have been made in Shanghai since 1993, but planning still requires greater professionalism and increased intervention powers in the development process. Increased co-ordination is required between sectoral and spatial planning, between different reform policies, and between the government authorities responsible for land, urban planing and economic development. Shanghai's central role is likely to be constrained by the recent decentralisation of decision-making authority from the State to the local level, despite generally positive economic trends. This suggests the development of effective regional planning mechanisms in order to provide a framework for the efficient allocation of resources and a base for national and international competition.

At a more general level, a balance should be struck between the needs of Shanghai as an international centre, with economic and urban development being oriented to making Shanghai into a 'world city,' and local needs for a stable and healthy living environment. This relates to the extent to which development based around external capital can create grass-roots development, and the nature of partnerships and/or competition between local and international business interests. Development processes and outcomes typify the policy dilemma facing governments in developing countries, where shortages of domestic revenues for urban and industrial development force a subjugation of national interests to those of international capital. China is experiencing this basic policy dilemma in all sectors and at all spatial levels: externally-initiated growth is required to kick-start a lagging economy although the long-term affects of such an approach may not be desirable.

Shanghai stands at the interface of the world economy and the Chinese nation state. This influences its economic and political imperatives, and its spatial and social structure. The discussion in the previous sections has shown how Shanghai has since Economic Reform and opening been increasingly influenced by her participation in the international market-place; at the macro level, this is in terms of increasing levels of foreign trade and FDI receipts, at the meso level, in the locational choices of multinational companies and by the activities of overseas real estate developers, and at the micro level in the internationalisation of the socio-cultural sphere, including the visual image of the city, symbolism, consumer opportunities, and architecture.

Acknowledgements

I am extremely grateful for Dr Fulong Wu's comments on the first draft of this chapter and his suggestions for amendment. I would also like to acknowledge the input of Dr Bernard Fingleton, Dr Peter Nolan and Dr Roger Chan. Responsibility for the opinions expressed however rests entirely with the author.

References

Castells, M (1996) The Rise of the Network Society *Oxford: Blackwell.*

Chan, RCK, Hsueh, TT and Luk, CM (eds.) (1996) China's regional economic development *Hong Kong: The Chinese University Press.*

Chen, XM (1991) "China's city hierarchy, urban policy and spatial development in the 1980s" in Urban Studies *28:3 pp. 341-363.*

China Daily, *various editions.*

Eng, I and Lin, Y (1996) "Seeking competitive advantage in an emergent open economy: foreign direct investment in Chinese industry" in Environment and Planning A *28 pp. 1113-1138.*

Gong, HM (1995) "Spatial patterns of foreign investment in China's cities 1980-89" in Urban Geography *16:3 pp. 198-209.*

Hall, P (1984) The World Cities *London: Weidenfeld and Nicolson (Third edition).*

Harvey, D (1989) The Urban Experience *Oxford: Blackwell.*

Hein, S (1992) "Trade strategy and the dependency hypothesis: a comparison of policy, foreign investment, and economic growth in Latin America and East Asia" in Economic Development and Cultural Change *40:3 pp. 495-521.*

Hs,, ICY (1995) The Rise of Modern China *New York: Oxford University Press (Fifth edition).*

Hu, ZL and Khan, MS (1997) "Why is China growing so fast?" Washington DC: International Monetary Fund, Economic Issues Series, Number 8.

Kim, WB (1991) "The role and structure of metropolises in China's urban economy" in Third World Planning Review *13:2 pp. 155-175.*

Knox, PL (ed.) (1993) The Restless Urban Landscape *Englewood Cliffs, New Jersey: Prentice Hall.*

Leung, CK (1996) "Foreign manufacturing investment and regional industrial growth in Guangdong Province, China" in Environment and Planning A *28 pp. 513-534.*

Ng, MK and Wu, FL (1995) "A critique of the 1989 City Planning Act of the People's Republic of China" in Third World Planning Review *17:3 pp. 279-293.*

Romer, PM (1986) "Increasing returns and long run growth" in Journal of Political Economy *Vol.94 pp. 1002-1033.*

Romer, PM (1993) "Two strategies for economic development : using ideas and producing ideas" in Proceedings of the World Bank Annual Bank Conference on Development Economics *Washington DC: World Bank.*

Sassen, S (1991) The Global City: New York, London, Tokyo *Princeton, NJ: Princeton University Press.*

Sassen, S (1994) Cities in a World Economy *Thousand Oaks, CA: Pine Forge Press.*

Schachar, A (1994) "Randstad Holland: A 'World City'?" in Urban Studies *31:3 pp. 381-400.*

Shanghai Municipal Statistics Bureau (1986) Shanghai Statistical Yearbook 1986 *Shanghai: China Statistics Publishing House (in Chinese).*

Shanghai Municipal Statistics Bureau (1996) Shanghai Statistical Yearbook 1996 *Shanghai: China Statistics Publishing House.*

Shanghai Star, *various editions.*

Sit, VFS and Yang, C (1997) "Foreign-induced Exo-urbanisation in the Pearl River Delta, China" in Urban Studies *34:4 pp. 647-673.*

South China Morning Post, *various editions.*

State Statistics Bureau (1996) China Statistical Yearbook 1996 *Beijing: China Statistics Press.*

Thrift, N (1994) "Globalisation, regionalisation, urbanisation: the case of the Netherlands" in Urban Studies *31:3, pp. 365-380.*

Todaro, MP (1994) Economic Development *London: Longman.*

Wong, SY (1994) *"Globalisation and regionalisation: the shaping of new economic regions in Asia and the Pacific"* in Regional Development Dialogue *15:1.*

World Bank, The *(1993)* The East Asian Miracle *New York: Oxford University Press.*

Yan, XP *(1995) "Chinese urban geography since the 1970s"* in Urban Geography *16:6 pp. 469-492.*

Yao, SM and Liu, T *(1995) "The evolution of urban spatial structure in the open areas of Southeastern coastal China"* in Urban Geography *16:7 pp. 561-574.*

Yeh, AGO and Wu, FL *(1995) "Internal structure of Chinese cities in the midst of economic reform"* in Urban Geography *16:6 pp. 521-554.*

THE URBAN BASE:
THE RUSSIAN FAR EAST

Editors' note:
With the demise of central planning and the anarchic implementation
of a 'market economy', the Russian Far East is having to survive on its
own. Some of the immigrant Russian population is leaving, some of
the indigenous population are rediscovering older ways of living. With
a low population density and a location on the Pacific Rim, this region
is turning to its neighbours, Korea, Japan and China. One of its attrac-
tions is its great wildernesses, in some ways comparable to Alaska's,
but not yet as well developed for tourism.

6 Economic Development, Tourism, and Urbanisation in the Emerging Markets of Northeast Asia

Brian J. Sommers and Dallen J. Timothy

6.1 Introduction

With the opening of its borders and the emergence of a market economy in Russia, there are growing possibilities for Northeast Asia. Economically, the natural resource wealth of the Russian Far East (RFE) has provided regional balance to the technological wealth of Japan and the Republic of Korea (South Korea), and the labour wealth of the People's Republic of China. This has led to development opportunities within Northeast Asia which were impossible under state socialism in the former Soviet Union.

The opening of Russia's borders has also provided development opportunities through trade. In the RFE, this trade-related potential is not just linked to the region's resource wealth. It is also a factor of relative location and the region's transportation system. Although it provides a poor level of service within the region, the existing transportation system does provide for shipment through the RFE. With updating and expansion, this system could make the ports and rail cities of the RFE key transhipment points for trade between the economic powers of Europe and East Asia.

The nascent market economy in Russia and the development of regional economic linkages based on Russian natural resources is also having an impact on tourism. Tourism is emerging in Northeast Asia as another potential tool for economic development. While tourism is limited at present, especially when compared to resource and trade-based development, there is a great deal of potential. The size of the regional tourism markets alone makes cross-national tourism linkages a potentially lucrative source of income. Besides supporting growth in the region's transportation and communications infrastructure, tourism could also play a part in the region's development simply by making people aware of the RFE and what it has to offer.

While resource-based development, either through extraction, harvesting, or eco-tourism, is concentrated in often isolated locations of the RFE, the result of that development is increasing regional urbanisation. In spite of very low population densities across the RFE, rates of urbanisation are as high as anywhere in Russia. This is reinforced by the transportation system and centralised resource-refining infrastructure. Thus, while the resources may be in rural locations, the existing network of railroads, airports, roads, and refineries ties the resource-based development to a series of urban nodes. This pattern is further reinforced by trade through the region. The urban centres which serve as major transportation and industrial nodes within the RFE, also serve the trade and transport system through the RFE. While it may seem rather contradictory, this makes one of the world's most isolated and peripheral regions an area of great potential urban growth. It is this pattern of economic and infrastructure development and its potential impact on urbanisation in the RFE that is the subject of this research.

6.2 Understanding the Russian Far East

To understand the Russian Far East, there are a number of important factors which bear consideration. Some factors such as the region's isolation, its natural environment, and its resource endowment are inherent to the region and are thus separate from any consideration of past development strategies. Other factors such as the region's transportation and communication systems, and its economic and political environments can, however, be linked directly to past regional development strategies and to the wider concerns of the Soviet state.

6.2.1 Location, Environment, People

The one constant factor in evaluating the RFE as an arena for development is its location. The RFE has always been isolated from the core areas of Russia (Map 6.1). Even with air transport, travel times and distances involved are great. It is over 7200 km (4500 miles) from Vladivostok to Moscow. It also takes longer to fly between Khabarovsk and Moscow (6+ hours), than from Khabarovsk to Seoul or Tokyo (1 hour), to Beijing (1.5 hours), or even to Anchorage (5 hours). The absence of reliable transportation in much of the Far East has only served to accentuate this isolation from the rest of Russia. The region's isolation from the core of Russia, its proximity to the countries of Northeast Asia and the Pacific Rim, and the availability of ocean transport have led to a more outward looking attitude amongst the people of the RFE. It has also led to increased ties between the RFE and Japan, China, Korea, and the U.S.

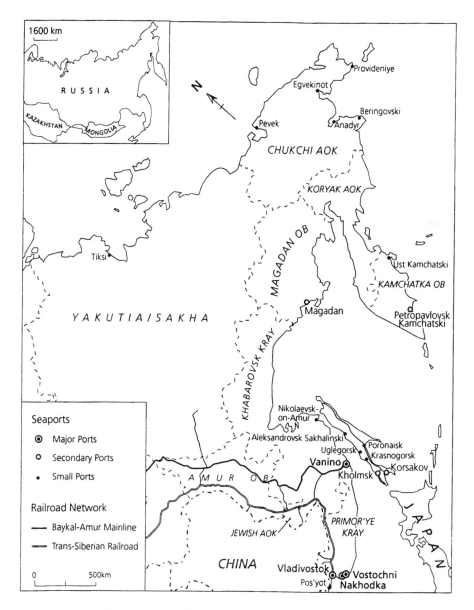

Map 6.1 The Russian Far East

The other constant for the RFE has always been its environment. Few areas can boast either the variety and quantity of economic resources that the Russian Far East can. The region is rich in non-renewable resources such as precious metals (gold) and stones (diamonds), ferrous and non-ferrous (copper and tin) metals, and hydrocarbons. The Far East also has a rich storehouse of renewable timber and fish resources.[1] While poorly developed and often inaccessible, these resources constitute an important part of the region's economic future.

The RFE is also rich in wildlife and scenic beauty. The intrusion of the occasional Siberian tiger into the region's urban areas in search of food (i.e. dogs) underscores the abundance of wildlife in the Far East (Klebnikov 1995, 87). The abundance and variety of wildlife is, on the one hand, a factor of the sparse human population and, on the other, of the variety in the region's biomes. The scenic beauty of the region is evident in the picturesque coastline of Primorye, the volcanoes of the Kamchatka peninsula, and of Lake Baikal (Findlay and Edwards 1990, 207; Klebnikov 1995, 90).

While the resources of the Far East provide it with a wealth of opportunities, these opportunities are checked by climatic constraints. While some areas (primarily on the east coast) have only limited climatic constraints to human activities, areas of the interior have some of the world's harshest environments. Extreme cold, wind chills, permafrost, flooding, extremely short to non-existent growing seasons, and long periods of darkness make many areas exceedingly difficult if not impossible to develop. Not only does this impact new development, but it also dramatically increases the frequency and cost of basic maintenance (Grace 1995, 71; Khartukov 1994, 26; Mote 1983).[2]

The isolation and environment of the Far East have a considerable impact on the quality of life experienced by those living in the region. In most areas, climatic constraints limit the ability of the land to support large populations and severely impact the region's desirability as a place to live. As a result, the Far East is sparsely populated with only about 7.8 million people in an area the size of China. This gives the region as a whole an average population density of less than 3 people per square mile. With over 5 million people living in the oblasts along the Chinese border, population den-

---

1 *The timber reserves of only a part of the region (Primorye, Khabarovsk, and Amursk) are almost half that of the entire U.S. (Klebnikov 1995, 88). In addition, the Sea of Okhotsk, once the private domain of the Soviet Pacific Fleet is one of the world's most productive fisheries (Klebnikov 1995, 87).*

2 *Victor Mote's chapter, "Environmental Constraints to the Economic Development of Siberia," in Robert Jensen's Soviet Natural Resources in the World Economy, presents an extremely detailed study on the extent of the environmental constraints within Siberia and the Russian Far East.*

sities average well below 1 for most of the Far East (Miller 1996, 79).

In spite of the fact that it has no cities of more than a million persons, the region is heavily urbanised. As a whole over 76% of the region's population lives in urban areas with some jurisdictions reaching rates over 85% (Minakir 1994). While rates of urbanisation did dip in the 1970s, this was a factor of increasing resource exploitation in the region and not an indicator of urban decline. In fact, the drop in the urbanisation rates was a relative one as all major cities in the region experienced population growth over the period (Minakir 1994). Then, as now, the bulk of the population resides near the Chinese border and near ocean ports (Klebnikov 1995, 87; Miller 1996, 79).

The urban populations of the RFE are primarily European, while those of the region's rural areas tend to be Turkic (Yakut) or Asiatic (Evenki, Eveni, Chukchi, Koryak). While literacy in the population is considered to be universal, education levels do vary significantly within the region and between ethnic groups. Many of the major non-Russian ethnic groups form titular nationalities of autonomous oblasts and okrugs in the RFE. Despite the political status given to some of these groups, their numbers are small. Over 90% of the population of the RFE is made up of ethnic groups derived from European Russia (Russians 80.0%, Ukrainians 8.0%, Belorussians 1.5%, Tatars 1.0%) (Miller 1996, 79).

6.2.2 The Soviet and Post-Soviet Context

While some elements key to our understanding of the RFE are specific to the region's locational and/or environmental contexts, others are tied to the history of Russian national politics and economics. In *Soviet Asia: Economic Development and National Policy Choices*, Leslie Dienes (1987) explored the prospects and policies behind Soviet Far East development into the late 1980s based on the region's historical context. Within this text, Dienes postulated three models for Far East development. Common to the models were problems and limitations of politics which were inherent in the system of the day. Primary among these were;

1) Limitations in local autonomy which required extensive interaction with central authorities.

2) Problems of uneven development and a poor mix of economic activities.

3) Lack of infrastructure.

The politics and economics of the day also produced very formidable hurdles to both foreign investment and development which were common to areas outside the Far East. Beyond facing the more localised problems listed above, outside investment faced;

1) Poor political, economic, and social linkages to the USSR in general and the RFE in particular.

2) A one-way relationship which provided the Soviet Union with hard currency for resources, but which failed to open Soviet markets to foreign products.

3) A business environment stifled by poor quality products and low worker productivity.

4) An investment environment far from conducive to outside intervention in these areas of concern.

These development concerns of the Soviet era, form the context of current Russian Far East economic and political issues.

6.2.3 Economics

In terms of economics, the RFE constitutes the furthest periphery of Russia. Isolated by distance from the major commercial and industrial centres of the European core of Russia, the RFE has in many respects been both out of sight and out of mind when it has come to the Russian economy.

In spite of its remoteness, the Far East did experience above average levels of state investment under a Soviet system which stressed a more balanced approach to economic development, possible because of strategic reasons. Consequently, peripheral regions like the Far East, while lagging behind the rest of the country in development, did experience more investment in relative terms than would have been expected in a market economy.[3] While the Far East produced 1.29% of the national income in 1991, its share of national consumption was 1.65% (Kirkow and Hanson 1994, 66-67). One example of how this somewhat parasitic relationship to the centre has aided development in the RFE is in state investment in heavy industry.

As a result of regional economic strengths and the development strategies that existed within the Soviet system, the RFE has a fairly significant heavy industrial base in the production of machine equipment and fertilisers. As is the case in the rest of the former Soviet Union, however, this heavy industry has not fared well in the transition to a market economy. Factories geared to the Soviet system face heavy internal and foreign competition due

---

3 *In addition to the ministerial structure of the economy and regional dependency concerns, the strategic value of the Far East also influenced state investment. The region's warm water ports on the Sea of Japan and its long border with China were all factors in its development under the Soviet system. As a result, the strategic value of the region led to higher levels of government investment than might otherwise have been predicted.*

to perceptions, just or unjust, of low quality and outdated technology. Coupled with low output and ongoing labour disputes, this presents problems for industrial expansion. While its products may not be good competitors in the world market, within Northeast Asia the products of the RFE have faired better. This is especially true in its trade dealings with China. The sheer size of the Chinese market for heavy equipment and fertilisers means that even a small presence in that market can pay significant dividends. While Far East industrial products are also competitive in North Korea, they are far less so in South Korea and Japan.

The RFE has, over past decades, developed as a source of natural resources meant for use elsewhere in Russia and the former Soviet Union. Many of the economic activities in the region are based on resource exploitation and are linked to end users outside the region (Swearingen 1987). This has created an economy which produces little in the way of consumer goods. Along with the small size of the internal market, this has created an economy dependent on heavy industrial exports and on light industrial imports. In a study of regional economic dependency, the RFE was found to have a regional dependency coefficient of 0.71, placing it in the middle of the eleven economic regions of Russia.[4]

Also important in an examination of regional dependency is the extent to which the products of the more independent regions are competitive in a larger world market. Those regions which have products that are competitive on a world market, primarily regions rich in natural resources, stand a better chance of being able to take advantage of free trade as a basis for economic development. For older industrial regions, the absence of strong economic dependency is not as important since it cannot compete as well in a world market. They are relatively independent, but cannot take advantage of this (Matsnev 1994).

6.2.4 Politics

The Far East has traditionally been in the furthest periphery of Russian politics as well. However, the political relationships between the Far East and the centre are quite dynamic and do directly impact development in the region. While the RFE may be inherently weak in some economic sectors, it is in an advantageous position with respect to Russia's political ambitions in Asia. It is the window to any Russian policies dealing with Northeast Asia. It

---

4 *The dependency values ranged from 0.49 in the Urals to 0.92 in the Central Chernozem region. While just an economic model, the variations in region dependency predicted by the model are supported by the political and economic realities in the regions (Matsnev 1994, 4).*

can also turn to outside sources of assistance in its dealings with the political centre of Russia. This provides the region with leverage in its internal affairs with Moscow and in trying to overcome its own weaknesses. The frailty of the Russian political centre only strengthens the position of the Far East to the extent that the region can strong-arm Moscow as a payment for co-operating with Russia's Asian policies (Christoffersen 1994, 516-517).

Outside of Russia, the RFE's political situation is also potentially strong. In its dealings with the countries of Northeast Asia and the Pacific Rim, the RFE can use its natural resource base and transportation access to European markets as bargaining chips in its political dealings (Christoffersen 1994, 516-518).[5] However, the RFE's need for outside support in exploiting its natural resources and in improving its transport system does draw significantly from its bargaining power.

An additional political complication for the region is its history of border conflicts. Ongoing border disputes between Russia and China as well as between Russia and Japan add to the complexity of the region's political environment. These disputes, along with increasing Russian nationalism, have also had the effect of producing ill will between ethnic Russians and non-Russian immigrants (Christoffersen 1994, 519-520; Klebnikov 1995, 89). As the region's economies grow together the environment for resolving territorial issues may improve (Carlile 1994, 427).[6] In the meantime, ethnic tensions are only intensified by border disputes which have had an unfortunate tendency to end in military or border guard actions (Christoffersen 1994, 520). Despite these border issues, ethnic tensions between the peoples of the RFE are as quiet and peaceful as anywhere in Russia.

With a relatively weak central government in Moscow, and with its political isolation from the Russian core, the RFE is for almost all intents and purposes autonomous. Although in a practical sense autonomous, there has to this date been no concerted drive for complete autonomy.[7] This may be the result of ethnic ties to the Russian heartland. It may also be a rationality which recognises that such a small and isolated population cannot support a

---

5 *This is especially true with respect to the Tumen River Project and expanding Russian linkages to Asia. Far East authorities, especially Primorye, have independently pursued political allies and trading partners in the region with little or no regard for the political intentions of the central authorities in Moscow (Christoffersen 1994).*

6 *For an in depth examination of the political economy of the relationship between Russia and its Northeast Asian neighbors see Carlile 1994, Christoffersen 1994, and Milner 1990.*

7 *This is in spite of a poor economic and social environment, and a political system dogged by allegations of corruption. In part, this may be a practical acknowledgement of the region's dependency on the Russian center. This may also be due to fears that corruption could increase with more regional independence. Regional drives for autonomy could also have been dampened by Russian government's response to the situation in Chechnya.*

developed, self-sufficient economy. An additional problem is that the econo-mies of the Far East are in competition with one another because they pro-duce the same slate of products. The virtually non-existent linkages between them also prevents integration even if they did have something to offer one another (Aliev 1994, 253). Consequently, any region-wide bid for complete autonomy would likely be only a first step toward autonomy for all of the political units of the region.

6.3 Transportation

One of the Far East's most valuable resources is its transportation sys-tem. Via ports in Primorye, the Trans-Siberian Railroad connects markets in Asia and Europe at comparable costs but faster than sea-based transport (BISNIS 1995i; Klebnikov 1995, 88). Important nodes in this system are the cities of Vladivostok and Khabarovsk. Vladivostok, the home of the Pacific Fleet and a closed city until 1990, is the major trans-shipment point for the Trans-Siberian Railroad (BISNIS 1995i; Hoff 1994, 19). Khabarovsk is the Far East's primary rail centre and boasts the largest international airport in the region (Hoff 1994, 19; Paisley and Lilley 1994, 41) (see Fig. 6.2).

In addition to Vladivostok and Khabarovsk, the port cities of Nakhodka, Vostochni, and Vanino are also major transportation centres. Nakhodka is the largest Far East port and is the home to a self-declared free trade zone (duty free zone). Vostochni, just north of Nakhodka is an important point of trans-shipment for natural resources bound for Pacific Rim ports. Both Nakhodka and Vostochni are linked, via Vladivostok, to the Trans-Siberian Railroad. Vanino is also an important export centre with links to the interior via the Baikal-Amur Mainline (BISNIS 1995i). Although there are a number of other ports in the region, the export capabilities of these are limited to non-existent. In fact, most are used solely as local points of supply and exchange for areas without rail or road access to other points in Russia.

While the RFE's natural resources and industrial products may be sub-ject to strong competition on world markets, transportation could become an area of specialisation for which the Far East would have few regional peers. This seems rather incongruous as the transportation system, today, is in some ways a liability. In the RFE, the cost of transport limits the extent to which firms in the region can compete in the markets of European Russia. How-ever, if one looks beyond the borders of Russia, the Far East can be an impor-tant land bridge between the markets of Europe and East Asia. Such a bridge could have a potentially huge impact on local development.

The problem with viewing the RFE as a transportation bridge between

Europe and Asia has to do with the condition of that bridge (Paisley and Lilley 1993, 41). Despite the growth of the region's transport system, its transportation infrastructure is still rather poorly developed. This is especially true when it comes to linkages extending beyond the borders of Russia to Northeast Asia and Pacific Rim countries (Khartukov 1995, 26). The capacity of the system is limited by seaports and railways strained under their current loads (Paisley and Lilley 1993, 41). This leads to slow-downs and bottlenecks at the break-of-bulk points. These problems have already grown acute in some areas, especially in cross-border trade with China where the system is taxed to its limits.[8] Improvements are also needed in the pipeline systems and export facilities. This would alleviate problems of rail bottlenecks and improve the flow of resources to consumers throughout the Pacific Rim (Khartukov 1995, 31). Due to the lack of development capital within Russia, however, these improvements are reliant on heavy foreign investment from Japan, South Korea, U.S., and international sources (Khartukov 1995, 31; Yevtushenko 1991, 7).

One problem resulting from the breakdown of the Soviet system and the market transition has been the decline in the region's defence-related industries (Hoff 1994, 18). Due to strategic concerns in the region, between 1/4 and 1/3 of all Soviet military forces were stationed in the area (Zaitsev 1991, 37). This military presence in the region will likely further decline through changes in the political mood of the country and as military cutbacks begin to take hold. While the de-militarisation is creating short-term economic difficulties, it is also creating long-term benefits. With demilitarisation, key defence-related facilities such as the dockyards and depots in Vladivostok and military airfields in Khabarovsk are being made available for redevelopment. These military facilities are key elements in improving the region's trade and transport systems.

6.4 The Russian Far East and Resource-Based Development

Even with heavy outside competition, any framework for future development in the Russian Far East is likely to be in some way export-oriented. This is due to the strong complementarity of the Northeast Asian economies (Findlay and Edwards 1990, 210-211). This is especially true if

---

8 *The Grodekovo railyard in Khabarovsk is a good example. The yard overflows with loaded rail cars, many of which are used to store materials awaiting shipment. Passenger traffic at the terminal has more than doubled even though the facility was inadequate to start with. As passengers have to pass through customs to enter, people using the restrooms in a neighboring building are escorted there by border guards (Martinov 1991, 2).*

one looks at the relative strengths of the regional economies in the three types of exports.

Labour Intensive: People's Republic of China
Resource Intensive: Russian Far East
Technology Intensive: Japan and South Korea

In terms of potential development, the RFE's greatest strength may lie in its abundance of natural resources. Considering the weak position of the RFE's trading partners, the region's abundant supplies of oil and natural gas, timber, ferrous and non-ferrous metals provide it with potentially lucrative sources of investment (Findlay and Edwards 1990, 207). The existence of resources and the exploitation of those resources, however, are two separate issues and it is here that the Far East faces difficulties. Nowhere is that more readily evident than in the problems facing the development of the region's hydrocarbon resources.

Due to its isolation and weather conditions, the Far East has always had difficulties developing its oil and natural gas deposits and in creating an effective distribution system for those resources.[9] The small size of the local market has also made resource exploitation difficult. In spite of the region's reserves, most of the oil and gas used locally comes from deposits in Siberia over 3,100 km away (Khartukov 1995, 26). The small size and relative insignificance of the local market means that, despite the wealth under their feet, the people of the Far East are subject to often severe fuel shortages (Christoffersen 1994, 527-528; ITAR-TASS 1994).[10] These fuel shortages are found even in areas of petroleum and natural gas production.

While one would expect that the region's dependency on the centre would foster stronger internal ties, the region's isolation from the industrial core of Russia and high transport costs actually favour ties to the RFE's Asian neighbours; China, Japan, and South Korea. An additional advantage of the

---

9 One example is on the island of Sakhalin. The island has become the focus of a number of high-profile oil and natural gas development efforts. Those efforts are hampered by the islands isolation and by an almost complete lack of infrastructure related to hydrocarbon development and exploitation. The island has only one existing port for hydrocarbon exports and that port is situated for exports to Russia rather than regional markets. Most of the island's ports are open only half the year due to the weather problems. Sakhalin also has only the most rudimentary pipeline and rail system. Many of its communities are subject to lengthy fuel shortages despite the sizeable deposits of oil and gas that surround the island (Khartukov 1995; Lilley 1994).

10 The City of Ussuriysk in Primorye exemplifies the severe energy crisis that faces the region as a whole. Due to fuel shortages, power can be shut off to residences for up to 6 hours a day. Some of the cities industrial enterprises commonly go without power for days at time (BISNIS 1995g, 2).

existing import dependency is that it has created an environment of ready acceptance of imports from countries outside the FSU (Hoff 1994, 18). These products often have a more positive image and a higher level of acceptance than products from inside Russia (Hoff 1994, 18).

The Far East's resource endowment and the competitiveness of those resources on world markets provides it with not only hard currency, but also with a means of dealing with its shortcomings. The region's resources also open avenues of trade to countries outside Asia, Australia, Finland, and the U.S. in particular, which have expertise in developing and marketing natural resources on world markets (Findlay and Edwards 1990, 207). This application of resources as a development tool may, in fact, be the only means for the RFE to deal with its economic problems.

6.5 Recent Trends in Trade and Economic Development in the RFE

One of the most important avenues for development in the Far East is development based on foreign trade. Foreign trade is a source of hard currency for investment and has the potential to break the region's dependency on European Russia as a source of goods and supplies (Findlay and Edwards 1990, 208). For the Far East, leading exports are fish and seafood, timber and wood products, and chemical products (Primarily Fertiliser). Imports consist primarily of high-tech equipment and consumer goods.

Table 6.1 Foreign trade of the RFE, in percent

Exports	%	*Imports*	%
Agriculture & Food Products	29.4	Consumer Goods	41.0
Fuel, Minerals, & Metals	26.7	Agriculture & Food Products	27.1
Machine Equipment & Transport	17.8	Machine Equipment & Transport	24.0
Wood Products	13.4		

Source: (BISNIS 1995i; BISNIS 1995a; Kirkow and Hanson 1994, 75).

Since the fall of communism, the primary partners in this trade have been Japan, China, Korea, and the U.S. Other significant trading partners include Australia, Germany, and Vietnam (BISNIS 1995a, 1; BISNIS 1995f. 2). The patterns of export/import activity for the Far East as a whole are reflected in the patterns for the individual oblasts and krays. This is evident in the following information on trade for Primorye and Khabarovsk krays.

Primorye - Primary exports from Primorye are timber, fish and fish products, iron, steel, non-ferrous metals, and nitrogen fertilisers (BISNIS 1995h; BISNIS 1995f; BISNIS 1995b; Kirkow and Hanson 1994, 74). Of total exports in 1994, 33% were with Japan, 21% were with China, and 11% were with South Korea. As with the Far East as a whole, exports to Japan and Korea have grown while exports to China have declined since 1991 when the export totals were 16%, 6%, and 68% respectively (Aliev 1994, 247). Primary imports are consumer goods, industrial machinery, commercial and transport services (BISNIS 1995f, 2). In total, over 80% of the kray's turnover was with Japan, China, and Korea (BISNIS 1995f, 1; BISNIS 1995c, 3).

Khabarovsk - In Khabarovsk, timber, ferrous metals, oil products, and non-ferrous metals account for 54%, 14%, 9% and 7% of total exports while foodstuffs 31%, consumer goods 30%, machinery and equipment 30% are the primary imports (BISNIS 1995a; BISNIS 1994b). 1994 imports and exports were as in Table 6.2.

Table 6.2 Trading partners of the RFE, in percent

	Japan	*Korea*	*China*	*USA*
Exports	67.9	7.4	7.7	1.9
Imports	11.0	26.6	17.2	13.3

Source: (Miller 1996; BISNIS 1995a).

Overall, the RFE has maintained a trade surplus with its trading partners outside Russia (Miller 1996, 161; Kirkow and Hanson 1994, 75; Nezavisimaya Gazeta 1994; Tregubovo 1994, 2). Although there are sources which show trade deficits for some oblasts, such deficits are based on accounting practices which show Russian gold exports coming through Moscow rather than from the source areas (Miller 1996, 162).

The value of RFE resources and trade has spurred significant foreign investment and large numbers of joint ventures. In fact, RFE foreign investment and joint ventures are comparable to those for the core economic regions of Russia. Many of these are resource-based, with foreign development capital being exchanged for Far Eastern natural resources. This pattern alleviates the problem of limited internal capital, while addressing regional needs for infrastructure improvements (BISNIS 1995e; BISNIS 1995d; BISNIS 1994a; Peitsch 1995).

6.6 Tourism in the Russian Far East

In many parts of the world, tourism is a key to economic development. While off to a slow start in the RFE, the development of tourism could help to build on the gains of resource-, trade-, and transport-based development.

Although Russia was a significant tourist destination during the Soviet era, especially for Germans, Finns, and Eastern Europeans, tourism was never considered a priority for economic development. The government's fear of outside ideologies and their influences on Soviet citizens hindered the development of international tourism on a large scale except for group tours and holidays in a selected number of urban areas and at several Black Sea coastal resorts. Domestic tourism focused on health seeking and group holiday camps in areas near warm seas with mineral springs and medicinal muds. This group-style tourism was largely a result of the government's distrust of foreigners, individualism and privatism, and the high regard paid to teaching nationalist ideals to children and young adults (Shaw 1991).

During the Soviet period, travel by foreigners to the eastern portion of Russia was limited to the Siberian cities of Novosibirsk and Irkutsk, as well as other communities located along the Trans-Siberian Railway (Hudman and Jackson 1990). During the 1970s and 1980s, Irkutsk and nearby Lake Baikal became rather important destinations for domestic tourists and some adventurous international travellers. The dearth of destinations in eastern Russia was in large part a result of low population densities, distance from the core and related travel costs, and a general lack of large urban tourist centres (Shaw 1991). Strained political relations with the West, the desire to monitor the travel and actions of foreign visitors, and a general mistrust and fear of outside influences kept most of the Russian Far East, especially rural areas and coastal cities, closed to foreigners until the 1990s. Together these issues isolated the RFE in terms of both physical and market access.

The shift to a market economy is also opening Russia to tourism. In the RFE, much of the tourism that has ensued, as with the region's economic development in general, has been tied to the region's rich resource endowment. The RFE possesses a great deal of natural beauty, including high mountains, active volcanoes and geysers, scenic river valleys, various species of wildlife and vegetation, and rugged coastlines. In fact, as of 1995, 19 nature reserves were located in the RFE, totalling more than 6,500 square hectares (Miller 1995). In addition to nature, the area is rich in cultural traditions, including nomadic herding, traditional celebrations, and village life.[11] One

---

11 *Several communities near Lake Baikal, for example, are in the beginning stages of developing a tourism industry, based largely on the area's rich cultural heritage and natural beauty (Poltoradyadko 1996).*

competitive advantage for the RFE is its claim to be the "greatest expanse of virgin wilderness left in the world" (Klebnikov 1995). This theme is a major selling point in the marketing campaigns of international companies who have begun to offer extensive adventure tours throughout the region.

Improvements to Accessibility

After the break up of the Soviet Union in 1991 and the subsequent change towards a capitalist economy, the RFE began to open up further for international tourists. Improvements to physical accessibility, which was not a high priority under the communist administration, are now being made. Although the purpose of such improvements is largely to enable more efficient transportation of freight between the western and eastern parts of the country, it is also facilitating the expansion of tourism into the region (Kirillov 1994).[12]

Several efforts are under way in parts of the Russian Far East to assist the growth of tourism in the region by improving both physical and market accessibility.[13] Of these, physical accessibility has seen the most pronounced gains. Direct flights to many parts of the RFE have been introduced in recent years. In 1996, flights to Petropavlovsk-Kamchatskii (PK) from Alaska by Alaska Airlines and Aeroflot were introduced as part of a Seattle-Anchorage-PK-Khabarovsk-Vladivostok route (Russian Far East Update 1996a; 1996b). In February 1997, Alaska Airlines announced its new service to Yuzhno-Sakhalinsk, which will open up yet another part of the RFE that has been closed for many years. Aeroflot, Asiana, Korean Airlines, Alaska Airlines, and Japan Airlines all now offer regular, direct flights to the RFE from the Chinese cities of Harbin, Shengyan, Dalien and Chunchung, from cities in Japan, as well as from Anchorage, Seattle, Los Angeles, San Francisco, Singapore, Seoul, and Pusan (Miller 1995). Aeroflot also recently announced its plans to turn the Khabarovsk airport into a major international and domes-

---

12 *One example is a highway linking Chita and Khabarovsk that is currently under construction. It was actually initiated in 1976, and by 1993, 453 km of the road had been constructed. Recent measures have been taken by the new Russian government to accelerate the construction of the highway, so that all 2,150 km will be completed by the year 2006 (Kirillov 1994). When completed, the new highway will enable people to travel from the western borders of the country to the east coast.*

13 *Physical accessibility refers to the degree of ease with which a traveler can physically get to a place. Causes of physical inaccessibility include harsh climates and physiographic barriers, lack of roads and railroads, and political isolation, such as that resulting from wars and boundary restrictions. Market access refers more to perceptual issues such as cost, time, and cultural differences. For some travelers, the cost of going somewhere or the fear of being surrounded by an unfamiliar culture, makes the place inaccessible even if physical accessibility is adequate.*

tic hub to stimulate more foreign travel to the RFE. These developments have been supplemented by improvements in sea-based transport from Japan and by Discovery Shipping which recently received permission to conduct cruises from Alaska to Kamchatka and Russia's Commander Islands (Russian Far East Update 1996b).

Market accessibility has also improved in the RFE. Several regional parks have been established in Kamchatka to help preserve the natural environment and to serve as bases for tourism. New tourist-class hotels are being built, and private investors are being sought for yet more. Arrangements are also being made to facilitate the widespread use of credit cards and travellers checks by foreign tourists (Russian Far East Update 1996b). In addition, the government has opened up for tourism and is promoting rural areas and cities, including Vladivostok, that were once closed to all outsiders. Since 1991, Vladivostok has been developed into an international tourism destination, multi-national manufacturing centre, and commerce hub. As a result of this designation, Seattle-based Alaska Airlines now operates regular flights into Vladivostok (Russian Far East Update 1996b).

Although many of the improvements discussed above have been brought about to stimulate forms of economic development, such as international trade and manufacturing, they have also been instrumental in initiating the growth of tourism in the region. Although data are not readily available yet for tourism in the RFE, the growth of tourism can be seen in the development of new tourism destinations and new types of tourism.

-The Lake Baikal region has developed rapidly as a tourist destination since the break up of the Soviet Union. A scan of internet advertisers and travel brochures reveals a great deal about the types of trips and growing number of travel companies offering tours to the region. Lake Baikal's main activities include windsurfing, sailing, and ecological safaris.

-Several tour outfitters have begun to offer hiking and camping ecotours in the Kamchatka region, focusing largely on the rugged coastlines and the volcanic landscape. Most of the ecotour destinations can only be reached by helicopter and hiking trails (Miller 1995; Russian Far East Update 1996b).

-Hunting is becoming a popular outdoor activity for many international travellers and the number of tour companies offering hunting expeditions has expanded a great deal in recent years. For hunters, the local brown bear, the red deer, and the wild boar are the main attractions of the region. Fishing too has become a significant form of tourism in the RFE with people fishing for trout, salmon, and more exotic species such as taimen and kunja (Klebnikov 1995). Several tour companies based in Alaska, Washington, and California have begun offering guided expeditions for hunters and fishing enthusiasts.

A new form of RFE tourism that has developed in recent years is known as barter tourism. After nearly 30 years of strained relations, travel by Chinese and Soviet citizens began to be normalised across their common boundary in 1988 between the border towns of Heihe, China, and Blagoveshchensk, Russia. Rather than part with valuable foreign currency, each party received the same number of tourists as its counterpart, and provided the same services with comparable quality, including food, hotel accommodations, and sightseeing (Zhao 1994b). The three main purposes of these trips were to sightsee, simply to experience being abroad after so many years of isolation, and to shop. In 1988, each side sent 520 people across the border. In 1989, this number increased to 8,000, and in 1992, the number had grown to 49,000 (Zhao 1994a).[14] Barter tourism has spurred the growth of shopping tourism in several other border communities (Kotkin and Wolff 1995; Rozman 1995). However, the concept has also expanded beyond its original location to include more distant cities, such as Khabarovsk and Vladivostok where many Chinese now travel for week-long shopping and sightseeing trips (Zhao 1994a).

Most travel to the RFE is still business-oriented and is dominated by Japanese, Korean, Chinese and American business representatives. Nevertheless, this form of international travel is an important factor in the development of other forms of tourism in the region. As trade and commerce increase in the RFE, the tourism-related infrastructure, such as hotels and restaurants, will be improved. This will stimulate the growth of large numbers of pleasure-seeking tourists as it reduces the effects of market and physical inaccessibility.

6.7 The Urban Aspect of Development in the RFE

While economic development based on resource exploitation, trade, transportation, and/or tourism may seem to push growth in many different directions, in the RFE the common result is likely to be continued urbanisation. It may seem incongruous to think of one of the world's most isolated regions on the far periphery of world economics and politics as being "urban." However, the history of industrial development in the region, the evolving patterns of trade, and the existing transportation infrastructure all focus growth on the major cities of the RFE.

Resource extraction and harvesting activities in the Russian Far East take place in some of the world's most isolated and inhospitable locations. However, the network of processing and refining centres is centralised with

---

14 *Most Chinese cross-border shoppers favor Russian furs, hats, electronics, leather goods, and watches. In China, the Russians buy cosmetics, sportswear, and cigarette lighters (Zhenge 1993).*

respect to the urban-based transportation system. This is largely a factor of Soviet-era development wherein transportation of resources was a factor of logistics more than it was weight-based costs. As a result, weight reducing activities, rather than being at the point of extraction or harvesting, took place at sites hand picked for their logistical or strategic importance. The difficult environmental conditions and the transitory nature of such extraction activities also played a part in the selection of centralised refining sites in the southern parts of the RFE. The primary basis for the selection of centralised urban sites was that they also provided transportation access to end-users of the region's resources via the Trans-Siberian Railroad, the Baykal-Amur Mainline (BAM) or via ports on the Pacific Ocean and Sea of Japan.[15]

Foreign trade, whether dealing with resource exports from the RFE or with the trans-shipment of materials through the RFE to points in Europe and East Asia, is also focussing growth in the region's urban areas. Transportation facilities, transportation linkages, transportation services, communications, financial services, and other activities necessary for trade in the region are urban. Business centres, consulates, and trade delegations in the region are also urban. Not only are these functions urban, but they are also concentrated in and around the cities of Khabarovsk (Khabarovsk Kray) and Vladivostok (Primorye Oblast).[16] These cities are key rail/air and rail/seaport hubs respectively and would seem to be vying for urban supremacy in the RFE. Rather than financing the added costs required of completely new development, joint ventures, foreign investment, and internal investment has focussed infrastructure investment (especially in transportation) on the major urban areas of the RFE (Miller 1996; Rozman 1995). Cross-border trade with China is also fostering the development of smaller urban centres in a "twin cities" format at key border crossings (Rozman 1995, 286).

Because cities often receive government priority in terms of infrastructure development and public services, regional tourism in most parts of the world begins to develop in cities. Tourism requires an established transportation infrastructure, attractions, and establishments where travellers can eat and sleep. While the attractions in the RFE are primarily rural, the transportation infrastructure and tourist facilities are primarily urban. As such, the urban areas where these facilities are present, however rudimentary they might

---

15 *It could also be argued that the refining centers led to the development of many of the region's cities and influenced the pattern of transportation. For the purpose of this research, the arguement over chicken versus egg will be left unanswered.*

16 *Khabarovsk and Vladivostok had 609,00 and 637,000 people respectively based on 1994 statistics. They are more than double the size of any other cities in the RFE. They have become regional banking and trade centers, they are home to regional stock markets, and are the dominant economic centers of the RFE.*

be, become the "jumping off points" for regional tourism (Law 1993; Page 1995). This reinforces existing patterns of business travel which are based on the major urban economic centres of the RFE. Some cities are also becoming tourist attractions in their own right, both for foreign visitors and for workers in primary sector activities looking to get away from the isolation of small mining and/or lumber settlements.

Politically, urban development is also a keen interest in the region. With the restructuring of the Russian economy and its effects on heavy industry, job growth is an important issue both in the RFE and throughout the former Soviet Union. With an overwhelmingly urban population in search of economic opportunities, and with democracy in local government, the possibility that development might be pushed into more rural locations is very doubtful. With the social infrastructure of the region concentrated in its urban centres, a pattern of continued urban development is likely.

6.8 Conclusion

With the opening of the Russian Far East's borders with its neighbours in Northeast Asia, new patterns of trade are reinforcing old patterns of development. Although current conditions in the RFE may not be inspiring, the regional development potential is great. Prospects for development through primary sector activities, trade, and tourism that have been made possible by the region's relative location, natural resource endowment, and transportation system have the potential for fuelling growth in the region well into the future. While the natural resources for the primary sector and for tourism are in often remote locations, the impact of resource-based development will likely be felt in the region's urban areas. This is a heritage of both Czarist and Soviet-era strategies of development which saw the growth of an entire regional economic and transportation system built on a few key centres. Over the years these centres have become cities that will continue to dominate the economic landscape of the Russian Far East.

References and Bibliography

Note : *BISNIS* is a Department of Commerce information service for the Newly Independent States

Aliev, R. (1994) "Russian Far East: Strategy and Tactics of Development in Primorye. *International Studies,* June 27, 241-261.
BISNIS. (1994a) "Russia Foreign Investment Statistics." Washington, DC: U.S. Department

of Commerce, Market Research, September 16.

BISNIS. (1994b) "Khabarovsk Territory Overview of Foreign Economic Relations." Washington, DC: U.S. Department of Commerce, Market Research, June 21.

BISNIS. (1995a) "Russian Far East Customs Department Report on Foreign Trade." Washington, DC: U.S. Department of Commerce, Market Research, July 28.

BISNIS. (1995b) "Russian Far East Commercial Tidbits-No 4." Washington, DC: U.S. Department of Commerce, Market Research, July 5.

BISNIS. (1995c) "Russian Far East Commercial Tidbits-No 3." Washington, DC: U.S. Department of Commerce, Market Research, May 31.

BISNIS. (1995d) "U.S. Trade and Investment Days in Nakhodka." Washington, DC: U.S. Department of Commerce, Market Research, May (19.

BISNIS. (1995e) "Russia Economic & Trade Overview." Washington, DC: U.S. Department of Commerce, Market Research, July 17.

BISNIS. (1995f) "Russian Far East Commercial Tidbits-No 2." Washington, DC: U.S. Department of Commerce, Market Research, May 15.

BISNIS. (1995g) "U.S. Trade and Investment Days in Ussuriysk." Washington, DC: U.S. Department of Commerce, Market Research, April 25.

BISNIS. (1995h) "Russian Far East Commercial Tidbits-No 1." Washington, DC: U.S. Department of Commerce, Market Research, April 6.

BISNIS. (1995I) "Russia Far Eastern Port Development." Washington, DC: U.S. Department of Commerce, Market Research, March 21.

Carlile, L.E. (1994) "The Changing Political Economy of Japan's Economic Relations with Russia: The Rise and Fall of *Seikei Fukabun.*" *Pacific Affairs*, 67(3), 411-432.

Christoffersen, G. (1994) "The Greater Vladivostok Project: Transnational Linkages in Regional Economic Planning." *Pacific Affairs*, 67(4), 513-531.

Dienes, L. (1987) *Soviet Asia: Economic Development and National Policy Choices.* Boulder, CO: Westview Press.

Findlay, C. and A. Edwards. (1990) "Assessing Economic Opportunities in the Soviet Far East." *The Pacific Review*, 3(3), 207-213.

Grace, J.D. (1995) "Russian Gas Resource Base Large, Overstated, Costly to Maintain." *Oil & Gas Journal*, February 6, 71-78.

Hoff, R. (1994) "Russian Far East an Attractive Market for Nearby U.S. Exporters." *AgExporter*, November, 18-19.

Hudman, L.E. and Jackson, R.H. (1990) *Geography of Travel and Tourism.* Albany: Delmar.

ITAR-TASS. (1994) "Maritime Kray Faces Acute Fuel Shortages." October 26.

Khartukov, E.M. (1994) "Oil and Gas in the Russian Far East: Its Impact on the Asia-Pacific Region." *Petromin*, October, 58-71.

Khartukov, E.M. (1995) "Foreign Funded Pipelines Key to Russian Far East Oil, Gas." *Oil & Gas Journal*, May 15, 26-31.

Kirillov, A. (1994) "Chita-Khabarovsk auto road will link Moscow with Vladivostok." *Rossiyskaya Gazeta*, July 5, 1.

Kirkow, P. and P. Hanson. (1994) "The Potential for Autonomous Regional Development in Russia: The Case of Primorskiy Kray." *Post-Soviet Geography*, 35(2), 63-88.

Klebnikov, P. (1995) "Russia-on-the-Pacific." *Forbes*, 155(7), 85-90.

Kotkin, S. and Wolff, D. (eds) (1995) *Rediscovering Russia in Asia: Siberia and the Russian Far East.* Armonk, NY: M.E. Sharp.

Law, C. (1993) *Urban Tourism: Attracting Visitors to Large Cities.* London: Mansell.

Lilley, J. (1994) "Boom-in-Waiting: Sakhalin fights to turn its oil and gas into a better life." *Far Eastern Economic Review*, September 1, 28-29.

Martinov, V. (1991). "Rail Freight Problems at PRC Border." *Gudok,* December 20, 2.

Matsnev, D. (1994) "Strategies for Economic Policy at Regional Levels Examined." *Delovoy*

Mir, March 31, 4-5.

Miller, E. (1995) *The Russian Far East: A Business Reference Guide, Second Edition.* Seattle: Russian Far East Update.

Milner, B.Z. (1990) "Multinational Economic Cooperation in the Soviet Far East." *Sino-Soviet Affairs,* 14(5), 43-52.

Minakir, P.A. (1994) *The Russian Far East: An Economic Handbook.* London: M.E. Sharpe.

Mote, V. (1983) "Environmental Constraints to the Economic Development of Siberia." In *Soviet Natural Resources in the World Economy,* Robert Jensen, Theodore Shabad, and Arthur Wright (eds) Chicago: The University of Chicago Press, 15-71.

Nezavisimaya Gazeta. (1994) "Joint Ventures Boost Imports, Exports 10-Fold." September 21, 2.

Page, S. (1995. *Urban Tourism.* London: Routledge.

Paisley, E. and J. Lilley. (1993) "Bear Necessities." *Far Eastern Economic Review,* July 8, 40-43.

Peitsch, B. (1995. "Foreign Investment in Russia." *OECD Observer,* No. (193, 32-34.

Poltoradyadko, V. (1996) "Transbaykal's economic potential." *Delovoy Mir,* 29 March: 6.

Rozman, G. (1995) "Spontaneity and direction along the Russio-Chinese border." In S. Kotkin and D. Wolff (eds), *Rediscovering Russia in Asia: Siberia and the Russian Far East,* pp. 275-289. Armonk, NY: M.E. Sharp.

Russian Far East Update. (1996a) "Alaska Airlines' Kamchatka service." *Russian Far East Update,* 6(12)

Russian Far East Update. (1996b) "Prospects for Kamchatka's tourism business." *Russian Far East Update,* 6(11), 9.

Shaw, D.J.B. (1991) "The Soviet Union." In D.R. Hall (ed), *Tourism and Economic Development in Eastern Europe and the Soviet Union,* pp. 119-141. London: Belhaven.

Swearingen, R. (1987) Siberia and the Soviet Far East: *Strategic Dimensions in Multinational Perspective.* Stanford, CA: Hoover Institutional Press.

Tregubova, Y. (1994) "Shumeyko Favors Maximum Powers for Local Authorities." *Segodnya,* September 20, 2.

Yevtushenko, I. (1991) "Zones for Entrepreneurs." *Ekonomika I Zhizn,* November, 7.

Zaitsev, V. (1991) "Problems of Russian Economic Reforms and Prospects for Economic Cooperation Between the Russian Far East and Northwest Pacific Countries." *Journal of Northeast Asian Studies,* 10(4), 35-42.

Zhao, X. (1994a) "Barter tourism: a new phenomenon along the China-Russia border." *Journal of Travel Research,* 32(3)): 65-67.

Zhao, X. (1994b) "Barter tourism along the China-Russia border." *Annals of Tourism Research,* 21(2): 401-403.

Zhenge, P. (1993) "Trade at the Sino-Russian border." *China Tourism,* 159: 70-73.

THE URBAN BASE: INDIA

Editors' note:
The chapters that follow have a very different 'feel' from the earlier ones on China. This may be for a variety of reasons – which could easily be seen as interrelated. Firstly, more temporal data is available for India from the Census, making it possible to conduct more space-time statistical analyses than for China. Secondly, there is no monolithic state keeping the urban and rural inhabitants apart with a passport system, so that there have been no official barriers to 'surplus' rural population moving to urban areas. Thirdly, given that both are regional societies, there has been more devolution, longer, in India, and a greater diversity of local cultures and languages democratically expressing their identity. True there has been national planning – mostly of a sectoral sort and more influential in the 50's and 60's and 70's than now, but always this has been within a mixed economy. In sum, it is less easy to talk about Indian urbanisation from the point of view of policy pronouncements and political ideology, but easier to talk about it from empirical evidence which gives a varying story over the country.

Where both China and India do appear similar is that both initially went for autarkic development. Both have more recently liberalised their overseas trade and their equity markets - China slightly ahead of India. China attempted to industrialise without urbanisation. India is often accused of having achieved the opposite – but this simple view will be elaborated in what follows.

7 Indian Urbanisation and the Characteristics of Large Indian Cities Revealed in the 1991 Census

Graham P. Chapman and Pushpa Pathak

7.1 Introduction

According to the census, out of India's total population of 846 million, 218 million, or only 26%, were classified as living in urban areas in 1991. Thus India's level of urbanisation is quite low compared with many developing countries, and the level has hardly gone up at all in the decade 1981-1991. This reflects the changing rate of growth of the urban population. In the decade 1961-71 the growth of the urban population was 38%, to be followed by the highest figure so far, 46%, in the next decade, 1971-81. But in the most recent decade 1981-1991 the growth of the urban population has declined to 36% (Pathak and Mehta, 1995a). The general slowing down of urbanisation has generated intense debate among scholars. Some scholars attribute it to under-enumeration of urban population while others present a wide range of plausible explanations. The most significant of these explanations are: a decline in the rural-urban migration, identification of relatively fewer new towns, and an increasing concentration of population in the rural areas adjacent to large urban centres (Premi, 1991; and Krishan, 1993). However, there are wide regional variations in the level of urbanisation and the growth rate of urban populations, as well as in the development of regional urban conglomerations - as distinct from conurbations (Pathak and Mehta, 1995b; and Jain, Ghosh and Kim, 1993).

The Indian census has routinely since its inception placed towns and cities in 6 size classes, labelled I to VI. The class intervals used have retained the same absolute population figures, and currently, as in the past, those cities with more than 100,000 population (1 lakh) are designated Class I. Census publications have usually tabled more extensive data in convenient form for these cities on an all-India basis, whereas data for smaller urban places is locked up in individual state volumes. Previous analyses of Class I cities in the 1961, 1971 and 1981 Census reports have revealed some striking patterns (Misra and Chapman (1991), Chapman (1983), Chapman and

135

Wanmali (1981), Dutt (1988)). Perhaps principal among these has been the distinction between large, fast-growing industrial urban centres with adverse sex ratios on the one hand, and more stagnant, smaller, less industrial cities with more balanced sex ratios, and a great predominance of trade and domestic industry. This distinction in crude and general terms mirrors the distinction between generative and parasitic cities established in literature. The spatial pattern of urbanisation has given credence to the correlations too: the more generative cities occur in clusters in the north west, west and south of India, the more parasitic ones in the East, particularly in Bihar. A feature which has in the past few decades marked Indian urbanisation as different from that in the west has been the positive correlation between the size of cities and their growth rates. In the 1960s and 70s the largest cities grew proportionately fastest. There has also been a correlation and between size and specialisation in employment. At what might perhaps be called a more mature stage of urbanisation in the west, the largest cities have less, rather than more, specialised employment characteristics.

7.2 The 1991 Census and Data on Urban Areas

In this chapter we report on some of the findings of analyses of data from the 1991 Census. For the first time the results of this census have been available on computer discs, thus enabling researchers to use more of the data in their analyses for any given investment of time, although of course requiring them to aggregate consistently from the different tables. Four principal data sets were available for the analysis reported on here. These are

A) for each state and union territory, a tabulation of the total for the rural areas of each district of persons in each of 36 categories.

B) for each state and union territory, a tabulation of the total for the rural areas of each district of persons in each of 36 categories.

C) for each state and union territory, a tabulation of the same 36 categories for each individual city, city component or town

D) the male and female populations of all class I cities in the 1991 census, together with the same figures for the same cities in the 1971 and 1981 census, even if they were then below the 100,000 threshold. This data set was not presented in this form in the 1991 census, and was collated in the National Institute for Urban Affairs in New Delhi.

From individual state tables in C) we have abstracted the set of variables for each of the cities listed in D), so that as well as the growth of the cities, we have the data for their employment characteristics and other social indicators such as sex ratios and literacy. There are 296 cities in D), contrasting with the 5,600 units for which data is provided in the total of state tables C). This does not mean that there are 5,600 urban places in India, since for larger urban places the tables provide breakdowns by internal subdivision of large urban areas. It is clear that C) is a very large data set warranting much more detailed study than we have managed so far, but that is beyond the scope of the current chapter.

No census was held in Jammu and Kashmir in 1991, so the data in A) and B) covers 449 districts (we have omitted the offshore island territories.). In the earlier works using the 1971 and 1981 census, cited above, equivalent district data was not incorporated, hence in this chapter we have for the first time the opportunity not only to analyse the characteristics of the Class I cities in 1991, but also to relate these if possible to the regional context revealed by district level data.

The aim of the chapter is to analyse the spatial variation of urbanisation in India, to some extent to contrast urban and rural variables, and to examine in more detail the characteristics of the large cities. Correlations within and between the data sets are explored to see what further light these throw on the processes of urbanisation and on urban structure. The chapter is exploratory and descriptive in nature.

District Level Data

Some of this data was not used because:- 1) the data on area was inconsistent between data sets and quite often absent; 2) the definition of occupied residential houses was thought to be open to variable interpretation; 3) and similarly we were not certain of the consistency of the definition of household; 4) we noticed that there were very few persons counted as marginal workers, and that the definition was again questionable; 5) rather than use Non-workers, we preferred to use Employment Participation Rates based on the number of people actually defined in work categories.

Most of the data was consistent between the different sets, although there were two very minor discrepancies between the Class I city population figures and figures derived from the state lists of urban areas. There were also inconsistencies in the spelling of some town and district names, which required careful attention and adjustment.

7.3 Spatial Averaging and Mapping at the District Scale

Although the geographical surface of India is continuous, the data reflects in part the arbitrary way that it is aggregated into dichotomous spatial units - such as districts and urban areas. This means that, for example,

The following data are available in the district level data.

Table 7.1   District level census data

Not used
(see Text)

1	Geographical Area
2	Number of Occupied residences
3	Number of Households

for remaining variables, the number of persons:-

Population
Male Population
Female Population
Male Population <7 years
Female Population <7 years
Male Scheduled Castes
Female Scheduled Castes
Male Population Scheduled Tribes
Female Population Scheduled Tribes
Male Literates
Female Population <7 years
Male Workers
Female Workers
Male Cultivators
Female Cultivators
Male Agricultural Labourers
Female Agricultural Labourers
Males in Livestock, Fishing, Forestry
Males in Livestock, Fishing, Forestry
Males in Mining and Quarrying
Females in Mining and Quarrying
Males in Manufacturing and Processing in Household Industry
Females in Manufacturing and Processing in Household Industry
Males in non-household Manufacturing and Processing
Females in non-household Manufacturing and Processing
Male Construction Workers
Female Construction Workers
Males in Trade and Commerce
Females in Trade and Commerce
Males in Transport, Storage and Communication
Females in Transport, Storage and Communication
Males in Other Services
Females in Other Services

4	Male Marginal Workers
4	Female Marginal Workers
5	Male non-workers
5	Female Non-workers

two adjacent districts might have a very high and a very low measure of percent urban, although they share a single urban area spreading from one to the other, so that the overall figure for that region ought to be somewhere between the two. To overcome these problems the district data has been subjected to spatial averaging. A more sophisticated method would be to use population potentials, as used by Chapman and Wanmali (1981), but on this occasion the resources were not available to make an equivalent analysis.

Table 7.2 Derived averaged district variables

Derived Variables
% of population urban
Sex ratio: females per 1000 males
% population < 7 years old
% of population Scheduled Caste
% of Population Scheduled Tribe
% Male Literates
% Female Literates
Male Cultivators as % Male workers
Female Cultivators as % Female workers
Male Agricultural Labourers as % Male workers
Female Agricultural Labourers as % Female workers
Males in Livestock, Fishing, Forestry as % Male workers
Females in Livestock, Fishing, Forestry as % Female workers
Males in Mining and Quarrying as % Male workers
Females in Mining and Quarrying as % Female workers
Males in Manufacturing and Processing in Household Industry as % Male workers
Females in Manufacturing and Processing in Household Industry as % Female workers
Males in non-household Manufacturing and Processing as % Male workers
Females in non-household Manufacturing and Processing as % Female workers
Male Construction Workers as % Male workers
Female Construction Workers as % Female workers
Males in Trade and Commerce as % Male workers
Females in Trade and Commerce as % Female workers
Males in Transport, Storage and Communication as % Male workers
Females in Transport, Storage and Communication as % Female workers
Males in Other Services as % Male workers
Females in Other Services as % Female workers

Instead the averaging has proceeded in a very simple way. All the districts have had their geographical centre recorded as x and y co-ordinates. Then for each district the original value of any variable is replaced by an average value of the district itself and its six nearest neighbours. There is nothing special or magic about six, except that it is the average of the number of contact neighbours in a random point pattern. The transformed data produces a smoother surface, and is the data used for most of the maps of district variation shown here.

For a correlation analysis of the district data, the unaveraged original data is used because by definition the averaging process introduces spatial auto-correlation.

From both the averaged data and the original data, new variables have been derived as listed in Table 7.2, all of which are available for the urban and rural values of each district separately, except of course a variable such as "% urban.".

In order to examine the broad spatial patterns, the averaged data is

Table 7.3 Percentage urban and Z-scores

Average % Urban = 25, Standard Deviation = 11

% Urban	Z-score
46	2
36	1
25	0
14	-1
3	-2

further manipulated. Firstly, highly skewed data is transformed by taking either log or power functions as appropriate, and secondly this data is re-expressed in standardised scores (often known as 'z' scores or sometimes 't' scores) with mean of 0 and standard deviation of 1.

As an example, Map 7.1 shows the Urban Population as a percentage of the Total. The symbols range from those districts below -2 Standard Deviations, through the ranges -2 ~ -1; -1~ 0; 0 ~ 1, 1~ 2, to those > 2. The absolute values are not depicted (but are shown in Table 7.3), but the result shows clearly the variation within India. In a sense, since relative values are being used, the procedure cannot fail to show variation: what is significant is whether the spatial pattern reveals new understanding. In this case it clearly does. The less urbanised areas of Arunachal Pradesh, Assam, Bihar, Eastern Uttar Pradesh and Orissa are expected. The higher levels in Tamil Nadu are expected: what was not expected to be so clear is the way in which two of the

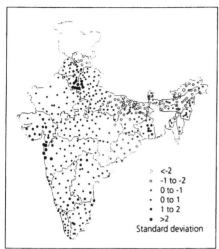

Map 7.1 India: Urban population as a percentage of the total

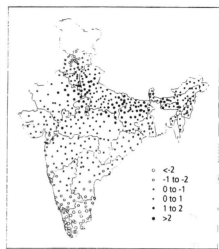

Map 7.2 Variation in the percentage of the rural population below the age of 7 years

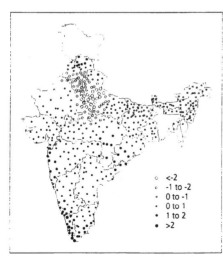

Map 7.3 Variations in the rural sex ratio

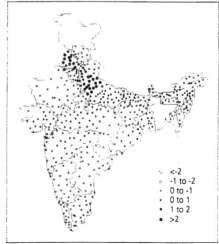

Map 7.4 Variation in male/female urban literacy rate

four metropolitan areas are so clearly much more intensely urbanised regional systems (Delhi and Bombay (Mumbai) than the other two (Calcutta and Madras). Though Calcutta is a huge metropolis, it is a more primate urban area within a more rural hinterland. In the case of Bombay, the urban system clearly stretches north into Gujarat - a point which will be picked up again below in a slightly different context.

Map 7.2 shows the variation in the percentage of the rural population below the age of 7 years. The south-north distinction is extremely clear, with the more youthful population stretching down nearly all of the Ganges Valley, and into Assam. Exceptions in the north are the areas of Punjab and Himachal Pradesh, which are consistently different on map after map, the Calcutta hinterland , and the southern areas of the north-eastern states. Again in map after map, Nagaland and Manipur, and sometimes Mizoram and Tripura too, are distinct from Assam and Arunachal Pradesh. In the south of India, the two states of Kerala and Tamil Nadu form a complete contrast to the northern states. The map showing the percent of urban population under 7 years (not shown) is almost identical.

Map 7.3 shows the variations in the rural sex ratio. The north south gradient is known and expected, as also the peaking of the highest values in Kerala. What is perhaps less expected is the pattern within the north, where the region of very low values stands out so clearly within western Uttar Pradesh and close to the Delhi conurbation. The urban sex ratio (not shown) is similar, though not picking out western Uttar Pradesh quite so emphatically.

Literacy is examined here by comparing the male and female literacy rates for both rural and urban areas. The two maps, 7.4 and 7.5, are essentially maps of discrimination by sex. The discrimination in favour of males in rural areas reveals a concentration in the 'cow belt' of Hindu orthodoxy from Rajasthan to Bengal. Apart from the areas of the North-eastern states, this map shows a close proximity to the percentage of rural population under the age of 7. It is a fairly graphic association of the implicit link between low levels of female education and high birth rates.

The next set of four maps (not shown) concern Employment Participation Rates -the percentage of male or female population recorded as employed. On the whole the rural maps show a lower rate of employment in the Gangetic North, with the exception of high male rates in Punjab and Haryana; a higher rate in the whole of the Deccan with the exception that the participation rate in Kerala is low - which is something that is quite well known. The urban maps display more marked regional contrasts - in Nagaland, Manipur, Tripura and Mizoram the female urban participation rates show strongly, as indeed they do in Kerala. By contrast in Punjab, Himachal Pradesh and the capital region, male urban participation rates are high, while female ones are low.

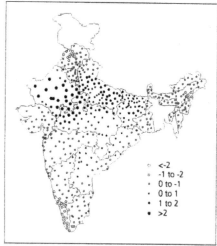

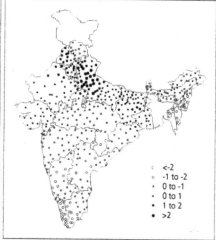

Map 7.5 Variation in male/female rural literacy rate

Map 7.6 The ratio of male urban participation rates/female urban participation rates

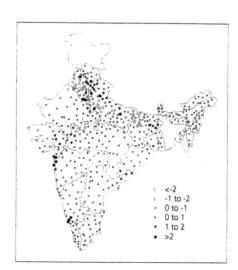

Map 7.7 The ratio of the proportion of tribal people in rural areas of each district compared with the proportion of urban people who are tribal

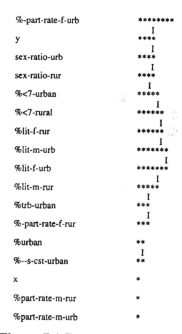

Figure 7.1 Structure vector of correlations of Indian district variables

Map 7.6 shows explicitly the ratio of male urban participation rates/female urban participation rates. The greatest 'imbalance' in favour of males is in western Uttar Pradesh, but Punjab and other parts of the north also feature. The least 'imbalance' occurs in the south in Kerala and southern Tamil Nadu.

Other kinds of 'imbalance' can be shown for the difference for any variable between urban and rural areas. Map 7.7 shows the ratio of the proportion of tribal people in rural areas of each district compared with the proportion of urban people who are tribal. The high values here indicate towns where the representation of the tribal people is significantly less than in the rural areas around. These are the towns often referred to as centres of 'internal colonialism' – around which tension builds up as the wealth and political power of the towns swamps the interests of the surrounding country. The map clearly picks up the towns of South Bihar and a line along the foothills of the Himalaya. The 'provocation' for producing this map was the recently announced intention of the company P&O to build a new port at Vardhavan north of Bombay, in Thane District and close to the Gujarati border, in an area inhabited by the Warli tribe. Opponents of the plan highlight the impact that such developments would have on Warli tribal culture. In South Bihar in the 1960s and 1970s the influx of Biharis and Bengalis into new industrial towns in tribal areas caused political tension and the reinvigoration of the demand for a tribal state, Jharkand. Map 7.7 highlights several areas in India where such tensions may be more probable. In the case of the area north of Bombay it is also worth again considering Map 7.1. The two together show how the area is well urbanised, but the tribal people are not represented in the towns.

7.4 Correlation Analysis at the District Level

The spatially-averaged data cannot be used for a correlation analysis, since by definition it has had auto-correlation built into it. Here we perform an analysis using data which has not been averaged. The question still remains open however, as to whether this data is still spatially auto-correlated, since neighbouring districts may be more likely to reflect each other than distant ones. However, if such auto-correlation is present, this does not mean that we cannot find patterns that are interesting, but rather that it is not possible to apply the usual tests of significance. In fact it is very easy to show that auto-correlation is present, since several variables correlate with the south-north co-ordinate (and some, to a lesser degree, with the east-west).

Skewed variables were transformed using either log or power functions, which eliminated most but not all of the skew from most of them. The data was then analysed using a Pearson product moment correlation coeffi-

cient. The exercise was repeated using a rank correlation coefficient, to eliminate the effects of whatever skew remained. The two correlation tables agreed
with each other loosely, varying a little in some cases in the strength of an
indicated correlation. The results of the more robust (assumption free) rank
correlation exercise are shown in Table 7.4 . (Given the number of observa-
tions (449) any correlation with an absolute value in excess of .15 is theoretically statistically significant at the 0.1% level. However, auto-correlation
means this value cannot be used with authority, and in any event the variance
explained remains low.)

The variables clearly mesh with each other over a network of direct
and indirect correlation. The next step in orthodox procedure would be to
use Factor Analysis, to impose orthogonal axes in place of the original variables. This procedure is not very satisfactory, because in a network of connections orthogonality seems an odd imposition, and because some variables
may load partially on more than one Factor. Here two different steps are followed. Firstly, and totally arbitrarily, correlations of less than Á.5Á are dropped
from the analysis; and secondly the remaining correlations are analysed in
terms of Q-connectivity - a device which considers multi-dimensional and
indirect connectivity. Q-analysis was developed by Atkin (1974), and is further exemplified by Chapman (1981 and 1984). The structure vector is shown
in Fig. 7.1. The vector shows how one variable - the Female Participation
Rate - is connected to many others, but stands in a group somewhat on its
own. On the other hand, both male and female literacy, both urban and rural,
form a high dimensional cluster with the percentage of children under 7 years
in both rural and urban areas. This information enables us to plot quite quickly
the correlation diagram shown in Fig. 7.2, where the direction of the correlations can also be shown. The correlations show both urban and rural characteristics vary together. Hence both urban and rural literacy rates inversely
connect with both urban and rural populations under the age of 7. This does
not mean that there are no differences between urban and rural areas - in
absolute terms the levels of literacy and the percentage of children under 7
may differ quite widely in many districts. But it does say that variation across
India in the absolute levels or urban areas will tend to mirror variations in
rural areas, and vice versa. Hence variation within India will be found to be
regional, rather than at the urban/rural interface.

The diagram also highlights the pivotal nature of the Female Urban
Participation rate. It is connected both to urban and rural sex ratios, and is
higher wherever these are more 'favourable' (more females per male.) These
three together are all negatively correlated with the North-South co-ordinate
'y'. The value of y decreases with decreasing latitude, so again what is
displayed is not urban-rural differences, but regional differences, again high-

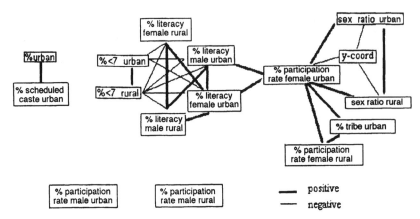

Figure 7.2 Diagram of correlations of Indian district variables

lighting the higher southern sex ratios. The Female participation rate also correlates with Percent Scheduled Tribe in urban areas and the Female Rural Participation rates. Inspection of the maps shows that this again is a regional result, reflecting in particular the combination of values in the North-eastern States.

In the diagram the variable Percent Urban is strongly associated with Percent Scheduled Caste urban. This in a sense implies that there may be more work opportunities for the lower strata of society where urbanisation rates are higher, but it may also reflect some of the regional impacts evident in the Punjab and Northwest. Two variables remain disconnected - the urban and the rural Male Participation Rates. Inspection of the correlation table shows that the rural rate has no particular high correlations - suggesting that it will be the result of local circumstance rather than any pan-national associations. The urban rate has quite strong negative association with the Percent of Rural Population under 7, and with urban Female Literacy Rates. One can only speculate in very general terms why this should be so: that the Male Participation Rate is like the Female Participation Rate a sign of 'modernity', but less markedly so.

7.5 Analysis of Large Cities

This is conducted in two stages. First there is an analysis of the class I cities(over 100,000 population) and their variables on their own in this section. Then an attempt is made to analyse the cities in relation to their regional settings.

7.5.1 Correlation Matrix for Large Cities

Table 7.5 shows the complete correlation matrix for the selected variables. Again, the approach adopted is not to use factor analysis to force orthogonal vectors onto this material, but to investigate the connectivity through a Q-analysis and a descriptive diagram. The value chosen for slicing the correlation matrix is arbitrary: in this case it is for absolute 'r' values greater than .35 - a value at which a clear structure emerges in the Q-analysis. At low values, e.g. 'r' = >|.10|, many variables are related to many others, and at high values many variables and many connections between variables are excluded.

The analysis shown in Figs. 7.3 and 7.4 immediately highlights one point by comparison with the analysis of the 1981 data set. In that set the growth of cities over the previous decade 1971-1981 was correlated quite

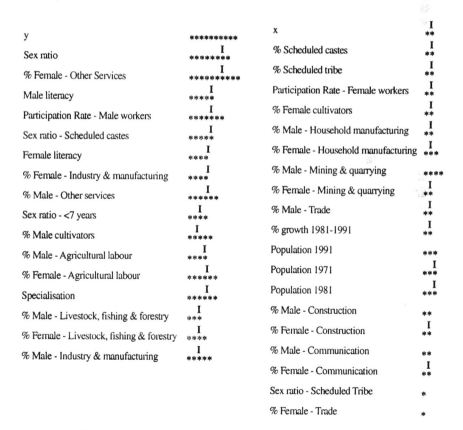

Figure 7.3 Structure vector of correlations of Indian city variables

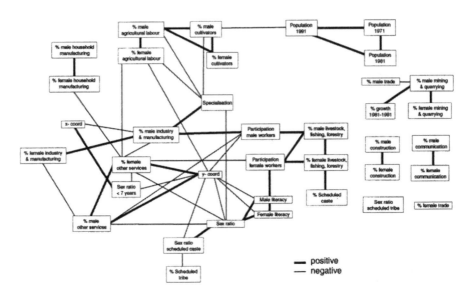

Figure 7.4 Diagram of correlations of Indian city variables

strongly with several variables, including size, the presence of organised industry, employment rate, and specialisation. This time growth hardly seems correlated with anything: in the diagram it appears to be correlated with Mining and Quarrying, but closer inspection of the data reveals this relationship to be overstated, because the data remains skewed even after a log transformation, and one city, Ramagundam in Andhra Pradesh, has had by far the highest rate of growth and has high levels of employment in mining. That growth has no particularly strong correlates in the more recent decade is interesting: a more diverse set of causes of the growth of cities perhaps indicates a different stage of urbanisation – perhaps one in which there are differing reasons for growth within regional clusters of specialising cities.

Many of the variables have been calculated separately for the two sexes. It is apparent that there is a close correlation between the two sexes for nearly all of these - for example the % of Female and Male employment in Mining and Quarrying, or in Industrial manufacturing. This does not mean that the employment levels of the two sexes are the same for any category of employment - but it does mean that given whatever absolute levels occur, then variations from one city to another in one variable is reflected by variation in the other. This in turn suggests that different types of city economy do not have a discriminating effect on male versus female employment which is additional to discrimination from other sources - e.g. cultural or demographic variation.

Three variables stand out in terms of the number of other variables with which they are correlated, and in terms of their total 'r' values. These are Sex Ratio, North-South Co-ordinate (y), and % Females in Other Services. The first is commonly referred to, as there appears to be such a regional difference between high Female/Male ratios in the South, and low ones in the North. The correlation of this with the North-south co-ordinate confirms this. Given such a crude way of measuring location in a single dimension, the relationship is extraordinarily strong.

The Other Services category is interesting too - it conjures up an image of pressure to find work in the informal sector. For both Females and Males it is negatively correlated with the respective workforce Participation Rate, and both are negatively correlated with levels of Industrial manufacturing employment. The Female Other Services is also negatively correlated with Household Industry, and even with Female Agricultural Labour.

Overall, the patterns of connection suggest the continuation of some distinctions seen in the previous analysis - between specialised industrial centres with higher employment rates, and less specialised centres with more diverse employment at lower participation rates.

All of these variables have been plotted on maps. Five these are reproduced here in 7.8-7.12, two of which will receive specific comment. The variable Specialisation is an entropy statistic derived from all the employment categories together, and is a measure of how much employment is con-

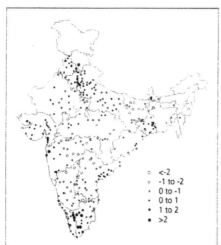

Map 7.8 Variation in male city participation rate from locally predicted value

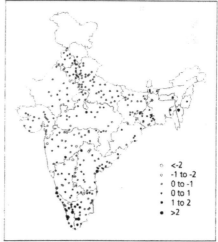

Map 7.9 Variation in female city participation rate from locally predicted value

centrated in one or a few categories. It does not differentiate between concentrations in two different categories - i.e. one city could be specialised in Industry and one in Household manufacturing, but both be equally specialised. The variable Percent Males in Construction emphatically and curiously does not correlate with urban growth. There does appear to be some geographical pattern worth explaining though, since the low values occur in Eastern Uttar Pradesh, Bihar, West Bengal and Orissa. These states collectively make up the stagnant part of two-speed India (Chapman, 1992) - not in the sense that population is not growing, but in that incomes per head in this zone have not improved for at least 30 years. This reminds us that a key variable which one would like to include in further analysis would be average incomes per head in these cities.

7.5.2 *Large Cities in their Regional Context*

Here we make an attempt to relate cities to their own regional context. Each city is linked to the district within which it is located. For these districts we have the average values of variables from the six nearest surrounding districts and the district itself. It is thus possible to calculate the local average value for any rural variable, in the proximity of any city. It is also possible to calculate the urban value from the surrounding six districts and the district itself, but in this case omitting the value of the Class 1 city included in the district (to avoid including this data on both sides of a correlation.) Table 7.6 shows the results for values > .|3|. The table invites almost endless speculation, and we do not presume to be able to pronounce on all the associations revealed. Some of the results are simple and straightforward and much as to be expected. For example the Percent Scheduled Caste and the Percent Scheduled Tribe in Class I cities both on average correlate with their respective regional trends, both urban and rural. It is for this reason that departures from the trend as revealed in Map 7.7, are to be noted. Literacy rates in Class I cities correlate with their local regional trends in literacy, again both rural and urban. Male Class I Participation Rates correlate with local regional Male Participation Rates, and Female with local Female Participation Rates. But then there is a distinction between Males and Female. The Female Class I Participation rate also correlates with regional Female Literacy Rates, but for the males this is not so. Presumably this is a hint that the characteristics of the male population of large cities are less likely to reflect local trends, perhaps because males migrate further to large cities than females. However, the Sex Ratios of the Cities are correlated with the urban and rural local Sex Ratios. Class I Female Construction workers are correlated with Rural Scheduled

Tribes, and both Urban and Rural Female Participation rates. This relationship depicts well the use of such labour in the arc of 'new' industrial towns in South Bihar and West Bengal, in the area known as 'Jharkand' (see above).

A last example of the possibilities of manipulating this data is a consideration of the extent to which particular values of variables for Class I cities differ from those that would be predicted by a local regional trend. In this case we are specifically interested in Participation Rates in Class I cities, and the local trend in Participation rates in Rural areas. The logic of this is that where there are large differentials, then we might expect local rural-urban migration rates to be higher too. The analytical procedure is as follows.

The average values for the rural Male and Female Participation Rates for the district containing the Class I city and its six nearest neighbours have already been calculated. The actual values of the participation rates in the Cities are then regressed against these local rural averages. The correlation between the average rural values and the individual city values is high - in the order of .5 - so we are right in supposing that cities on average do show participation rates which reflect local circumstance. The last stage is to map the residual between actual urban participation rates and the expected participation rate predicted by the regression. Two maps are shown here 7.8 and 7.9 - for the Male and Female Participation rate residuals - again scaled in standard deviations of the residuals. The map of Male Residuals emphasises the Delhi , Calcutta and Bombay metropolitan areas to some extent, but not Madras. In the South the industrial cities of inland Tamil Nadu make a very distinct pattern on the map. Rural male participation rates in this area are not particularly low by national standards. In one of the lower ranked districts of Tamil Nadu - Coimbatore - they are about national average. But the male participation rates in town are clearly much higher than expected from these values. It would be exciting to say that this demonstrated a dynamic and growing urban area. However, the figures for urban growth do not show this area to be expanding fast, and indeed growth rates for the Class I cities remain the hardest to correlate with any other variables. We have yet to agree on the true significance of this pattern.

7.6 Conclusions

In a previous paper (Misra and Chapman 1991) as a result of an analysis of data on the Class I cities of the 1981 census, the authors concluded by showing a regionalisation of India's major cities. The fact that they produced such a regionalisation was indicative of the belief that urbanisation in India has distinct regional characteristics, as well as also having some national trends.

Prominent among the national trends was the ability to explain growth rates of Indian cities in terms of their size and their manufacturing base. One implication of the correlation was the unwelcome one that already very large cities could grow even larger.

In this chapter we have not found it possible to explain the growth rates of cities in the last decade in any simple way. Specifically, it is no longer true that the largest cities are more likely to grow fastest. On the other hand, we have shown that regional systems of cities are important, and that in some of these the growth is dispersed throughout the city region. These regions - as with Bombay, Pune and parts of Thane District, are too big and too dispersed to be called simple conurbations, but the reality of their regional performance is clear. We have also been able to go one step further, and to show how many of the characteristics of cities are strongly correlated with their regional hinterlands, even if the absolute values of many variables do vary between urban and rural areas. In this sense, urbanisation in India represents a major change, but one which is mostly rooted in and develops out of the local context. We have also shown exceptions to this statement – in particular that urban areas in tribal zones may specifically NOT represent the local cultural context.

References

Atkin, R.H. (1974) *Mathematical Structure in Human Affairs* , Heineman, London

Chapman, G.P. and Wanmali, S. (1981) 'Urban-rural relationships in India: a Macro-scale approach using Population Potentials'. *Geoforum* Vol. 12 No 1 pp 19-44

Chapman, G.P. (1981) 'Q-analysis' Chapter 23, pp 235-247, in R. J. Bennet and N.Wrigley (Eds). *British Quantitative Geography, Retrospect and Prospect_* Routledge, Kegan and Paul

Chapman, G.P (1983) 'The Growth of Large Cities in India 1961-1971' *Geoforum_*Vol 14 No 2 pp 149-159

Chapman, G.P. (1984) 'The Structure of Two Farms in Bangladesh'. In *Understanding Green Revolutions* Ed T. P. Bayliss-Smith and S. V. Wanmali, C.U.P.

Chapman, G.P. (1992) ' Change in the South Asian Core: Patterns of Growth and Stagnation in India' Chapter 2 pp10-43 in *The Changing Geography of Asia*, Chapman, G.P. and Baker, K. (Eds), Routledge, London

Dutt, A. (1988)

Jain, M. K., Ghosh, M., and Kim, W. B. (1993), "Emerging Trends of Urbanisation in India : An Analysis of 1991Census Results", *Occasional Paper* No. 1 of 1993, Census of India, New Delhi.

Krishan, G. (1993), "The Slowing Down of Indian Urbanisation", *Geography*, Vol. 78, Part I, pp. 80-84.

Misra, H.N. and Chapman, G.P. (1991) 'The Growth of Large Cities in India 1971-1981" *Geoforum_*Vol 22 No 3 271-

Pathak, P. and Mehta, D. (1995a), "Recent Trends in Urbanisation and Rural-Urban Migration in India: SomeExplanations and Projections",*Urban India*, Vol XV, No. 2, pp. 1-15.

Pathak, P. and Mehta, D. (1995b), "Trends, Patterns and Implications of Rural-Urban Migration in India", in, *Trends, Patterns and Implications of Rural-Urban Migration in India, Nepal and Thailand*, Economic and Social Commission for Asia and Pacific, Asian Population Study Series No. 138, United Nations, New York.

Premi, M. K. (1991), "India's Urban Scene and its Future Implications", *Demography India*, Vol. 20, No.1, pp. 41-52.

8 Patterns of Urbanisation and Development in India in the 1990's

Swapna Banerjee-Guha

Editors' note:
One of the variables used in this analysis refers to crime (as one indicator, inversely, of the quality of life). Crime was touched on in Nigel Harris' opening chapter, and is touched on again in volume II, where one chapter is devoted to an analysis of the geography of crime in India. It is a topic which perhaps warrants more research.

8.1 Introduction

Independent India inherited from its colonial past a spatial structure in which agglomeration economies had favoured core cities, while giving them ambivalent roles in regional development. The Sixth Plan document in 1980 identified regional imbalance as a continuing characteristic feature in the Indian economic process. Standard core analysis undertaken by Dasgupta (1984) on India's socio-economic well-being using 1981 data reiterated these views. In the nineties with the official initiation of liberalized policies this pattern has become associated with the spatial dynamics of international capital and made unevenness a premise for capital accumulation. Against this backdrop, the following sections attempt to analyse the patterns of urbanisation and development that have obtained in India in the early nineties.

8.2 Pattern in Early Eighties

In 1981,the spatial pattern of urbanisation was of a few dominant cities followed by a large number of sub-regional and smaller centres and a vast impoverished periphery. The curve of socio-economic well-being peaked at the metropolitan centres and select pockets, such as Punjab, and declined steeply at the level of smaller towns and interior areas (Dasgupta, 1984), even in areas rich in resources such as Bihar or northeastern states. Intra-regional disparity existed in all regions. The northern region was marked by the affluence of Delhi and Punjab at one extreme and poverty and deprivation of western Rajasthan, eastern Uttar Pradesh and Bihar on the other. The Western region accounted for a very affluent core of Greater Bombay and simultaneously the poorest districts of the country in Eastern Maharashtra. Urban-rural and intra-rural disparity in the western region is corroborated by the fact that per capita net product in urban and rural areas are not correlated, reflecting disjunction between industrial and agricultural growth. While Maharashtra had the highest rank in urban consumption, it was way down in rural consumption (Bharadwaj, 1982). Dynamic and modern, the urban industrial core of Bombay with an extended umland up to Gujarat, however, did not show much sign of integration with the region with city-hinterland forward and backward linkages. The sugar-rich villages of south eastern Maharashtra stood out as deprived areas in terms of industry, infrastructure and quality of life. The eastern region of the country exhibited an expanded poverty surface with Calcutta displaying acute primacy. In contrast to these three regions, the southern region was found to suffer the least from intra-regional disparity with a number of regional and sub-regional centres emerging within a small area. Although in the western part in Kerala it primarily reflected a strong social infrastructure without a weaker production base, in the eastern part in Tamilnadu, Karnataka and Andhra Pradesh a considerable economic spread effect is identified with medium cities, such as, Tiruchirapalli or Kodagu as viable diffusion centres. However, placement of metropolitan cities like Bangalore, Hyderabad or Madras at a much higher level simultaneously indicated the officially endorsed policy promoting metropolitanisation and expansion of metropolitan space. In overall terms the 1981 Indian space economy was marked by a synchronisation of high level well-being and high disparity reflecting both a rural-urban gap and intra-regional disparity. A correlation matrix of development variables shows high positive values between metropolitan cities and level of urbanisation, improved infrastructure, non-agricultural workers and cognisable crime indicating a distinct urban bias in the development process till early eighties.

8.3 Early Nineties: Factor Analysis

The spatial pattern of urbanisation and development in 1991 reinforces the very process of centralisation and metropolitan hegemony despite agricultural improvement in some districts of Maharashtra, Karnataka, Tamilnadu and Andhra Pradesh. A factor analysis with 23 variables (district level data) has been used to examine the nature of development as a geographically variable condition. Twenty three variables were selected for analysis, representing the best compromise between coverage of the states and the social and economic conditions we wished to examine. The variables have been selected under the heads of basic ingredients of development in the following manner.

A factor analysis reduces the 23 variables to two factors. Factor I,

Table 8.1 Variables used in the analysis

A. <u>Status</u> :
1. TOTLIT   % of total literates
2. FEMLIT   % of -female literates
3. SCHECA   % of scheduled castes
4. SCHETB   % of scheduled tribes
5. MARWOR   % of -marginal workers

B. <u>Urbanisation:</u>
6. URBPOP   % urban population
7. URBCIT   % of urban population in class I cities.
8. URBGTH   % decennial urban urowth
9. NONAGR   % of non-agricultural workers

C.<u>Industrial Development :</u>
10. WORFAC   Average daily workers in registered factories
11. HOSIND   % of workers in household industries
12. CRDIND   Per capita bank credit to industries

D. <u>Agricultural Development :</u>
13. AGRLAB   % of agricultural labourers
14. LNDSIZ   Average size of operational holdings
15. CONFER   Per capita consumption of fertiliser
16. CRDAGR   Per capita bank credit to agriculture
17. TRALAC   Number of tractors per lakh hectare of gross cropped area
18. IRICRP   Percentage of irrigated area to gross cropped area

E. <u>Quality of Life :</u>
19. VILELC   Percentage of villages electrified
20. HOSBED   Number of hospital beds per 1,00,000 people
21. BNKDEP   Per capita total bank deposit
22. BNKCRD   Per capita total bank credit
23. COGCRM   Number of cognisable crime per 10,000 people.

Note: Data on the above variables have been collected from the Centre for Monitoring Indian Economy (CMIE), Bombay; Statistical Abstracts, Handbooks; Digests of Statistics and Ecnomic Reviews of various States; Bureau of Economics and Statistics, Government of Maharashtra; Times of India, Bombay (Reference Section) and Census of India,1991. The entire data is not of 1991. It varies between 1988 and 1991. The emerging pattern hence has been considered as of early nineties.

explaining a remarkable 63% of the variance loads high on urban population (URBPOP), total bank credit (BANKCRD) and deposit (BNKDEP), bank credit for industries (CRDIND), population in class I cities (URBCIT), average daily workers in registered factories (WDRFAC), total literates (TOTLIT), female literates (FEMLIT), number of hospital beds (HOSBED), cognisable crime (COGCRM) as well as consumption of fertiliser (CONFER), number of tractors (TRACLC), bank credit to agriculture (CRDAGR) and villages electrified (VILELC). The first six variables reflect the association between industrial and large urban populations. The association with crime is also of interest - it suggests either that there is more crime in wealthier and more modern urban centres, or that there is more reported crime. It is certainly tempting to believe it shows a degree of social deviance associated with urbanisation. The last four variables indicate the role of increased agricultural productivity and surplus in the recent urbanisation trends, for example, in Punjab, exemplified by factor scores of Chandigarh (12.63, Ludhiana (5.71) and Patiala (3.98). In this case urbanisation in some regional contexts is interconnected with rural growth. Factor I can be interpreted as the factor of industrialisation and urbanisation and modernisation. It is negatively associated with variables like percentage of agricultural labourers (AGRLAB) or percentage of scheduled tribe population (SCHETB).

The districts are ranked by Factor 1 scores. The ranking runs from the highest group, the single district unit of Bombay 17.39, through metropolitan Calcutta 14.57 (but the outer urban hinterland has much lower values), Chandigarh 12.63, Delhi 11.00, Madras 11.10, Bangalore 8.90, Hyderabad 8.08, Ahmedabad 6.58, Bhopal 6.60, Lucknow, Ernakulam 6.17, Ludhiana 5.71, Kanpur 5.11 Pune 4.45 Thane 4.18 and so on to the poorest districts - 1.00 to -2.90 in Rajasthan, eastern Karnataka, western Andhra, almost entire Orissa and the northeastern districts, and then less than -2.9 in Bastar areas of Madhya Pradesh, some parts of eastern Uttar Pradesh and eastern Bihar (Map 8.1).

The above factor also exemplifies the process of how capital operates in India (Smith, 1984), produces uneven development and capitalises space, especially of the largest cities. Once growth starts taking place in these locations along capitalist lines, it tends to attract resources even at the cost of depriving other regions (Bagchi, 1989). Funds have been diverted to these points for developing basic and social infrastructure and a skilled labour force. Multipliers being an important factor contributing towards uneven distribution, other areas remain out of the minimum threshold line and do not generate growth or associative factors of reproduction. Thus the process of accumulation operating in large urban or industrial centres turns out to be essentially a cumulative causation process generating backwash and outmigration

(Myrdal, 1957) from other regions. This is proved by the expenditure of large cities on consumer durables and infrastructure or by their share in bank deposits compared to medium towns and villages. For example, in 1991 Bombay accounted for 61% of the total deposits in the scheduled commercial banks of the western region while the four combined largest cities (Bombay, Calcutta, Delhi and Madras) accounted for 34% of the national total. Even in mid-eighties Delhi cornered 40% of the national urban planning budget. This predatory and expansive nature of large cities is an expression of their dominance through their expanded urban space (Castells, 1977). The recent con-

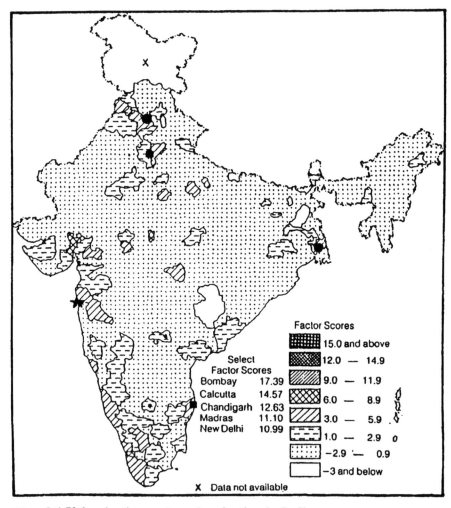

Map 8.1 Urbanisation and modernisation in India

centration of State investment of Maharashtra in Raigad district - the backyard of Bombay and an officially recognised backward region of the State, part of which is located in the newly developed metropolis of New Bombay - is a more contemporary example of expansion of a dominant metropolitan space. The very same process thwarts movement of capital and development beyond the limit of this expanded space in other needy regions and reconstructs a primacy of agglomerated urban space over single urban units. The order of control over capital and space also gets reconstructed contributing to regional disparity at various scales and orders.

The second order placements of Calcutta and Chandigarh show a two directional and temporal pattern. On the one hand, it distinctly distinguishes Calcutta as a declining core (albeit having some recent efforts towards capital accumulation) and on the other, it identifies the emerging trend of a rising capitalist class in Punjab and of a highly differentiated class structure in rural areas. A new trend of a symbiosis of agricultural surplus and urban growth in a number of states, besides Punjab, is also brought out by the factor. Placement of Delhi and Madras at the subsequent level, however, represents two divergent processes operating in north and south respectively. Delhi had a much later start in industrial expansion compared to Calcutta, Bombay and Madras. During the planning era especially since sixties Delhi and its surrounding areas got privileged patronage for industrial development, thanks to the making of the National Capital Region (NCR). The newly emerged capitalist class of Punjab, Haryana and nearby areas have invested large proportion of their agricultural surplus into these industrial estates to make the regional investment factor significant. The declining supremacy of Madras is, however, associated with a positive trend of spread during the plan periods, exemplified by the development of centres like Bangalore, Hyderabad, Ernakulam, Tiruchirapalli, Kodagu, Trivandrum etc. The primacy index in the prevailing urban system of southern India is thus lower (Banerjee-Guha, 1985).

Factor II (explained variance 21%) loads high on fertiliser consumption (CONFER), number of tractors (TRACLC), irritgated area under cultivation (IRICRP), agricultural bank credit (CDRAGR), agricultural labourers (AGRLAB), electricity in villages (VILELC), and average size of operational holdings (LNDSIZ). Because of the combination of these variables factor II can be interpreted as the agricultural development factor. Punjab stands out singularly having the highest scores (districts ranging from 3.5 to 7.00) while rice regions of Andhra Pradesh have the second highest scores ranging from 3.5 to 6.00. This region covers districts of West Godavari (4.17), Nizamabad (3.64), Nellore (3.5) and East Godavari (3.18). While in Punjab variables of agricultural and industrial activities have shown synchronisation making higher total score, variables contributing to high scores in the second group of districts represent only agricultural development. Punjab's prominence in both Factor I and II, as mentioned earlier, is an expression of convergence and symbiosis of agriculture and industry in contrast to their earlier divergent status with a focus on agriculture. Pockets of agricultural development with factor scores ranging from +1 to 3.5 located in south-eastern Maharashtra in the sugar belt districts, such as, Kolhapur (2.07) and central Gujarat also are indicators of the trend of 'pocketed' growth. On the other hand, districts in Tamilnadu with higher scores, such as, Chengal Pattu (2.80), North Arcot

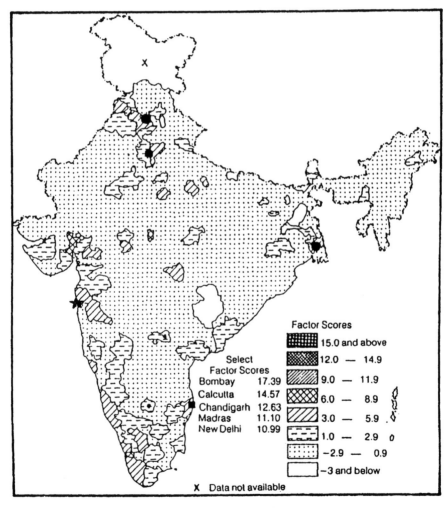

Map 8.2 Agricultural modernisation in India

(2.74), Thanjavur (2.73),South Arcot (2.34), Periyar (1.71), Dindigul (1.41) and Madurai (1.36) point at a sub regional process of local resource utilisation for raising both food and cash crops aided by the State Government through price subsidisation and supply of inputs.. Districts of Chikmagalur (1.42) and Chitradurga (1.27) in south Karnataka forming parts of the old Mysore State, have been agriculturally and industrially developed for a long time. Coorg had well-developed plantation economy with a favourable man-land ratio. Most of the districts have nearly 100% villages electrified. In Andhra Pradesh, districts in Krishna and Godavari deltas stand out. While Kurnool benefits from the Tungabhadra Multipurpose project, Nellore and Chittor get water of lower Pennar. Other districts harness water from the Nagarjunasagar Project. In Uttar Pradesh mainly the districts of Meerut (2.08), Jaunpur (1.92), Agra (1.58) in the west, Varanasi or Allahabad (1.87), Gorakhpur (1.33), Mirzapur (1.26) in the east and a few districts of Bihar such as Gaya (1.43) and its adjoining districts (1.43) have factor scores between 1.00 to 3.5. Interestingly, all districts in this part located close to large urban centres have higher scores in agricultural bank credit indicating urban-biased agricultural credit operations. In West Bengal significant districts are Bardhaman (2.24) and the adjoining Birbhum (2.21) and Malda (1.20) - the only region in the state where input-based agriculture partially succeeded. The latter also shows high scores in bank credit to agriculture (Fig.3). Huge contiguous areas in Rajasthan, Madhya Pradesh, Orissa, Maharashtra and Gujarat are found to have negative factor scores.

8.4 Summing Up

The above pattern of urbanisation and development brings out the basic characterisitic features of the prevailing socio-economic process in India. Industrial growth and the related urbanisation process are found to have a very limited spatial expansion as shown by the very high factor scores of a few large urban centres. Similarly the post liberalisation investment concentration in less developed areas having proximity to these centres irrespective of their regional, ecological or resource considerations reinforces metropolitanisation in an expanded form without any link with proper regional development (Banerjee-Guha, 1997). The latter is also explained by the slowing down of the outmigration rate of many backward districts that lie close to and far from metropolitan cities during 1981-91 suggesting that the policy of liberalisation in India which only promotes greater movement of capital and natural resources. Effectively it gets associated with the growing immobility of workforce and population. Next comes the import of capital intensive agriculture and associated rural disparity characterised by a mass of poor and a small minority of rich peasants and landlords.The control of land, private capital goods, credit, public infrastructure and government patronage by rich peasants and landlords who gain through increasing commercialiasation of agriculture and use of non- labour inputs in land (Bagchi, 1984) is also explained by the higher score areas of Factor II. Finally, Factor I and II together represent the recent trends of convergence of industry, agriculture and infrastructure in 'nouveau riche' areas and transformation of urban landscapes towards international capital requirements in established metropolitan cities as well. Increasing marginalisation of the organised workforce and informalisation of the economy especially in the urban areas also can be explained as major features through the variables of household, industrial and marginal workers. Spatial inequality has been increasing. Measured in terms of co-efficient of variation (cv) in per capita Net State Domestic Product, and taking the average value for five years the cv increased from 22.6 per cent during 1970-75 to 29.1 per cent during 1980-85. In 1993 the cv again went up to 32.2 percent (Kundu, 1996). Similar increases in regional disparity is noted in case of per capita consumption expenditure, infrastructural facilities, power consumption, transport, health services, etc.

All these are explicit despite the avowed objective of official planning to reduce spatial disparity and poverty.

References

Bagchi, A.K. (1984): 'Towards a Political Economy of Planning in India',*Contributions to Political Economy*, Vol.3, pp 15-38.

Bagchi, A.K. (1989): *Political Economy of Unerdeveloped Regions*, Charles Duckworth, U.K.

Banerjee-Guha, S. (1997): *Spatial Dynamics of International Capital: A Study of Multinational Corporations in India*, Orient Longman, Hyderabad, India.

Banerjee-Guha, S.(1985): 'Colonialism, Neo-colonialsim and RegionalDisparity: The Indian Experience', Unpublished Monograph, Department of Geography and Environmental Engineering, The Johns Hopkins University, Baltimore, U.S.A.

Bhalla, G.S. and Chadha, G.K. (1983): *Green Revolution and the Small Peasant : A Study of Income Distribution among unjab Cultivators*, Concept Publishing Co., New Delhi, India.

Castells, M. (1977): *The Urban Question*, Routledge, U.K.

Dasgupta, B. (1971): 'Socio-economic Regionalisation of India', *Economic and Political Weekly*, August 5, 1971, Bombay.

Dasgupta, N. (1984): A Geography of Socio-economic Well Being in India, Unpublished M.Phil Dissertation, Department of Geography, University of Bombay.

Kundu, A. And Gupta,S.(1996): 'Migration, Urbanisation and Regionalinequality', *Economic and Political Weekly*, `Vol.XXXI (52), pp 3391-3398, Mumbai, India.

Myrdal, G. (1957): *Economic Theory and Underdeveloped Regions*,Charles Duckworth, U.K.

Planning Commission (1980): *Sixth Five Year Plan* Government of India,New Delhi.

Smith. N. (1984): *Uneven Development*, Basil Blackwell, U.K.

9 The Information Technology (IT) Industry in Bangalore: A Case of Urban Competitiveness in India?

Sampath Srinivas

Editors' note:
Bangalore was the fastest growing city in India 1971- 1981, the chosen location for many of the technologically advanced public sector industries in India, such as aeronautics. By the 1990s it was emerging as a centre for India's private sector IT industry, which has helped it to maintain its high growth rates.

9.1 Introduction

Bangalore today has the largest concentration of Information Technology (IT)[1] industries in India. The Bangalore situation (as it may be described), is a result of a spatial reorganisation of the global IT industry and the dynamics of growth offered to the global IT industry by India. This chapter examines the causal reasons for Bangalore's dominance in the IT industry in India. To understand why, within India, its is Bangalore in particular, it is necessary to attempt to define the notion of urban competitiveness.

9.2 Understanding the Theme of Urban Competitiveness and its Importance

Why does a city become the base for successful international competitors in an industry? Why is a city often home for so many of an industry's world leaders? These questions often come to the minds of present day urban managers. The influence of the city on the pursuit of competitive advantage in a particular field is of central importance to the level and rate of productivity growth available. These concerns have led to the identification of a series of issues[2] all of which can be bundled together under the theme of urban competitiveness.

Studies related to urban competitiveness or territorial competition have

---

1 *For the purpose of the present paper, IT is defined as a group of information and electronics based technologies that includes computers and computer software.*
2 *Research have focused on different issues like territorial competition (Cheshire et.al, 1996), regional advantage (Saxenian, 1995), competitive advantage of cities (Porter, 1995), and urban competition (Kresl, 1995).*

largely been conducted in Western Europe and in North America. In the first half of the 1980s, both Europe and North America turned their attention to regional liberalisation, 'for which additional gains appeared to be more promising'(Kresl, 1995:46). The completion of the single internal market in the European Community, which was soon followed by the North American Free Trade Agreement, led to the realisation that the ability of individual nations to manage the international economic relations was constrained (Kresl, 1995).

Thus according to Kresl, as nations moved with 'slow but sure steps towards the sidelines, regions and, particularly cities became both significant as economic actors and more aware of the greater burden on them to plan strategically for their economic futures in an environment in which they were increasingly exposed to challenges to existing economic activities and to opportunities for growth into new areas emanating from the international economy' (Kresl, 1995:46). In the developing world, there are different reasons for studying local competitiveness. The economic liberalisation that is now taking place in many countries, including India, has increased the importance of the urban areas as recipients of Foreign Direct Investment (FDI). In many developing countries, new growth has concentrated in only a few large cities – the phenomenon known as 'urban primacy'.

9.2.1 What is Urban Competitiveness?

There is no universally acceptable or available definition of urban competitiveness. The definition adopted here is very specific to the current paper, and should be seen only in the context of present research. For the purpose of the paper, urban competitiveness is defined as a process by which one urban area or one region dominates a particular economic activity more than other urban areas or regions in a geographical area (a country or a group of countries).

More important than the definition of the concept are the elements that can actually be used to explain the concept of urban competitiveness itself. Singh and Kresl (1994) use a number of elements to work out the competitive ranking of forty US cities. Their study focused on three indicators of urban competitiveness. According to them, each of these variables 'captures an important aspect of the performance of a city's economy' (Singh; Kresl, 1994:429). The variables chosen by Singh and Kresl (1994) and their role in indicating competitiveness are as follows:

The growth of Retail Sales. Relatively rapid growth of retail sales will be a function of growth of the city's population, of rising income of its inhab-

itants and of the degree to which it is an attractive location for non-inhabit-
ants to come to for shopping, recreation, cultural events and dining. Again
each of these components will be indicative of competitiveness.

The growth of Manufacturing Value Added. Relatively rapid growth of
value added in manufacturing will be reflective of investments in plant and
equipment, in human capital and in infrastructure. It will give an indication of
the overall competitiveness of the city's manufacturing sector.

The growth of Business Services. While services as a category in
genereal include several items, such as amusement, auto repair, and personal
services, which have little or nothing to do with economic competitiveness,
business services are essential to any expansion of economic activity and of
any transformation of economic activity

9.3 Determinants of Urban Competitiveness

The determinants are the economic and strategic variables used to measure or assess competitiveness. Economic determinants can be divided into
factors of production, infrastructure, location, economic structure and urban
amenities. The strategic determinants are mainly governmental effectiveness,
urban strategy, institutional flexibility, and private-public sector co-opera-
tion. Kresl and Singh consider that economic determinants will mostly be
expressible quantitatively, whereas strategic determinants are more usually
qualititative.

9.3.1 Economic Determinants

Factors of Production
It has been long recognised that high and increasing productivity are
keys to making an economy more competitive. This, of course, means one
must pay attention to the quantity and quality of the labour force, the capital
stock, and production and office sites (Kresl, 1995: 52). Mullen and Williams
(1990) argue that it is the newness of the capital stock that is important because newer investments are endowed with the most recent technological
advances. 'The productivity gains that result from newer technology generate output growth and thus an increase in labour efficiency, in part because of
the in-migration of skilled workers who are attracted to these better jobs'
(Kresl, 1995: 53).

Beeson (1990) gives backhanded support for the importance of capital
in her study of the decline of manufacturing in large cities. The primary cause

of the slow relative growth rate of large standard metropolitan statistical areas (SMSAs) in the USA between 1959 and 1978 was found to be low relative growth rates in labour and productivity, in part related to a lower rate of capital accumulation.

In getting access to capital for new investments, it is not necessary for the accumulation process to be local. Florida and Smith (1992) show the importance of co-investment in attracting capital from a distance, and Kresl shows that local governments should ensure 'an adequate and growing high-tech infrastructure [which] is needed by venture capital' (Kresl, 1995: 53).

The quality and the quantity of labour in enhancement of a city's international competitiveness is significant (Pompili, 1992). Beeson (1992) stresses the impact that large agglomerations have in developing a high quality labour force through their impacts on learning. Her research leads her to conclude that: higher output per worker gives rise to more opportunity for learning by doing, producing high-technology goods does the same, and because "an individual's ability to learn depends on the average knowledge of those around him/her".

Cities facilitate this process by placing workers in close proximity with a large number of other workers from whom they can learn. Kresl (1995), suggests that a city can enhance its competitiveness by introducing subsidies to improve the average level of education and training through an industrial policy that subsidises high tech industries. Bangalore has subsidised hi-tech and electronic forms, and its development plan is biased in favour of low-pollution industries.

Infrastructure

The infrastructure needs of a city trying to enhance its competitiveness are dependent on the specific role that the city seeks to create for itself. Some cities may try to become point-of-access or bridge cities, whereas others may seek to develop a national, or international vocation. They may wish to become export bases, to develop niches in certain industries, or to build on an existing capacity to become a centre of research and development (Kresl, 1992).

Some strategies will require investment in transportation infrastructure to move goods or people, or in communications to facilitate information flow and the quality of decision making. A study by World Bank revealed that Bangalore has by far the best quality data communication facilities among the larger Indian cities (IBRD, 1992). The relationships between universities, firms, and research labs throughout the economic space may be essential to success. Moomaw and Williams (1991) argue that the most important determinant of variation in total factor productivity growth (which they link to rapid output growth) are investments by the state in education and transport and communications infrastructure.

Location

The rapid development of new technologies of production, transportation, and communication are causing a profound reconsideration of the basic tenets of traditional location theory (Kresl, 1995). 'The advantage of the centre over periphery is now greatly reduced, and goods and services, information, and decisions can be transported longer distances with no loss in time, efficiency, or cost effectiveness' (Kresl, 1995: 56). A fine example of this change in the location theory can be explained by the global software industry. Many companies, all over the world, use the fibre optic networks, and telecommunications links to control and decentralise the production process, thus reducing the cost and time in producing a product (software in this case).This then places a considerable burden on the local government to implement the correct set of policy initiatives that "will in a sense, defeat distance" (Kresl, 1995: 57).

Economic Structure

The economic structure of an urban economy is important for that city's competitiveness because it is one of the key determinants of the degree to which that economy will be able to (a) take advantage of the opportunities that present themselves, (b) gain access to foreign markets and technologies, (c) establish necessary linkages with others, and (d) provide growth in employment, production, and exports. The aspects of economic structure that seem to be most significant are the distribution of firms by size, the role played by foreign-owned firms, and the richness of the complex of firms providing business and financial support services.

It has been argued that smaller firms are better able to position themselves to take advantage of niche markets, to exploit new technologies, in production and, in general, to react quickly to rapidly changing competitive conditions (Holmström, 1994; Humphrey, 1995; Humphrey and Schmitz, 1996). But smaller firms often lack the resources and experience with exporting to foreign markets that are common to most large firms. Nevertheless, 'some of the literature contends that cities and other level of government can assist with export financing; first encounters with foreign exchange markets, foreign legal systems, and foreign languages; and contacts with the appropriate providers of the wide variety of business services that exporting requires' (Kresl, 1995:58). In the last few years, however, increasing number of researches have demonstrated that many small firms, in fact, perform work given to them by large firms[3], so that a healthy small-business sector may in some instances depend on the existence of a set of healthy large businesses.

The presence of foreign-owned firms is supported because many innovations in technology, new products, and advances in management are kept within the originating firm. If this is not the case, it is usually true that it is relatively costly to gain access to them in open market transactions. Thus Kresl argues that 'an economy with a high number of foreign-owned firms may then be one with privileged access to elements that are crucial to competitiveness' (1995: 59).

The last aspect of economic structure, the adequacy of the complex of providers of business and financial support services, is easier to resolve. 'All studies now demonstrate that this is an important item for a city that seeks to enhance or just to retain its international competitiveness' (Kresl, 1995: 60). In the case of Bangalore, Holmström mentions the role played by various state and federal institutions. 'cheap loans, advice and assistance from a number of central or state government institutions, particularly a Small Industries Service Institute and an Industrial Training Institute, and the Central Machine Tools Institute' (Holmström, 1994: 18), all of which provided assistance and consultancy services to the upcoming and established small firms in Bangalore.

Previously, it was perceived that the larger firms were quite indifferent to the services provided in a city in which they were located because much of the work that service firms would provide was done internally. Now, however, large firms are paring their staffs and the functions they perform and are contracting these tasks out to other, usually smaller firms. This accounts for much of the rapid growth in recent years in virtually all aspects of the services sector in virtually all locations (Kresl, 1995). Benería and Roldán's (1987) study suggests that among the reasons given by firms for sub-contracting, the most prevalent was the lowering of labour costs. However, as other studies (for example, McFarlan and Nolan 1995; Rothery, 1995) indicate, lower labour cost may not always be the most determining factor in outsourcing or subcontracting. This is especially true in the case of IT industries, where quality has a far more important role to play than price or the cost of labour.

Urban Amenities
Urban amenities are those aspects of human society that can be realised only in cities, where there sufficient people to support a variety of serv-

---

3 *The literature uses terms like "sub-contracting" and "outsourcing" to explain such a phenomenon. Some of the finest written articles on this issue include Mcfarlan and Nolan (1995), Rothery (1995), Storper and Scott (1992), Sabel (1984), Holmström (1993 and 1994) Ernste and Meir (1992) and Benería and Roldán (1987).*

ices. 'The provision of 'high culture' (museums, orchestras, operas, and dance companies, galleries (or their regional equivalent), and other performance and exhibition spaces) rank high on this list, along with historic districts for residences, shopping and recreation, educational institutions, parks, and so forth' (Kresl, 1995). Though these amenities may appear to have more relevance to a healthy tourism industry, they are perhaps more important for the effect they have on the local labour supply. Other variables considered by researchers to be amenities, such as temperate weather and closeness to mountains or an ocean, may also be significant determinants of the ability of a city to advance itself as a prime location for desirable economic activity . Indeed, Van den Berg and Klassen (1989) argue that given modern transportation and communications, firms can locate their plant in any city or town and that this 'calls for greater economic competition among the towns' As a consequence 'cities need to develop a strategy of supplying better-quality housing, shops, cultural and leisure time provisions' (Van den Berg; Klassen 1989: 57).

9.3.2 Strategic Determinants

Effective Governance

An effective and responsive urban local government can contribute in making the city internationally competitive. The need for metrowide governance is given by Blackley and Greytak (1986) in their study of economic activity in Cincinnati, USA. They highlight a pattern of location that is heavily influenced by the on-going substitution of capital intensive production, and changes in the technology of transportation and communication. The result is a reduced incentive for firms to cluster in the traditional core and a positive incentive to spread their activities throughout the metropolitan space. Grasland (1992) in his study of Montpellier, France (where an attempt to create a technological centre has been made by the French government) shows that it was Montpellier as the centre of a regional technological space that was important. This involves both physical structures and institutions, as well as budgetary considerations and creation of regional networks that tie the central city together with its outlying towns, universities, and firms.

Urban Strategy and Public Private Co-operation

The goal of any urban strategic plan is to mobilise local resources, to create an effective structure for their interaction, to precisely identify end objectives, and to give specific tasks to each of the relevant entities. Porter (1991), and Krugman (1991) also stress the importance of going beyond the traditional focus on technology and factor endowments, to include such concepts from economic geography as core and periphery, activity clustering, and agglomeration. Shachar and Felsenstein (1992) point out that many firms do not choose between places, they are born into one place. These firms will be primarily interested in survival and markets, and this suggests supportive policies from the local government. Later it is imperative that the city officials work to assist these firms in establishing international linkages. Both the public and the private sectors must be involved, and the coordination of their activities must be closely monitored

Institutional Flexibility
The foregoing discussion on the determinants of competitiveness highlighted how various economic activities and industries tends to get concentrated in a particular urban location. A number of studies[4] have highlighted how high technology firms get concentrated in certain urban locations. Almost all of these studies have been carried in the industrially advanced economies.

9.4 Urban Competitiveness in India

How far are these ideas relevant to India? The technology available to the vast majority of the Indian population is nowhere near that available in the western. Nor is the infrastructure in the country anywhere near that of international standards. In such a circumstance, why are the IT companies making a beeline to India, and more so prominently to Bangalore? Empirical evidence suggests that Bangalore has the highest proportion of the IT industry in the country.

The literature available on the location of the IT industry suggests that IT and other high technology firms concentrate in regions that offer high quality work force, interaction with research laboratories, availability of state of art infrastructure, and favourable state or federal laws to promote such industries.

9.5 Brief Review of the IT industry in World and in India

Over the last two decades, the price of computer hardware has been falling, while the demand for software products and documentation has been escalating at 12% per year. In relative terms the software component of any

---

4 For examples see Saxenian (1995), Castells and Hall (1994), Segal and Quince (1985),

IT system has become more and more important and expensive, and the hardware relatively cheaper. The number of trained software analysts in developed countries has been rising by only 5% per year, so that there has been a bottleneck in software development. Together these conditions have provoked IT companies to seek global solutions.

In many cases, software professionals are invited from another country to work on a client site (which has been very commonly referred to as body shopping).[5] Table 9.1 shows the difference in monthly wages between software professionals in India and US in US $ terms. As is evident, companies looking for good quality software professionals can save enormous sums of money by hiring Indian instead of American professionals.

Table 9.1 USA and India: Difference in the salaries of software professionals (1995) in US$

Professional Category	USA	India
Junior Programmer	6000-7000	290-300
2-4 Years Experience	7000-8000	380-430
System Analyst	10,000-12,500	610-720
Project Manager	12,500 +	750 +

Source : Compiled from Heeks (1996 : 115 and 117).

Table 9.2 Unfilled vacancies

Country	Deficit in Software Professionals
UK	16,000 (1989)
USA	87, 000 (1992)
Japan	400,000 (1993)

Source : Heeks, 1996 : 108.

A major trend in today's software industry is to offer clients an integrated system, which involves integrating a number of different products from software vendors which are designed to work together as a single system. Now system integration services have begun to bring together different types of software and hardware and integrate them to provide a unique configuration more or less bespoke to end users' requirements.

The ever decreasing prices of computer hardware has enabled firms to

---

5 *Body shopping is very cost effective for companies, as well trained software professionals can be hired at almost a fraction of the cost one would have paid the local software professional. But since 1994 countries like the United States have imposed stricter work permit rules to reduce body shopping, but still the trade of body shopping continues unabated.*

take on system integration, "or the progressive assembling of system components into the whole information system" (IBM, 1994). A systems integrator combines standard hardware components with custom software- or certain standard software packages modified appropriately- in a unique configuration more or less tailor-made to end users' requirements. This is a common trend all over the world. Today large integrated systems are possible, using satellite communication, modern networking technologies, and multi-user, multi-tasking operating systems. This is significant, because the greatest gains from the use of IT, in principle, come when components are integrated into a single networking entity.

For a long time it has been argued that the only niche that Indian software companies enjoy is cost. But there have been increasing evidence to suggest that it is not the whole truth. The average wage of Indian software professional in US $ has actually at least doubled between, 1990 and 1996. By 1996, the World Bank Study, projected that India's revenue productivity will be the same as Ireland and Singapore (IBRD, 1992). To stay competitive in the software industry, country and companies have to move beyond cost factors alone, and emphasis has to be laid on the quality of the product, and the level of skills involved in software development.

9.6 Bangalore as a Premier Centre for IT Industry in India

There is very limited official statistics to show the spatial spread of the IT industry in India. The Annual Survey of Industries (ASI) primarily provides a brief state level analysis of a set of industries, of which the IT industry is merely a component. Therefore, year wise information of the IT industry for all the locations in India was collated from the various volumes of the IT Magazine -Dataquest (published from India), for two time periods to offer a temporal analysis. The results presented here are based on the analysis of data of firms from 1993 and 1996.

In 1993, more than a third of the IT firms in India were based in Bangalore. Although this proportion fell to under 23 percent by 1996 (Fig. 1), the city still has the largest concentration of IT firms in the country. Table 9.3 reveals that Delhi, Bombay and Madras have emerged as the strongest competitors to Bangalore for the location of the IT industries in India.

In quantitative terms Bangalore does enjoy a premier position among the Indian cities. How does it rate qualitatively vis-à-vis other cities? An analysis has been made to do a similar thing for individual cities. One way of understanding the productivity of an industry is to analyse the Net Value Added

Table 9.3 India: Location of IT firms in major urban areas (1996)

Urban Area	Number of Firms	Percent to Total
Bangalore	205	22.2
Bombay	172	18.6
Calcutta	28	20.2
Delhi	187	20.2
Hyderabad-Secundrabad	75	8.1
Madras	169	18.3
Pune	28	3.0
Others	61	6.6
Total	**925**	**100.0**

Source: Compiled from Dataquest 1996.

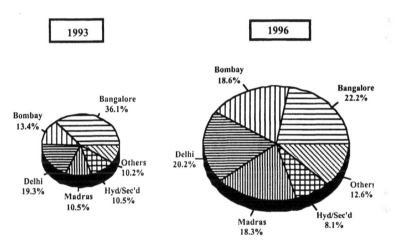

Figure 9.1   India: IT firms in major urban areas- 1993 and 1996
(proportion to the national total)
Source: Dataquest 1993 and 1996.

(NVA)[6] per employee in that industry category. Owing to the nature of data collection methods adopted by the Dataquest Magazine, the NVA (Net Value Added) per employee cannot be calculated. Instead, the Turnover per Employee (TOE) for each category and for each location has been calculated for

---

6   *The NVA needs to be interpreted carefully. High NVA per worker can be found in sectors that have high ratios of capital to labour. This is mainly because capital intensive industries must earn a normal return on large investments, they must charge prices that are larger markup over labour costs than labour-intensive industries, which means that they have high value added per worker.*

the 1993-1996 time period. The TOE is used as a proxy variable for NVA per employee. The Turnover Per Employee (TOE) in Bangalore is one of the highest in the country. Bangalore's TOE for all firms in 1996 was 30 percent greater than that of Bombay, and 16 percent higher than the all- India aggregate. However, Delhi's TOE is 30 percent higher in all firms category than Bangalore (Table 9.4). Thus it well established that the IT industry in Bangalore not only excels in quantitative terms, even quality wise, it one of the best in the country.

The analysis of data for all type of IT firms and those that are specialising in Software and services only show that, Bangalore has the largest concentration of IT industry in India. The comparative figures for 1993 and 1996 (Fig. 1) show that Bangalore's share of total number firms and employees has in fact fallen (from 39.9% to 27%), but the Turnover per Employee, has increased during the same time period , and is second highest after Delhi (Fig 2)

More than 80% of current IT firms in Bangalore started their operation between 1986 and 1996 (Table 9.5). One-fifth of all the IT firms started in India during 1986-90 period and a quarter of the firms that started operation in India during 1991-96 period established themselves in Bangalore (Table 5).

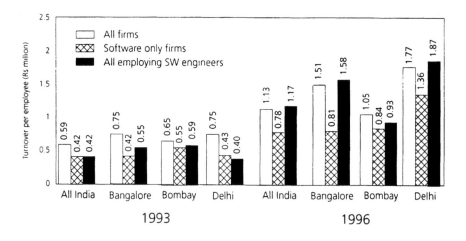

Figure 9.2   Turnover per employee in IT firms (1993 and 1996 current prices)
Source: Dataquest, 1993 and 1996.

Year	Bangalore			Bombay			Delhi			All Locations		
	All Firms	Software Only Firms	All Firms Employing Software Engineers	All Firms	Software Only Firms	All Firms Employing Software Engineers	All Firms	Software Only Firms	All Firms Employing Software Engineers	All Firms	Software Only Firms	All Firms Employing Software Engineers
1993												
Total Firms	127	69	88	47	20	30	68	28	41	352	173	233
Total Employees	8,896	3,641	5,263	14,793	5,932	9,998	10,818	3,654	5,619	48,880	16,365	29,343
Total Turnover (Rs Billion)	6.698	1.466	2.926	9.752	3.269	5.951	8.199	1.573	2.300	28.999	7.013	12.32
Average Turnover Per Employee (Rs. Million)	0.753	0.427	0.555	0.659	0.551	0.595	0.757	0.430	0.409	0.593	0.428	0.420
1996												
Total Firms	205	98	153	172	60	120	187*	62**	134***	925(742)	363(305)	694(560)
Total Employees	16,924	7,277	14,114	29,900	16,826	22,664	16,214	3,990	13,252	84,924	36,999	62,691
Total Turnover (Rs Billion)	25.714	5.960	22.367	31.690	14.266	21.170	34.954	6.945	26.647	107.733	33.19	79.262
							29.718	*5.451*	*24.901*	*96.019*	*29.05*	*73.938*
Average Turnover	1.519	0.819	1.584	1.059	0.847	0.934	2.150	1.740	2.010	1.268	0.897	1.264
							1.771	*1.366*	*1.879*	*1.130*	*0.785*	*1.170*

* 86 out of 187 Firms in Delhi have not disclosed the number of employees, which has skewed the Average Turnover per Employee in its favour

** 25 of the 62 have not given Employee figures

*** 62 of these 134 have not divulged employee figures

The Figure in Italics for Delhi and All Locations excludes the firms without information about employees

Source: Dataquest 1993 and 1996

Table 9.4 Turnover by sector, 1993 and 1996 - three major Indian cities

Table 9.5 India: Start up of IT firms by location

Year	Bangalore	Bombay	Calcutta	Delhi	Hyderabad	Madras	Pune	Others	Total
1950-59	1								1
1960-69		6		2	1				9
1970-80	10	18	1	9	3	2	5	3	51
1981-85	20	25	2	12	4	14	6	9	92
1986-90	71	64	10	82	26	60	8	26	347
1991-96	104	58	15	82	41	93	9	23	425
Total	205	172	28	187	75	169	28	61	925

Source: Dataquest, 1996.

Table 9.6 Bangalore: Turnover of IT firms (1996)

Rupees million 95-96	*Count*	*%*
Not Available	11	5.4
0-9	104	50.7
10-20	28	13.7
21-30	14	6.8
31-40	10	4.9
41-50	2	1.0
51-99	9	4.4
100-250	14	6.8
260-500	6	2.9
760-1000	2	1.0
1001-2000	2	1.0
2001-3000	1	0.5
3001-4000	1	0.5
7001-9000	1	0.5
TOTAL	205	100.0

Source: Dataquest, 1996.

Bangalore mostly has the smaller firms, unlike Bombay, which has all the large IT industries (i.e. more than 500 employee). One-third of Bangalore's IT firms employ 10 or less number of employees, and another quarter between 11 and 25. In fact over 70 percent of Bangalore's IT firms employ 50 or less number of people (Table 7).

Table 9.7 Bangalore: Employees in the IT firms (1996)

	Total Employees 1996	
Bangalore	*Count*	*%*
Not Available	12	5.9
0-10	60	
11-25	52	
26-50	33	
51-75	11	
76-150	15	
151-300	7	
301-500	10	
501-750	2	
1000-2000	2	
3001-4000	1	
Total	205	100

Source: Dataquest, 1996.

In Bangalore, software firms dominates the IT scene (Table 8).

Table 9.8 Bangalore: Disaggregation of the IT industry

	Category of the IT Industry	
Bangalore	*Count*	*%*
Information Technology	12	5.9
Hardware Only	18	8.8
Software & Services Only	98	47.8
Dealers	37	18
Peripherals & Others	34	16.6
Training including Software Training	6	2.9
Total	205	100.0

1 Includes firms doing business in Hardware, Software, and System Integration
Source: Dataquest, 1996.

Given the growth trend of Bangalore's computer software industry over the past decade, many business journals, even outside India, have started referring to it as "India's Silicon Plateau". 85 percent of new software companies that chose Bangalore as their headquarters do so because of the availability of "a large pool of low cost professionals" (Arthur Andersen Study- c.f. Business Standard, 1995, New Delhi). According to the Report in the Business Standard, Andersen has advised a number of multinationals on siting for their new Indian operations. Based on a range of factors including transport, power, telecommunications, labour availability and "livability", Andersen rated Bangalore as the first preference for locating software development. Bombay was ruled out because of its high property prices, while Andersen says, "Delhi and Madras could be considered as alternative, backup locations" (Business Standard, 1995). The capital of Karnataka (Bangalore), thus owes its success in attracting new IT investments to a combination of political, industrial and geographical factors (Financial Times, 1995).

9.7 Understanding the Urban Competitiveness of Bangalore

9.7.1 The Study Methodology

The empirical analysis has been conducted at two levels. One at the national level of policy making, and another at the city level. The research is based on both secondary sources of data and primary data collection. The study relied on two types of field surveys, a firm-level survey and a policy makers survey. An understanding of competitiveness of Bangalore is carried out using a set of indicators which include inter alia level of telecommunications infrastructure, government policies, availability of industrial/office space, skilled labour and specialised services. The field work for the research was carried out in Bangalore between July and October, 1995. In all 52 IT firms were contacted. Of these 20 were domestic, and 16 each were foreign owned and joint venture firms. These 52 IT firms can be broadly grouped into IT firms (those involved both in software and hardware), software only firms, hardware only firms, value added resellers, and peripherals manufacturers. Structured interviews were the main format of the investigation.

9.7.2 Summary of Findings

Table 9.9 Bangalore: Category of the surveyed firms

Category of the Firm	Number	Percent to Total Surveyed Firms
Software and Services Only	42	80.8
Information Technology	4	7.7
Hardware Only	3	5.8
Peripherals	3	5.8
Total	52	100.0

Source: Field Survey, 1995.

Table 9.10 Start up year of the surveyed IT firms in Bangalore

Year of Commencement Production	Number	Percent to Total of Surveyed Firms
1970-1980	3	5.8
1981-1985	7	13.5
1986-1990	20	38.5
1991-1995	22	42.3
Total	52	100.0

Source: Field Survey, 1995.

Table 9.11 Single most reason for any IT firm be interested in India

Characteristics	Category of Firm (% to the total responses)		
	Foreign Owned	Domestic	Joint Ventures
Best of Value and Quality	62.5	20.0	25.0
Lower Wages Compared to Competitors	6.3	35.0	12.5
English Speaking IT Professionals	6.3	25.0	0
Hardworking and Reliable	0	10.0	0
Large Untapped Domestic Market	25.0	10.0	62.5

Source: Field Survey, 1995.
Table 9.11 is based on the answers given to the question on why any IT firm should be interested in India.

9.7.3 IT Firms: Main Reasons for Choosing Bangalore as a Production Base

Some of the surveyed firms, also made reference to the 'technology and industry culture' that originated in the region in the early part of this century. In the early part of the twentieth century two Dewans (Prime Ministers) of the then Princely State of Mysore, Sir Viswesvarayya and Mirza Ismail, 'launched a remarkable programme of agricultural, industrial and social development. Viswesvarayya established a polytechnic in Bangalore, and government-owned factories there and elsewhere in the state' (Holmström, 1994: 16). India's leading industrial firm Tata endowed what is now the Indian Institute of Science, 'to be followed by a host of publicly and privately funded research institutes' (Holmström, 1976: 8). During the Second World War, 'India's first aircraft factory Hindustan Aircraft (now Hindustan Aeronautics) was founded in Bangalore. Thus at the threshold of India's Independence (in 1947), Bangalore had one of the most technologically advanced industries and work force of the time in India.

In the years after the Independence, the national government established some of the country's biggest public sector factories, notably Indian Telephone Industries, Hindustan Machine Tools (making machine tools and watches for export and domestic market), Bharat Electronics (mainly supplying the defence forces), and Bharat Earth movers. The private sector, according to Holmström, 'followed, taking advantage of the large number of engineers and skilled workers trained in the vast public sector factories' (Holmström, 1994: 18).

Table 9.12 Importance of locational decisions

Major Factors for Choosing Bangalore as a Production Base	*Very Important Locational Decision*	*Important Locational Decision*	*Less Important Locational Decision*	*Total (%)*
Government Related Factors				
Government Support	50	44.2	5.8	100
Software Technology Park	0	23.1	76.9	100
Electronic City	0	7.7	92.3	100
City Related Factors				
Availability of High Technology Professionals	82.7	11.5	5.8	100
Availability of Research Institutes and Laboratories	82.7	17.3	0	100
Already a Major Centre for High Technology Production	36.5	51.9	11.5	100
Cheaper Cost of Living than Bombay/Delhi	21.2	67.3	11.5	100
Favourable Physical Climate throughout the year	13.5	65.4	21.2	100
Others				
Sheer Convenience	13.5	11.5	75.0	100

Source: Field Survey, 1995.

Among the government related factors, more than 90 percent of the surveyed firms consider government support as an important locational factor (50 % of these find it as the most important locational decision). Availability of research laboratories and institutes scores higher than the availability of professionals. All the 52 firms find the research laboratories and institutes as an important locational factor (with 82.7% stating it as the most important factor). Interestingly, 5.8 percent of these 52 firms find availability of skilled professionals as a less important locational factor. This clearly demonstrates the importance attached by the IT firms in Bangalore to the research institutes and laboratories in Bangalore. Bangalore's position as a high technology centre, cheaper cost of living than Delhi and Bombay, and favourable physical climate all are important locational decisions (but not the most important ones).

9.7.4 Bangalore's Future and the Issue of Competitiveness

Most of the firms established a clear causal reason for being located in Bangalore. They were further probed on the sustainability of Bangalore, increasingly beset with pollution. In the set of questions that addressed this issue, the first one asked the firms if they still considered Bangalore as an attractive city for the location of the IT firms in India. More than three quarters (78%) of the surveyed firms still consider the city as an attractive location for setting up IT business in India. The most important reason to support that answer (in order of importance) are: better telecommunications and data communication infrastructure than any other city in India (35%), best city for IT business in India (28%), and more and more IT firms choose Bangalore as a production base (37), all of which according to the surveyed firms indicate Bangalore's continued competitive position. However, the overall infrastructure supply in the city has come under criticism. Only half of the surveyed firms (53%) feel that Bangalore's infrastructure is adequate to support further growth of IT industry in the city. This is a clear message to the urban administrators of Bangalore.

The surveyed firms point to three clear problems that needs to be addressed by the city's administrators immediately. The first and foremost problem is the woefully inadequate power supply situation not only in the city but throughout the state of Karnataka. This is notwithstanding the fact that most of the IT firms are low consumers of power, and many of them have their own captive power generating facility independent of the state power supply. However these firms cannot provide electricity to the homes of its executives! Increasing pollution levels in the city, and congestion on the roads of Bangalore is considered as the next major problem (34%). Absence of an international airport (20%), and very high turnover of professionals especially the software engineers (20%) are other problems largely worrying the surveyed firms.

9.8 Conclusions

It is only in the last couple of years that studies on the issues of urban competitiveness have started appearing in the field of urban and regional economics. It is a topic of great importance to the cities of the developing world. By specifically targeting a group of economic activities cities may be able to attract investment into a particular sector or an activity. They must identify the key sectors or economic activities that offer maximum leverage, so that the urban area can consolidate its position in that activity, and attract a larger share of investment than other urban areas. In this context the study of IT industry in Bangalore offers insights into how the city has been able to project itself as a major centre for IT industry in India, and thus seems to answer the question raised in the title of this paper. However, there are many facets that have performed poorly in Bangalore. Poor urban infrastructure in Bangalore, and absence of venture capitalists to fund further expansion are perhaps two of the most crucial factors that may prevent Bangalore from continuing to be competitive in the future IT sector in India.

References

Beeson, P. (1992) 'Agglomeration Economies and Productivity Growth', in E. Mills and J. McDonald (Eds.), *Sources of Metropolitan Growth,*(pp. 19-35), Centre for Urban Studies, Rutgers University, New York.

Beeson, P. E. (1990) 'Sources of the Decline of Manufacturing in Large Metropolitan Areas', *Journal of Urban Economics*, 28, 78-84.

Benería, L; Roldán, M. (1987) *The Cross Roads of Class and Gender- Industrial Homework, Subcontracting, and Household Dynamics in Mexico City*, University of Chicago Press, Chicago.

Blackley, P; Greytak, D. (1986) 'Comparative Advantage and Industrial Location: An Intrametropolitan Evaluation', *Urban Studies*, 23, pp. 221-230.

Calem, P; Carlino, G. (1991) 'Urban Agglomeration Economies in the Presence of Technical Change', *Journal of Urban Economics*, Vol. 29, pp. 82-95.

Castells, M; Hall, P. (1994) *Technopoles of the World: The Making of the 21ˢᵗ Century Industrial Complexes*, Routledge, London.

Dataquest (1993) *The Indian Top 20*, New Delhi.

Dataquest (1994) *The Indian Top 20, Vol. I, and Vol. II* ,New Delhi.

Dataquest (1995) *The Indian Top 20. Vol. I, and Vol. II* ,New Delhi.

Dataquest (1996) *The Indian Top 20, Vol. I, Vol. II , and Vol. III*, New Delhi.

Ernste, H; Mier, V (1992) (Eds.) *Regional Development and Contemporary Industrial Response : Extending Flexible Specialisation*, Belhaven Press, London.

Florida, R; Smith, D. (1992) 'Venture Capital's Role in Economic Development: An Empirical

Analysis', in E. Mills and J. McDonald (Eds.), *Sources of Metropolitan Growth,* (pp. 187-209), Centre for Urban Studies, Rutgers University, New York.

Grasland, L. (1992) 'The Search for an International Position in the creation of a Regional Technological Space: The Example of Montepellier', *Urban Studies*, 29, pp. 1003-1010.

Hansen, N. (1987) 'The Evolution of the French Regional Economy and the French Regional Theory', *The Review of Regional Studies*, 17, pp. 5-13.

Heeks, R. (1996) *India's Software Industry-State Policy, Liberalisation, and Industrial Development*, Sage, New Delhi.

Heenan, D. (1977) 'Global Cities of Tomorrow', *Harvard Business Review*, 55 (3), pp. 79-92.

Holmström, M. (1994) *Bangalore as an Industrial District: Flexible Specialisation in a labour-surplus economy?*, Pondy Papers in Social Sciences, Institut FranÁais De Pondichery, Pondichery.

Humphrey, J. (1995) 'Industrial Organisation and Manufacturing Competitiveness in Developing Countries- Editor's Introduction', *World Development*, Vol. 23, Number 1, pp. 1-7.

Humphrey, J. ; Schmitz, H. (1996) "The Triple C Approach to Local Industrial Policy", *World Development*, Vol. 24, Number 12, pp. 1859-1877.

IBRD (1992) *India's Software Services- Export Potential and Strategies*, International Software Studies, Presented by Maxi/Micro Inc., USA.

Klassen, L. (1987) 'The future of the Larger European Towns', *Urban Studies*, 24, pp. 251-257.

Kresl, P. (1995) "The Determinants of Urban Competitiveness : A Survey", in P. K. Kresl and G. Gappert (eds.) *North American Cities and the Global Economy*, Sage, Thousand Oaks, pp. 45-68.

Kresl, P. K.; Singh, B. (1994) 'The Competitiveness of Cities: the United States', in OECD Sponsored Conference on *Cities and the New Global Economy*, Conference Proceedings, pp. 424-446.

Krugman, P. (1991) *Geography and Trade*, MIT Press, Cambridge, Massachusetts.

Krugman, P. (1996) 'Competitiveness: A Dangerous Obsession', *Pop Internationalism*, MIT Press, Cambridge, Massachusetts pp. 3-24.

McFarlan, F. W; Richard Nolan (1995) 'How to Manage an IT Outsourcing Alliance', *Sloan Management Review*, Winter.

Moomow, R; Williams, M (1991) 'Total Productivity Growth in Manufacturing: Further Evidence from the States', *Journal of Regional Science*, 31(1), pp. 17-34.

Mullen, J. ; Williams, M. (1990) 'Explaining Total Factor Productivity Differentials in Urban Manufacturing', *Journal of Urban Economics*, Vol. 28, 103-123.

Nolan et al. (1988) *Managing Personal Computers in the large organisations*, Lexington, Massachusetts.

Paul Cheshire ; Gordon I. (eds.) 1995 *Territorial Competition in an Integrating Europe*, Avebury, Aldershot.

Piore, J; Sabel, C. (1984) *The Second Industrial Divide : Possibilities for Prosperity*, Basic Books, New York.

Pompili, T. (1992) 'The Role of Human Capital in Urban System Structure and Development: The case of Italy, *Urban Studies*, Vol. 29, pp. 905-934.

Porter, M. (1990) *The Competitive Advantage of Nations*, Free Press, New York.

Porter, M. E. (1995) 'The competitive advantage of the inner city', *Harvard Business Review*, May-June, pp. 55-71.

Rothery, B. ; Ian Robertson (1995) *The Truth about Outsourcing*, Gower, Hampshire.

Saxenian, Annalee (1996) *Regional Advantage: Culture and Competition in Silicon Valley and Route 128*, Harvard University Press, Cambridge, Massachusetts.

Schware, R. (1992) *The World Software Industry and Software Engineering*, World Bank Technical Paper, Number 104, Washington.

Shachar, A.; Felestein (1992) 'Urban Economic Development and High Technology Industry', *Urban Studies*, 29, pp. 839-855.

Singh, V.; Borzutzky, S. (1988) 'The State of Mature Industrial Regions in Western Europe and North America', *Urban Studies*, 25, pp. 212-227.

Storper, M.; Scott, A. (1992) *Pathways to Industrialisation and Regional Development*, Routledge, London.

Van den Berg, L.; Klassen, L. (1989) 'The Major European Cities: Underway to 1992', in M. Belil (Ed.) *Eurocities* (pp. 55-60), Organising Committee of the Eurocities Conference, Barcelona.

10 Sustainable Tourism: Resources and Strategies for Western India and the Case of the Shekhawati Painted Walls

Allen G. Noble and Frank J. Costa

Editors' note:
In Chapter 6, the extensive wilderness areas of the Russian Far East were described as an attraction for tourism. Here, unique folk culture, such as dance, dress, handicrafts, painting, architecture, and house types are attractions of a different kind. The elaborate paintedwalls of Shehawati pressent an opportunity for improving local income through tourism

10.1 Introduction

As world population continues to expand, countries are forced to search ever more diligently for strategies to support the growing number of their citizens. Because of its already large population, India is one of those nations most concerned about such problems. As long ago as the 1950s India began a national program to curtail population growth. A considerable part of its national resources has been devoted to family planning education, to the provision of contraceptive devices, to promotion of other birth control programs, and even to the encouragement of sterilisation as an effective limitation to human reproduction.

Despite the acceptance of these measures as national policy and the considerable expenditure involved, the Indian programs have been only partially successful. Some of the difficulties standing in the way of mounting an effective program are directly related to the fact that the Indian governmental system is basically democratic. Thus, the draconian measures adopted by China, for example, have been resisted in India as inappropriate.

Progress toward controlling population growth is likely to be quite measured and the problem of how a growing population can secure economic support which would result in rising living standards remains to be tackled effectively.

One strategy receiving growing support, not only in India, but throughout the world is an emphasis on sustainable development. Such an approach requires a more effective utilisation of existing resources, on a more or less permanent basis, without the depletion of those resources. The expansion of tourism is one of the most often cited types of sustainable development.

Western India appears to be a region where such sustainable tourism could be effectively employed to extend economic support. A unique range of structures and decorative features found there can form the basis for regional tourism. These features include the wall murals of Shekhawati (Rajasthan), the carved house facades (*haveli*) of Gujarat, the stepped wells of both Gujarat and Rajasthan, and a number of minor, but still distinctive features. Proximity to the already well-established tourist centres of Jaipur and Udaipur and the large urban centre of Ahmedabad is a further advantage.

10.2 Shekhawati Painted Walls

The Shekhawati area lies midway between the medieval kingdoms of Jaipur and Bikaner in the modern state of Rajasthan. It comprises three modern districts of that state — Churu, Jhunjhunu, and Sikar. Shekhawati emerged sometime in the 15th century from a much larger Rajput domain (Cooper 1994, 11). Beginning in the 18th century, prosperity and local peace encouraged a flowering of artistic endeavor. Particularly noteworthy are the painted wall murals.

Although wall murals decorate towns or cities in much of western India, it is in the Shekhawati area that this art form reaches its pinnacle of development. "Nowhere else in the world is there such a profusion and concentration of high quality frescoes as in Shekhawati" (Rakesh and Lewis 1995, 8).

One important reason for the concentration of painted walls in Shekhawati is the success which local merchants (Marwaris) found in their trading activities as their influence grew throughout the length and breadth of India. Each time their wealth grew, they constructed some sort of building in the hometown, often extensive and elaborate. Of necessity, the houses were not ornamental on the outside so as not to create envy from royal protectors or to excite the interest of tax collectors. Thus, one finds the earliest, and often the best preserved and most finely executed, murals on courtyard and interior walls. Only in the late 19th and 20th centuries did exterior wall decoration find favour.

Because wall murals continued to be painted until just before Indian independence in 1947, and fewer and fewer restraints on ostentation were imposed as time passed, the later paintings are often more openly flagrant displays of wealth. Family members in fancy dress, jewelry, and possessions such as motor cars are frequently seen, and exterior walls more often used. Although wall murals appear in almost every town and village in Shekhawati, a few communities are especially important centres of such decoration. Map 10.1 locates these places.

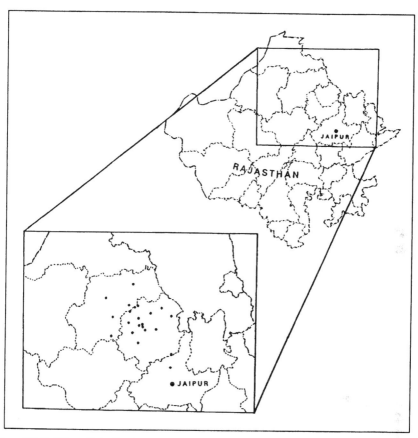

Map 10.1 Haveli in Rajasthan and Gujarat

10.3 Haveli

The word *haveli*, which originally meant an enclosed courtyard area, is widely used throughout western India to refer to substantial dwellings usually built around one or two courtyards. In Gujarat and southern Rajasthan, the term has a more specific meaning. It applies to dwellings with rich interior and exterior decoration. Especially characteristic are the intricately carved wooden facades. Map 10.2 shows the distribution in Gujarat of haveli as determined in a special publication by the Census of India in 1961 (*Census of India 1961* 1965). Ahmedabad and its environs, the southern part of the Kathiawar peninsula and the southern coast including the urban centres of Surat and Baroda (Vadodara), have the highest concentrations. In Ahmedabad and to a lesser extent in Mahesana, Baroda (Vadodara) and Surat, the houses

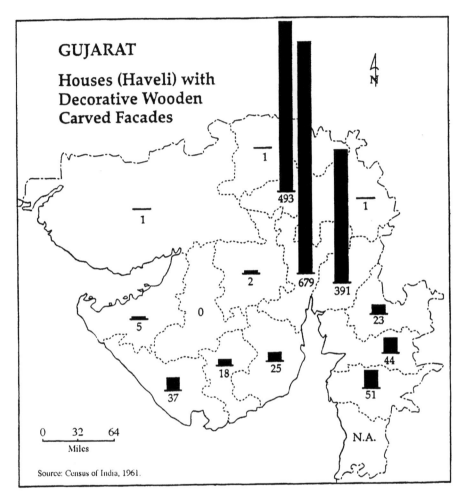

Map 10.2 Haveli in Gujarat

are contained within a settlement arrangement of *khadkis* and *pols*. The khadki is a collection of houses opening on to a cul-de-sac with a single outlet usually controlled by a massive gateway. Khadkis are grouped into larger neighborhoods, precincts, wards or quarters called pols which can extend over considerable urban territory. The pol normally has just one, or at most a very few entrances, also guarded by elaborate gateways. Although typically much smaller, at least one, in the 1960s the Mandvi Pol in Ahmedabad, claimed a resident population of about 10,000 (Spate 1967, 653). The pol houses closely associated castes or communities or even a single caste, whereas the khadki contains the dwellings of members of a kinship group. The Bombay

Gazetteer (1879, 294) suggested that the development of pols dates from Hindu-Muslim conflicts of the 18th century, but in truth their origin as a defensive feature seems much older.

Space within the khadkis and pol is almost entirely occupied by the compounds and dwellings of the residents. Entrance to the khadki is restricted. In some cases, not even residents of other khadkis within the same pol are allowed entry. One or two wells supply water to the inhabitants of the entire pol and public latrines, located close to the pol gateways for easy access by harijans for emptying and cleaning, complete the usual public facilities. Sweepers are permitted limited access to the pol to keep the street clean.

The evolution and the morphology of the Gujarat haveli have been extensively analyzed by Pramar (1989) and will not be discussed here. What is of interest is the utility of the haveli carved façade as a focus of sustainable tourism.

Because virtually all havelis are in an urban environment, they stand in danger from continuing population growth and consequent urban expansion. Perhaps the greatest threat is from government planners and engineers who desire to solve traffic congestion and circulation problems by connecting the khadki cul-de-sacs and removing pol gateways to facilitate vehicular movement. Such approaches rarely respect tradition or recognize the inestimable value of artistic resources.

The greatest treasure of the havelis, their intricately carved facades, unfortunately lie closest to the street and hence in greatest jeopardy. Minutely carved door and window frames, balcony porches and balustrades and pillars, brackets and struts combine to produce magnificent facades (Fig 10.1). Originally the carving was subsidiary to the structural use of the wood, but gradually the carving became a symbol of status and was created primarily for display.

10.4 Stepped Wells

A third distinctive vernacular feature which should be recognized and protected to promote sustainable tourism in western India is the stepped well. Variously referred to as a *baoli, bowri, baori, bavadi, baodi* or *wav*, they have a wide distribution in the dry landscape of Gujarat and Rajasthan, where obtaining a secure source of water is critical to survival. Pandit (1955, 97) suggests that virtually every town and village in Gujarat has at least one such well. However, only a handful ever possessed the elaborate architectural structure and ornamentation which would enable them to be resources for tourism. The most important of these are located on Map 10.3.

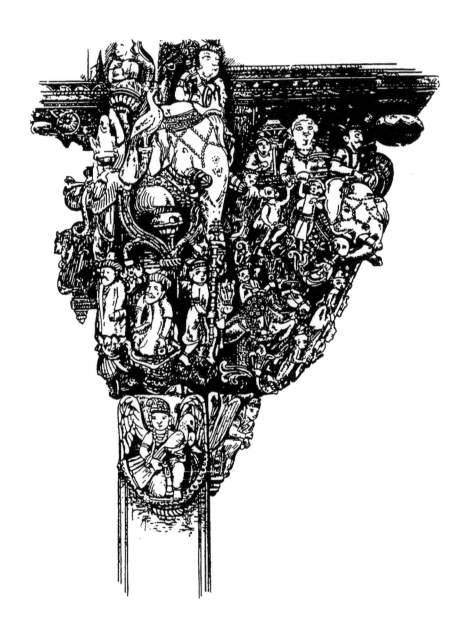

Figure 10.1 An ornamental bracket from a Haveli

The India Handbook 1999

Map 10.3 Location of stepped wells

The stepped well solved an important problem by reducing labour requirements. In western India, as elsewhere in much of the country, water to irrigate fields is drawn from shaft wells by using bullocks to lift the water. Water for household use drawn from the same wells, where the depth to the water table may be over one hundred feet, necessitates enormous human effort. To solve this problem, the stepped well evolved. In some cases a pavilion above ground marks the well; in other cases the structure is entirely below ground. Those wishing to secure water merely have to descend a long flight of steps to water level. A second advantage is that a stepped well permits access to the water of a large number of people at the same time. The stepped well is a modification of the shaft well. This latter part is still used to draw irrigation water by the traditional methods. The stepped well typically is decidedly elongated. The Dada Harir well, one of the best known, is 196 feet long by 40 feet wide (Bombay Presidency 1879, 282).

10.5 Minor Tourist Resources

A number of additional features occur in Rajasthan and Gujarat, which offer supplementary tourist sites. These include a large number of Hindu and Jain temples, Muslim mosques, forts, caravanserai and *chittai*. The latter, found mostly in Rajasthan, are ceremonial pavilions, often of exquisite design, erected to memorialize important personages. As befits a frontier region which saw the ebb and flow of various empires and kingdoms, western India possesses almost innumerable small forts and fortified palaces, the latter usually contained within the former. Many are impressive and could extend the tourist interest in this region. Caravanserai, while not numerous because of their unusual function, also may provide an additional tourist resource dimension. Finally, the religious structures add to the sites available for tourist development. Many are in close proximity to the more major facilities discussed above. Western India clearly has an abundance of resources to support sustainable tourism.

10.6 Sustainable Tourism Planning Strategies

Economic development through tourism is a policy filled with potential conflict for the physical and social environments of western India. Evidence from other parts of the world suggests that tourism development without adequate planning can result in the depletion and eventual loss of the tourism resource.

To avoid this circumstance, tourism development must provide for adequate protection of both natural and cultural resources for the present day, and for the future. In essence, this is the goal of sustainable tourism; protection and enhancement of the tourist environment for current and future residents and visitors.

Sustainable development in tourism strives to maintain and enhance a *symbiotic relationship* between *tourism* and *environmental protection* by maintaining and, wherever possible, enhancing the economic benefits of tourism without depleting the tourism resource, and by maintaining and, wherever possible, enhancing the natural, built and cultural environments of the tourist setting. The process should avoid all actions that are irreversible such as the cutting of old growth forests, the destruction of ancient monuments and artifacts, historic urban areas and irreplaceable natural landscapes (Cronin 1990, 12).

The most obvious beneficial impact of traditional tourism is as an impetus for economic development from foreign revenue infustion and local job creation. But the most obvious negative impact or cost is the premature and often unplanned development of tourist facilities which have damaged the physical assets that they were intended to exploit. The pattern of overdevelopment and environmental damage can be seen dramatically in the rapid and unplanned growth of Mediterranean coastal resorts of the 1950s and 1960s. Natural amenities were exploited and frequently destroyed in the rush for tourist revenues. Parts of the coastal regions of Spain, Italy, France and Greece provide visible testimony to the damaging effects of uncontrolled and unsustainable development. The dilemma that tourism destination areas must confront is how to take economic advantage of their resources without overusing and eventually destroying them. This specific dilemma and concern has given rise to the practice of sustainable tourism. Destination areas must plan for the use of their touristic resources so that the primary goals of sustainable development, resousrce conservation and intergenerational equity, are achieved.

10.7 Tourism Planning Levels

The national level of tourism planning is concerned primarily with establishing national tourism goals, formulating tourism development policy and broad implementation strategies. At this level, decisions are made concerning which regions should be developed for tourism (also an aspect of general economic development planning, how tourism planning should be coordinated with other planning sectors such as transportation planning and the financing and construction of major infra-structure to support tourism (airports, ports, road, rail). Other important national level tourism planning policies include the identification of gateway cities (where foreign tourists are received) and internal travel routes. Finally the national level is responsible for developing coordinating policies for environmental and socioeconomic development which should reflect sustainable development goals.

At the regional level, national policies are further developed and refined to reflect the particular assets or conditions of specific regions. In countries without national level tourism planning, the regional or state level must provide overall coordination of tourism development. The regional level is responsible for identification of appropriate regional level policies and strategies for the socioeconomic and physical development of tourism sites and resorts. Paraalleling the national level, the regional planning process should identify the major tourist access points (regional gateways) and an internal transportation network for regional tourism. The regional level is also responsible for providing greater detail and devising implementation strategies for environmental and socioeconomic development goals identified in general terms at the national level.

At the local level the focus of tourism planning becomes site specific. Local or district-wide land use plans which may include detailed development proposals or zoning of land should be prepared at the local level. Planning issues of importance include internal linkages between tourist attractions, the complementary and harmonious development of accommodations, restaurants and shops, and the appropriate integration of new development (either new tourist attraction or new support facilities) into the exisitng fabric of the tourist destination area. Concerns for the successful integration of visitors with residents also are addressed at this level through mechanisms to avoid congestion on overcrowding and to guarantee access for local residents to tourism attractions and other local facilities.

The first step in creating a sustainable tourism process at the local or regional level is the preparation of a tourism plan which includes a land use component, a heritage protection component, a set of design guidelines for new development, appropriate regulatory tools to guide the development process, and finally a management component to assure development is occurring according to the plan and its undergirding regulations. The entire process of plan development and implementation should reflect a recognition of development limits which are frequently expressed through the identification of the physical and social carrying capacities of the area or site. Physical carrying capacity refers to the ability of a natural system or environment to absorb population growth and physical development without significant degradation or breakdown. Social carrying capacity refers to the limits of local tolerance for new or increased development.

Specific sustainable tourism strategies for the areas under study in this chapter can be divided into two parts:(a) physical development and accessibility strategies and (b) socio-economic and cultural strategies. Physical development and accessibility strategies relate primarily to land development and transportation, while socio-economic and cultural strategies are concerned with a range of human-centered issues.

The most important physical development strategy is the identification of gateways or entry points into the tourist area which also serve as "immersion centres" from which tourists travel to sites in a controlled and directed manner to minimize intrusion into the tourist site or destination area. In essence this strategy is one of concentration of visitors and support facilities in areas capable of sustaining high density development, and then controlled dispersion to local sites with lower tolerance or threshold levels for tourist development as expressed in their physical and social carrying capacities.

Regional gateways perform several important functions in tourism. They serve as regional points of entry and connection for the comprehensive national network of air, rail and road transport and, thus, act as the focus of the regional transportation and accessibility network. Another important function for regional gateways is to provide accommodation and reception facilities for visitors destined to dispersed tourism sites within the region. In carrying out this latter function, regional gateways act as centres of concentration and information dissemination for visitors. Here should be located visitor centres, museums, cultural displays and other facilities designed to give a unified understanding and appreciation of the dispersed cultural and historic resources that tourists will visit in a guided route pattern and in mass transport modes such as shuttle buses, vans or rail when available. This approach is successfully employed in many areas. For example, this strategy of concentration and dispersion is used in the Amboseli National Park in Kenya.

Several social and cultural strategies can be applied to an area to maximize the prospects for sustainable tourism. The most important economic strategy is to minimize economic "leakage", or the loss of tourism revenues from the destination area. There are three ways in which this can be accomplished and economic "leakage" minimized. Provisions or policies on the part of government and incentives for private sector tourism operators to promote local ownership of tourism services such as reception facilities and transportation services within the regional gateway centre are critical. These would include hotel, restaurant, travel agency and private bus or van services. A second strategy is the promotion of education for local people to assure access to positions in the regional tourist industry. Employment in the hotel and restaurant industries can provide numerous, but primarily unskilled and seasonal positions for local people. However, skilled and permanent managerial positions should also be accessible to trained locals.

A third strategy, similar to the first, is the promotion of local entrepreneurship in the production of goods and services for the local and regional tourism markets. Provision of agricultural products, handcrafts, construction materials and skills together with services directly associated with tourism reception and transportation services are the most appropriate areas for local entrepreneurship. Shaw and Williams discuss a model originally proposed by Lundgren illustrating the progressive development of local entrepreneurship. Although they have utilized an island enclave, the model can be applied to any area where tourism is being introduced. The production and sale of local crafts, food products, and performance or exhibition of local art and dance can be the basis for entrepreneurial development at dispersed sites, while all of these with the addition of accommodations and transport services would constitute the basis for entrepreneurial development at the regional gateway location.

Finally the promotion of local folk culture as expressed in arts, crafts and performances has an economic benefit as discussed earlier, but of equal or even greater significance for sustainability, is its importance for the preservation of cultural authenticity.

In summary then, several key principles for sustainable tourism can be identified. These are:

*Sustainable tourism should subscribe to ethical principles that respect the culture and environment of the tourist area, the economy, traditional ways of life and political patterns;

*Sustainable tourism should involve the local population, proceed only with their approval and provide for a degree of local control;

*Tourism development should be undertaken with inteer-generational equity in mind. It should involve a fair distribution of benefits and costs among tourism entrepreneurs and host people, not only now but in the future.

*Tourism development should be planned and managed with regard for the protection of the natural environment for future generations;

*Development should be assessed on an ongoing basis to evaluate impacts andpermit action to counter any negative effects (Cronin 1990, 12-18).

10.8 Applying the Strategies to Western India

The most appropriate strategy to ensure sustainable tourism development in western India is to create a guiding mechanism which includes participants on all levels. A Gujarat-Rajasthan Tourism Council might be created as a coordinating body. It should be small and have a small professional staff. The membership of the Council should include individuals nominated by the central and state governments, from various segments of the tourist industry, environmental bodies and other groups as the need is perceived. Subsidiary advisory bodies could be constituted at more local levels.

Initially, funding would have to be provided from the central government, but a recognition of gradually increasing private sector funding should be recognized and clearly articulated from the outset. Funding should consist of central government grants, small state grants, through various regional and local planning bodies and, perhaps, membership dues from travel industry providers such as hotels, travel agencies, catering providers, private transport operators, and so forth. Tourist receipts, bed taxes and other similar revenue generating measures could be explored to provide additional funding. Financial resources should be used to support professional staff, to encourage development projects which contain sustainable elements, and to ensure environmental protection when tourism development occurs. A major responsibility of the professional staff would be to prepare and update tourism plans, and to monitor development projects. The Tourism Council needs to be given statutory responsibilities in order that it can enforce controls.

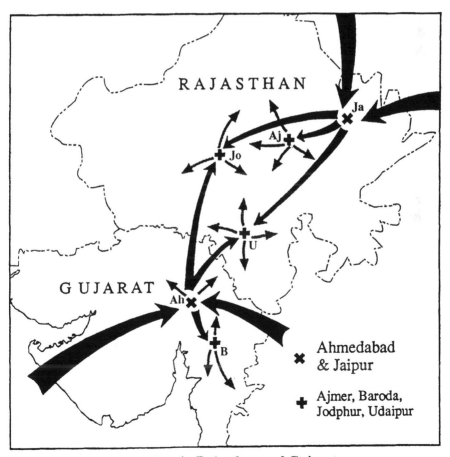

Map 10.4 Planned tourism in Rajasthan and Gujarat

Two gateway cities, Jaipur and Ahmedabad, and four regional tourist service centres, Ajmer, Udaipur, Baroda and Jodhpur should be designated (Map 10.4). Of these, Jaipur and Ahmedabad would have primacy since they are and will continue to be the major points of entry to the region. Both cities have important airway and railway connections, significant hotel facilities and other urban amenities to serve tourists. Also important is the fact that the two gateways are located in different provinces, which helps to ensure political balance. Development activities must never lose sight of the importance of political factors in order to ensure continued support from the entire region. Ajmer, Udaipur, Jodhpur and Baroda have more limited accessibility and will continue to function as service centres, but only for parts of the entire region. Because the area is large, each of these centres will function cooperatively rather than competitively. In each of these centres orienting museums or facilities need to be constructed. Each of the six gateways and regional centres can act as a centre for the organisation of visits to neighboring villages and towns which possess the actual resources. These locations can gain from handicraft activities, local catering and other specific tourist services.

10.9 Conclusion

As Indian population continues to grow, traditional settlements and architectural features are placed under the threat that their basic integrity will be damaged. The additional numbers of people in western India must also be provided with support, hopefully support which can be sustained into the future. Developing a sustainable tourism industry based upon existing architectural features holds promise of achieving both goals. However, the task will not be easy. Only a concerted effort by dedicated individuals and agencies has any possibility of success.

References

Bombay Presidency, 1879. *Gazetteer of the Bombay Presidency*. vol IV, *Ahmedabad*. Bombay: Government Central Press.
Bradnock, R and Roma Bradnock. 1998. *The India Handbook 1999*. Bath: Footprint Handbooks.
Census of India 1961. 1965. "Wood Carving of Gujarat," Delhi: *Census of India, 1961*, vol. 5, *Gujarat*, Part VII-A (2) pp. 1-45 plus plates.
Cooper, Ilay. 1994. *The Painted Towns of Shekhawati*. Ahmedabad: Mapin Publishing Pvt. Ltd.

Cronin, L. 1990. A Strategy for Tourism and Sustainable Developments. *World Leisure and Recreation.* 32:3:12-18.

Henry, W.R. 1980. Patterns of Tourist Use in Kenya's Amboseli National Park: Implications for planning and management, pp. 43-57 in *Tourism Marketing and Management Issues.* D. Hawkins, et al, (eds.) Washington: George Washington University.

Pandit, S.H. 1955. Stepped Wells of Gujarat - A Marvel of Subterranean Architecture pp. 97-103 *Souvenir Volume, Forty Second Session, Indian Science Congress.* Baroda: M.S. University of Baroda.

Pramar, V.S. 1989. *Haveli: Wooden Houses and Mansions of Gujarat.* Ahmedabad: Mapin Publishing Pvt. Ltd.

Rakesh, Pankaj and Karoki Lewis. 1995. *Shekhawati: Rajasthan's Painted Homes.* Delhi: Lustre Press.

Shaw, Gareth and Allan M. Williams. 1994. *Critical Issues in Tourism: A Geographic Perspective.* Oxford: Blackwell.

Spate, O.H.K. 1967. *India and Pakistan: A General and Regional Geography.* London: Methuen.

THE URBAN BASE:
SAUDI ARABIA

Editors' note:
Dr Mubarak reports on the process of urbanisation in Saudi Arabia in this century. In 1950 Saudi Arabia was the least urbanised country in West Asia, except for Oman and Yemen. By 1990 after the oil booms 80 per cent of the population was urban. The process of urbanisation has been part of the building of a nation state out of warring nomadic tribes, and part also of the modernisation of that state. Part of the melding of the tribes and their change from nomadic to sedentary existence is achieved by the reaffirmation of Islam. The process is in many senses ideologically driven, yet it could not be further from the ideology of China, where much of the stress has been on minimizing the growth of the towns, and keeping the peasants in the countryside. Another contrast is that whereas China has until recently maintained an autarkic economy, Saudi Arabia has remained ideologically closed while being hugely dependent on trading its oil resources internationally.

11 Nomad Settlements in Saudi Arabia: A Cultural Approach to Urbanisation in Developing Countries

Faisal A. Mubarak

11.1 Introduction

Saudi Arabia occupies an area of 2,149,690 square kilometers, slightly less than one fourth the size of the United States. One percent of the land is arable, while approximately 39% is meadows and pastures. The remaining area is an uninhabited sandy desert. It has no perennial rivers or permanent water bodies. When King Abdul-Aziz conquered and united what is now known as Saudi Arabia, he acquired a vast, rugged land, which was thinly populated and geographically isolated from the outside world by the forbidding expanse of deserts. In the early 1950s, it was estimated that less than two percent of the entire area was under cultivation.[1] The traditional and xenophobic society was afflicted with poverty, illiteracy, high mortality and endemic water shortages. The new nation lacked modern infrastructure, and technology and did not have an industrial base. The country had no social, medical or educational institutions in the modern sense.

The genesis of modern urbanisation in Saudi Arabia lies in the early battles fought by King Abdul-Aziz Al Saud, the founder of modern Saudi Arabia starting with the capture of Riyadh in 1902. Using the double-objective approach of religion and largesse to win the loyalty of the strong Bedouin tribes occupying the vast sparse terrain of the Arabian Peninsula, King Abdul-Aziz encouraged the formation of the *Ikhwan* movement in a successful attempt to turn their pugnacious independence into a sizable military force upon which he relied to expand his territorial gains. By doing so, he skillfully co-opted the tribal sheik, hence avoiding confrontation, while utilizing their latent energies and endurance to proceed with the consolidation of his realm.

Within a mere decade, King Abdul-Aziz originated the *hijar* program aimed at the settlement of nomads whose transient way of life and fickle allegiance had hitherto defied singular authority. The *hijar* program has since shaped the future of urbanisation in the fledgling nation-state and gave it a character of its own. This early formation of *hijar* foreshadowed the emer-

---

1 K. S. *Twitchell. Saudi Arabia: With and Account of the Development of Its Natural Resources (Princeton, NJ: **Princeton University Press, 1958**).*

gence of modern urban centres in the Kingdom. They were heavily dependent on the central government for financial support, a characteristic of their
urban economies which has since thwarted genuine decentralisation and local administration of the cities' economies (Mubarak, 1992).

Urbanisation studies in developing nations have overlooked essential,
particular socio-political developments occurring before, and parallel to urban development whose clarification helps explicate the context within which
the urbanisation process takes place. For example, in some political systems,
political factors such as national and regime security and stability may take
precedence over immediate economic goals as the major component of national policy. A traditionalist absolutist monarchy presiding over a traditional
society will most likely opt for a different economic development path as
opposed to a secular and decentralised democracy. I argue that generalisa-
tions of Western urbanisation must be taken with caution when applied to
developing non-Western societies undergoing rapid transition.

Still, the study of urbanisation in developing countries, especially in
the Middle East, is lacking when it comes to national, political exigencies,
which play a considerable role in the structuring of the modern urban network of settlements and sets the stage for future urban development. The
factors behind the creation and development of such growth centres constitutes a key to the understanding of the prevailing socio-political and economic factors. Careful analysis of these dimensions at the macro level will
contribute to insights crucial to policy decisions. Studies that address only
quantitative indicators of social transition such as demographics, economic
indicators and infrastructure risk overlooking the fundamental cultural aspects, resulting in a limited analysis and outcome.

11.2 Genesis: the Formation of the New Nation-state

Tilly (1975b), differentiates between traditional and modern states in
the following four aspects. First, the modern state manages to attain territorial expansion and consolidation; second, it exerts a great degree of control
over social, economic, and cultural functions within its confines, third, it encompasses mechanisms of governance and institutions that were previously
separate from other institutions within its territory; fourth, rulers of modern
states attain a greater capacity in the monopolisation of means of violence.[2]

---

2 C. Tilly, *"Reflections on the History of European State-Making,"* in the *Formation of National States in Western Europe. Princeton, N.J: Princeton University Press, 1975b): 3-83.
Reviewed in Hechter and Burstein (1980).*

To maintain its standing, the modern state forges bureaucratic and legal means to collect taxes, maintain armies, delimit the national economy, and manipulate economic growth, all created as sources for legitimate authority (Hechter and Brustein, 1980). The survival of a political unit is maximized by recognized territorial land holdings, 'cultural homogeneity,' and 'geographical advantages,' and the availability of extractable resources and a stable political environment .

The state is the consolidation of the dispersed authority of traditional society into a central power and source of control. Bendix (1980) states, "In Europe, the nation-state was a natural outgrowth of centralizing power of capitalist accumulation." And the industrial revolution flourished in the complex cultural and political developments in gestation since the middle ages. In Europe, nationalism was developed by kings to gain control over landlords and tribes and to abridge the power of the religious ecclesiastics. It has often been argued that nationalism makes for national integration, a prerequisite essential for modemisation which, according to Karl Deutsch, involves social mobilisation and change. According to C. E. Black (1967), central authority and its rationalized functions owe their existence to the rule of law, which is maintained by the bureaucracy specifically created to facilitate its implementation, as well as to the improved rapport between the state and the citizen. Yet, the emergence of industrialisation in the nineteenth century Europe ushered in rapid social, political and economic developments which spread beyond the continent in an unprecedented manner in history. Even countries that had been building their political institutions for centuries had to look for Westem inspiration and direct help to cope with unprecedented problems in the process of modemisation (Bendix, 1980). The aforementioned attributes and development phases of evolving modem states have helped set the stage for a theoretical background for the understanding of the emergence of the Saudi nation-state.

At the turn of the 20th century, the Arabian Peninsula was little more than a loosely knit patchwork of independent oasis towns, fishing hamlets and tribal grazing grounds ruled by feuding princes and tribal sheiks. Whenever there were severe water shortages and droughts, marauding tribesmen invaded poorly walled villages and looted caravans. Nomads' loyalties were seldom granted to central authority, and given the prevailing state of the economy, appropriate conditions for a modem state seemed centuries from reality. Warlords appeared in various parts of the Arabian Peninsula whose subordination to the Ottoman Caliph became nominal. The socio-political order of the Arabian Peninsula had changed little since the time of the Prophet Mohammed when the Arabian tribes extended their sway over the vast of the Islamic world as it is known today. With the disintegration of the Ottoman

Empire, the political climate in the Peninsula was ripe for strong leadership as political fragmentation created a vacuum.

In 1902 Abdul-Aziz, with a mere forty of his acolytes, invaded Riyadh killing its Rashid viceroy and bringing the pre-industrial settlement of approximately five thousand inhabitants under his family's control. From there with scattered battles he established control over the central and eastern parts of Arabia, and then mounted a full-scale invasion targeting the Hijaz (Western Arabia) region with its historical and religious centres of Makkah and Madinah, as well as the cosmopolitan port town of Jeddah, and the formidable mountainous region of Asir . After the first world war the climax came when the rival dynastic powers in Arabia in 1926 ceded Jeddah to Abdul-Aziz. In 1932, the country was officially named the Kingdom of Saudi Arabia, a hereditary monarchy which appealed to traditions and was committed to modernity.

Like other nation-states, the Saudi government chose to modernize its underdeveloped society, a process which threatened the integrity of the traditionally based society. According to Black (1967), economic and social transformation requires a range of resources and skill such as is rarely found within the limits of a single political organised society. "National unification, part of the process of nation building, normally constitutes a phase of the transfer of power from traditional to modernizing leaders."[3] Thanks to oil affluence, which was found in commercial quantities in the 1940s, the Saudi government opted for a policy of rapid development resulting in remarkable consequences causing dramatic shifts in the society. This decision to pursue modernisation hinged on the pressures to maintain internal integrity and to stave off external threats aimed at destabilizing the fragile Saudi political state in its early decades. Militarily, the state has followed a multi-objective policy, emphasizing the strengthening of the country's defense capabilities to thwart outside hostilities and, internally, to deter fragmentation of the nation's subregions, traditionally susceptible to segregation by tribal affiliations. The sedentarisation of nomads in Saudi Arabia was a manifestation of such objectives.

Intervention in the process of urban development has been a hallmark of the Saudi monarchy since before the promulgation of the new nation-state in 1932. The enlistment of national financial resources, however meagre, and man power in the consolidation of the new nation-state resulted in a rapid increase of population in cities and indirectly resulted in an increase in urban living standards of the population. In the civil domain, the Saudi government has tried to create a stable, modern economy centered around a central bureaucracy heavily dependent on oil revenues. Learning from the experience

---

3 *Black (1967), 149.*

of the previous two realms, King Abdul-Aziz introduced his creative *hijar* program to settle the nomads who comprised 75% of the population at the birth of the nation-state. The following decades witnessed a transformation from a traditional society marked by poverty and tribal warfare, into a nation with an urban population rising from 15% in 1950 to 77.2% in 1992 (Frisbie, 1996). A new political system emerged in which monarchical authority exercised the use of power and paved the way for the development of the modern state with its monopolisation of functions in the hands of the central government.

11.3   The Ikhwan and the Formation of Hijar: *The Politico-Military Phase, 1912-1930s*

King Abdul-Aziz recognized the danger of his fickle Bedouin subjects. He found a divergence between the Unitarian teachings of Islam and those of the nomads. Bedouins were largely ignorant of the pure principles of monotheistic Islam as introduced by Prophet Mohammed in the sixth century A.D. and reintroduced by Ibn Abdul-Wahhab twelve centuries later. He was also watchful of the fact that their transitory ways of life could not be counted upon for regular enlistment in a dependable military force. Using the double objective of settlement and religious indoctrination, he started to co-opt the powerful sheiks in his realm.

Bedouins were willing to give up nomadism by selling their livestock and joining in the newly introduced agricultural colonies. In exchange, the government assisted them in the new livelihood of subsistence agricultural production. Religious clerks advanced this transformation by likening the transfer to settled colonies as resembling that of Prophet Mohammed's night (*hijrah*) from Makkah (previously the land of unbelievers) to Madinah (*dar-ussalam*), or the home of peace, hence such colonies came to be known as *hijar*.

They were charged with the duty to bear witness to Allah, to uphold the true faith, and to instruct others in the ways of Prophet Mohammed. Allah alone is the head of the ummah (nation of Islam), a notion of expanding community organised around Islamic principles and unbound by any form of rigid physical or political boundaries. Philby wrote, "Villages sprang up in every suitable centre with surprising rapidity: each having a present stake in the land as well as contingent one in eternity."[4]

It was estimated that the new system of *hijar* settlements, throughout

---

4 Quoted in Gary Roeller. *The Birth of Saudi Arabia, Britain and the Rise of the House of Sa'ud.* (London: Frank Cass, 1976), 129.

the regions lying within King Abdul-Aziz control, put 76,500 Bedouin fighters as a standing army under the control of the King, a large number even by the standards of modern states of our time (Saleh, 1985).

Artawiyah rapidly grew to a town of 10,000 souls. Men who had been traditionally the scourge of any and every government, unpredictable and fickle as mayflies, settled down to a life of fierce dedication They abandoned their tents and their nomad life, sold their sheep and camels, and gave themselves up to farming, to prayer and the Koran.[5]

From 1927 onward, King Abdul-Aziz devoted his energies, to the extent that his resources would permit him, to the economic betterment of his nascent state. "It was, in short, the moment when Ibn Saud's new kingdom began to assume the outline of a contemporary nation-state in place of the amorphous structure of a traditional desert fiefdom;..."[6] King Abdul-Aziz embarked upon a structural consolidation and the establishment of law and order as a necessary prelude to the operation of foreign mining companies and other economic programs. By doing so, he moved from the phase of reliance on Bedouin support to the strengthening of an organised army as his coffers benefited from foreign aid and royalties and taxes trickled down from the state's diverse regions.

The promulgation of the state with its standing and mechanized army lessened the King's reliance on the *Ikhwan*. The kaleidoscopic incidents during the first three decades created a backlash by the recalcitrant *Ikhwan* when they realized that the King had a limited notion of *jihad*. They preferred and desired to continue their conquests, as did the Prophet before, outside the demarcated boundaries settled between the King and regional states and mediated by the British. To the *Ikhwan*, the notion of a state demarcated by political boundaries diverged from their goal of a boundless Muslim community. A series of violent confrontations settled in favour of King Abdul-Aziz brought the militaly-religious phase of the *hijar* prograrn to an end (Lebkicher *et al*, 1952).

11.4 The Tribal Centre: 1940s-1960s

The second generation of sedentarisation efforts by the Saudi government was commenced during the decade of the 1940s. The new state had safely passed through the 'red' formative decades marked by battles and treaties that resulted in the creation of the monarchic system under Abdul-Aziz's control. The previous attitude of cajoling Bedouins to win their sup-

---

5 D. Holden and R. Johns. *The House of Saud (London: Sidgwick and Jackson, 1981),69.*
6 Holden and R. Johns, *op.cti.*, 79.

port was substituted with a "negative" one which perceived nomadic life as anachronistic. From the arrny reserves in the conquest decades of the 1910s and 1920s, the King attempted to turn the latent energies of his nomad subjects into settled manpower that could be called upon to build the new nation.

With improved economic conditions brought by imposed security under the new state demand for urban functions grew. These demands which were stimulated by population growth, the settlement of nomads and other migrants along with rising living standards, allowed the central state to assume a large role in the everyday life of towns and cities. Some *hijar* were endowed with more auspicious ingredients for growth than others. *Hijar* that housed tribal sheiks, were located on tribal crossroads, or were bestowed with sufficient quantities of potable water or happened to be in the proximity of industrial or regional centres enjoyed relative prosperity and were bestowed with the status of administrative centre or regional hubs. Artawiyah epitomizes such a tribal centre.

Early estimates put the population of the Kingdom in 1944 at 5.2 million. In 1932 the population of nomads accounted for 70% of the national population. The first official census of 1974 put the percentage of nomads at 27%. The rapid improvements in economic conditions brought by the new nation-state were again expanded with the discovery of oil in 1938, and a concomitant increase in demand for oil following the end of WWII. The growing state bureaucracy and oil industrialisation offered an improved income to the traditional population and the demand for oil was sufficient enough to create the need for foreign labour and immigration. Both 'push' and 'pull' factors accounted for the historical flight from nomadism and rural areas to booming towns. The annual rate of urban population growth averaged 58.1% during the 1950s and 21.1% during the 1980-1985 period (Al Khalifah, 1996). This astronomical growth retlected high natural increases, internal and international migration and reclassification of settlement into urban centres. Natural causes played their role as well. The northern region of the Kingdom where the majority of nomads lived, suffered a drought that lasted six years. The 1958 drought destroyed 90% of the region's livestock.

Encouraged by optimistic reports by a number of American geologists and agricultural experts, the government embarked upon a major initiative to ameliorate the conditions of the various regions susceptible to drought and abject poverty. Government projects included drilling water wells, the introduction of mechanical means to obtain water and numerous agricultural programs and projects all resulting in unprecedented upward changes in living conditions in numerous villages and *hijar*. Such programs both accelerated the settlement of nomads and also triggered a larger movement throughout the chain of national urban hierarchy, nomads and ruralites alike flocked to the larger towns and cities where demand for a large labour force was generated. The percentage of urban dwellers increased from an estimated 5.9% to 15% in 1950 to 50% by the early 1970s, and urbanisation continued apace to reach 77.2% in 1992 (Al Khalifah, 1996; Frisbie, 1996). The combined developmental efforts shouldered by the ever expanding role of the national government included a number of programs and projects which focused on the improvement of agriculture and the settling of nomads.

The Wadi Sirhan project in the northern region of Saudi Arabia, south of the border with Jordan, offered *hijar* and village dwellers water pumps and technical support to encourage Bedouin settlement and lessen the negative impact of the vagaries of the meager waterfall on the Saudi population in general. The project was suspended and revived ten years later following the drought of the late 1950s. It placed more emphasis on pastoral grazing management than on agricultural development. The government distributed fodder, provided veterinarians, cleared old wells and drilled new ones and offered financial and technical aid to increase the number of livestock. The project, which covered an area of 40,000 hectares, represented a confrontation with reality, for practical imperatives clashed with previously held theoretical attitudes relating to the settlement of nomads at the expense of forgoing livestock production. Moreover, the government, partly due to limited resources and partly due to the cumbersome magnitude of transforming a large segment of the nomads who lived for millennia with no settlements, was increasingly confronted with the conclusion that such intervention in the lifestyles of a large segment of the national population must be approached with caution and at a slower pace than previously sought.

The Tabuk Basin Project (TBP) in the northern region bordering Jordan and Israel, on the other hand, placed greater emphasis on the settlement of nomads in the sparse and desolate region. Water, agricultural education and training was delivered by the government for the project which benefited 2,400 families. The area covered by the TBP amounted to 950 hectares, including 130 water pumps for irrigation. In addition, the government interspersed the region with 3,300 wells serving a total of 734 Bedouin clusters (*ezbah*). Another project espoused by the central government was in the area of Wadi Assabha. Its area was 40,000 hectares and was distributed to 1,000 recipients, each receiving four hectare agricultural lot.

The King Faisal Model Project (KFMP) targeted nomads inhabiting the southern region of the country, 240 kilometers south of Riyadh, the nation's capital. It covered an area of 40 square kilometers. The KFMP followed the (Saudi) classical approach to nomadism, that is the transformation of their transient lifestyle into a settled tribal labor force engaged in agricultural and other sedentary ventures. However, unlike the traditional approach in which the end-goal was the settling of nomads in easy-to-control and develop *hijar*, KFMP represented the first full scale project extending into the larger goal of national development. The KFMP was envisaged to meet the rising demand for foodstuffs nationwide brought by improved economic conditions; self-sufficiency in agriculture and other 'strategic' staples was the motto of the era. The KFMP was essentially a modern government-run farm with training facilities, experimental nurseries, and a settlement housing 1,000 Bedouin families.

Wadi Jebreen which is located 100 kilometers at the south of the KFMP, represents another endeavor to alleviate the austere living conditions of Saudi nomads. It was basically an agricultural-related settlement which included training and production facilities aimed at inducing the settlement of nomads at the furthemmost southem edge of the inhabited expanse abutting the formidable desert of the Empty Quarter. In essence, in addition to their long-term objective of settling the nomads and as drought emergency relief centers, these agricultural 'outposts,' also served ulterior political objectives, namely that of reinforcing the sovereignty of the central state especially in the disputed border regions and to instill national association and patriotism among its nomad population whose previous mode of production and allegiance had little do with any central authority. These projects typified the government's approach prior to the prosperous 1970s, and were aimed at engaging the nomadic population in national economic development programs, and through encouraging bedouin settlement, they would help establish the sovereignty of the central state in desolate and disputed border regions during the early decades of its formation.

11.5 The Bedouin Hijar as Urban Centers, 1970 to Present

The 1970s brought a tremendous increase in oil prices worldwide comprising a windfall for the Kingdom's treasury. As oil revenues rose sharply in the 1970s, government spending increased substantially. For example, annual oil income increased from $1,214 million in 1971 to $22.6 billion in 1974, to $101.8 billion in 1981. The programs which had been introduced in the 1950s and 1960s were substantially expanded to astronomical sums of capital which ultimately triggered more migration to the cities which, in the span of a few generations, has resulted in the concentration of the majority of the Kingdom's population in urban centers. Concomitantly, improvements in health provisions have decreased infant mortality and have prolonged life expectancy, causing a sharp upward kink in world populatlon.

The social unit of the agrarian society, the tribe, drastically mutated into the mobile nuclear family as the new social unit, was prone to migrate in pursuit of economic trends in the industrial age. Improved communication systems facilitated and increased access to information, increasingly suffusing the society, while technological breakthroughs in transportation shortened travel time, hence allowing fast and easy mobility between and within urban centers. All in all, these and other complex developments encouraged and reinforced the process of social mobilisation, which is spatially translated in the physical migration of the population away from its rural habitat. As illustrated in the following examples, these trends were also observed in the urban growth of the north-eastern region of Saudi Arabia, which was primarily inhabited by nomads.

To the south-east of the Wadi Sarhan project, Bedouins benefited from the nearby transnational pipe line (Tapline) carrying oil from the oil producing fields in eastern Saudi Arabia to Lebanon. The unprecedented venture of building the Tapline project required an enormous labor force amounting to 16,000 workers at its peak. The early structures of Arar housed Bedouins seeking jobs and benefits in the industrial carnp. In addition to the Arar industrial camp, three other major pump stations at Al Qaisomah, Rafha, 284 kilometers to the south, and Turaif, 238 kilometers to the north were added. The new industrial compounds serving the pump stations formed the "Tapline urban corridor" upon which future growth ensued. At the location of each pump station, an industrial compound was built to house a professional staff and support facilities to maintain these pump stations.

In line with the agreement between the Saudi government and the Arab American oil company (Aramco), Aramco built schools, health clinics, warehouses, airfields and some public-oriented facilities. Moreover, at the request of the government, the Tapline constructed a tarmac road paralleling the pipeline, connecting the various growth poles dotting the Tapline corridor, a project which was completed by the mid-1960s. Although the road was essentially meant to serve the trans-Saharan pipeline and its industrial compounds, it generated a substantial volume of traffic altering the regional urban network and creating considerable commercial activity of regional, national, and international significance. The combined effects of creating health services, providing water, as well as educational and transport facilities have since bestowed the Tapline corridor's towns with large numbers of migrants and brought modern civilisation to the desolate and arid region.

During the 1960s, Arar's population, in particular, grew at an annual growth rate of 4.6 percent, a higher growth rate per annum than its Tapline sisters (Turaif, 3.5%, and Rafha, 3.7%) a rate which was more than double the national population growth rate of 2.5 percent. Between 1962 and 1973,

Turaif's population grew from 7,000 to 9,500 and Rafha from 4,000 to 5,500. In comparison, Arar's population increased from 9,000 to 14,000. Between 1973 and 1986 Arar witnessed an increase in its population at the average growth rate of 24 percent, a growth which, by 1987, gave Arar a population of 65,000. During the same period Turaif's population reached 20,000 while Rafha attained 17,000, increasing at an annual rate of 7.3 percent and 13.9 percent respectively. Such growth rates reflected the acute urbanisation process associated with economic and administrative sector development brought by oil industrialisation and the growing influence and affluence of the central government (Mubarak, 1992).

Especially during its early years of formation, the transport of oil was the *raison d' être* for Arar's existence. Yet, oil served only as the starting point for Arar's subsequent development. Oil-related activity set in motion an urbanisation process that continued even with the suspension of most of the oil transported through the Tapline following the eruption of the civil war in Lebanon in the mid 1970s. Although the genesis and early growth of the Tapline's towns reposed on the economic principles of oil marketing, later urban development must be credited to the national government's interest in enhancing urban centers in the desolate northern region for several factors, one of which was the central government's concern for national security and sovereignty. Arar was also to benefit substantially from increasing expenditures of the military where an increasing number of troops were stationed.

The improved oil revenues of the 1970s fuelled the migration to urban centers at an unprecedented pace. Bedouins and town ruralites flocked to regional centers and to the emerging oil towns seeking better job opportunities and public services. For example, during the early 1970s, the Northern region's urban centers offered an average income of $334 per annum in comparison to $163 for those living outside cities (Doxiadis, 1974).

11.6 Assessment and Evaluation

In the light of the volatile political environment which prevailed during the first half of the century, the *hijar* program was a political success. Yet the *hijar* program must be evaluated in the overall socio-cultural and economic implications that resulted from such an intervention in the spatial system of the Kingdom. Given the scale and the shallow institutional development of the Saudi state during the first half of the century, shortcomings were expected in the transformation of Bedouins from nomads to urban dwellers. The major goals upon which the development oriented phase of *hijar* program predicated were as follows: (1) the redistribution of population to increase agricultural land and even development, (2) the settling of nomads in communities with basic public services whereby they could then contribute to the national economy, (3) the provision of public services and infrastructure in *hijar* that promised future growth as well as to stop the development of clusters that did not possess the requisite ingredients for future development. To attain such goals, a number of means were devised including the coordination between the various government agencies and sectors (national, regional and local) handling settlement functions; developing regulations and guidelines that governed positive development of *hijar* in line with the national economic goals of the Kingdom, and limiting the development of scattered fortuitous *hijar* and *ezab* which lacked growth potential; and the provision of public services and projects in *hijar* that enjoyed favorable conditions for future development (Al Musallam, 1986).

According to Al Musallam, the average *hijrah* contained 40 residential units housing 251 inhabitants. The average number of inhabitants varied between the different regions of the Kingdom. A 1982 study by the Ministry of Intenor covenng 4,020 *hijar* concluded that 68% were not suitable for (efficient) future growth. In addition, a survey by the Ministry of Municipal and Rural Affairs indicated that there were 4,636 residential clusters, *ezab* that did not mature to established communities. These clusters constituted a small number of shacks and ahwash - walled, irregular-shaped land lots containing a room or more, a yard and a shaded area for sheep, goat or camels. In many instances, small *hijar* and *ezab* were either deserted or kept for temporary use during the grazing seasons. Bedouins developed the custom of moving between the established towns and cities where they worked or were hosted by their relatives during the summer, and moving back to their *hijar* during the winter. In other instances, during the early decades prior to the prosperous 1970s, Bedouins squatted on public lands forming desultory communities called *hilal*, shantytowns on the outskirts of towns and cities. Some built their walled structures *ahwash*, which sprang up around highways branching from towns, hoping to later file a claim for the land and then sell it to another party or wait for land values to increase as development occurred near the highway.

Prior to the prosperous decades of the 1970s and 1980s, state intervention in urban growth and planning was limited. Subsequently, the layout of *hillal* (Bedouin shantytowns in urban centers) and *hijar* paid no attention to any geometric or other perceived logical organisation of its components.[7] Temporary and poorly built structures were typically made of mud or cement

---

7 *F. Mubarak. Urbanisation, Urban Policy and City Form: Urban Development in Saudi Arabia Ph.D. Dissertation, University of Washington, Seattle (1992), 71.*

blocks, cardboard, metal sheets, discarded water barrels, and cloth. Houses comprising a room or more were strewn about in space, without any logical relationship, ostensibly evoking the way Bedouins pitch their tents in the desert. Occasionally, substantial though tawdry houses were built by a prominent sheik, a retired or absentee government official or a livestock merchant. However, socio-cultural factors did play a role in the location of homes. Houses and tents were grouped in a fashion that attested to family ties; additionally, an appropriate distance was allowed to maximize visual and acoustic privacy between dwellings. Goats, sheep, donkeys and stray dogs and camels were tethered nearby. In these *hijar*, one could find a mosque abuttmg an incipient *suk* (market) lying on the road or at the nexus of desert treks. It was at a later stage (circa 1970s), and only in *hijar* that grew into sizable towns, that a municipality (*Baladiyyah*) was established which imposed administrative measures in the growth of the settlement and made some alterations in its organic structure.

In more substantial *hijar* such as these, people enjoyed a stable population and were bestowed with propitious factors for growth (such as relative amounts of potable water, laying on tribal desert tracks, a newly established regional road network, or located in the proximity of growing regional towns or military bases), the government introduced schools, a security apparatus and other bureaucratic facilities soon to be the very *raison de' etre* of future development. In these settlements, the urban economy became increasingly driven by exogenous influence, that is of central state support, rather than self-support.

In addition to geographic and demographic considerations, a number of problems thwarted healthy development of *hijar* . Tribal association and struggles and kin vendetta played a role in the location and development of some *hijar*. Lured by status and incentives provided by the state, young and nomadic Bedouin sheiks were encouraged to relocate to new sites without due appreciation for future growth prospects. Such communities proved to be a waste for government resources and if ever inhabited never grew further than small ezab. In other instances, tribal sheiks were allotted substantial areas to be distributed among their members only to lay waste due to inner-tnbal contentions. *Hijar* residents inflated their communities needs to government agencies while allocation of resources were affected by the relative power of some tribal sheiks, all confusing rational and equitable distribution of vital resources among competing Bedouin communities. Finally, administrative obstacles such as 'redtape,' the dispersed authority among the various government departments handling different aspects of *hijar* and lack of coordination between these departments all hampered positive outcomes. Consequently, the government halted the haphazard creation of new *hijar* unless warranted by sufficient supporting studies.[8]

More and more, Bedouins became urban dwellers as they engaged in the economic development programs of the 1970s and 1980s, which offered lucrative agricultural and real estate investment loans, which were governed by lax requirements, and small industrial establishments which provided windfall profits to recipients. In the burgeoning towns and cities, the emergence of loyalties to neighborhood, family, and occupation gradually overrode the traditional ties of kinship. The burgeoning oil industries in the Eastern oil producing region (Mubarak ,1995), as well as the civil and military establishments absorbed an ever increasing number of Bedouins and ruralites, thus migration to urban centers continued apace.

11.7 Conclusion

The kaleidoscopic events that engulfed the Arabian Peninsula at the turn of the century resulted in the creation of the new nation of Saudi Arabia, a monarchic state governed by tradition. A centripetal process ensued in which the fragmented power of tribal association weakened and diminishing territorial control of tribal turfs was paralleled by the rising political aplomb of King Abdul-Aziz, eventually to be hardened into an absolutist monarchy presiding over a bloated bureaucracy.

From a general developmental perspective, the *hijar* program constitutes a unique expenment in urbanisation. For such a relatively massive and visionary intervention involved not only the transition of the nomadic segment of the population to that of settled tribal villages, but also a profound social and economic transformation initiated primarily for political and later economic objectives. Such intervention in the national urban process, which dates back to 1912, was seen by King Abdul-Aziz as a necessary measure to defuse centrifugal tendencies that prevailed for millenriia as well as to convey legitimacy on the rising Saudi monarchy. Unlike most other urbanisation processes in which a predominately rural population is driven out of their agrarian settlements due to the simple socio-economic 'push' or 'pull' factors, the Saudi sedentarisation model exhibited a profound metamorphosis of the bulk of the population from that of a traditional nomadic life to that of settled peasantry and urban dwellers. Such a metamorphosis was made possible by sweeping cultural developments forged by King Abdul-Aziz's vision-

---

8 *For example, Royal Order no. 2219/8 dated 29/7/97 (1979) demanded strict observation of regulations that govern the creation and provision of services in hijar. Royal Orders no. 10607 dated 8/5/1401 (1981); 11398 dated 140501402 (1982); and RO 3/2/10251 dated 2/5/1403 (1983) all imposed detailed requirements concerning the creation of hijar.*

ary political determination amidst propitious International conditions.

From a socio-economic perspective, the urbanisation of nomads in Saudi Arabia, the new spatial system of *hijar* clusters, villages and tribal centers, however hard and unfamiliar offered a new chance to construct new lives with new sets of relationship, and where the residents grasped at the bright opportunities within their reach.

Bibliography

Al-Hathloul, S. and Narayanan Edadan (eds): The Role of State in Shaping Urban Forrns, in *Urban Development in Saudi Arabia: Challenges and Opportunities*, (Riyadh, Dar Al Sahan, 1996), 247-286.

Al Khalifah, Abdullah, H. "Urban Social Structure," in *Urban Development in Saudi Arabia: Challenges and Opportunities*, edited by S. Al-Hathloul and Narayanan Edadan (Riyadh, Dar Al Sahan, 1996), 89-112.

Al Musallam, H. S. "Al Istitan wal-Qawa'ad allati Tahkumu Nomou wa Takween al *Hijar*," (Sedentarisation and the Roles that Governs Hijar growth and Forrnation.) in proceedings for the Second Conference for Municipalities and Rural Communities, (Riyadh, National Guards Press, 1986), 457-485.

Al Rasheed, Madawi, "Durable and Non-Durable Dynasties: The Rashidis and Saudis in Central Arabia," *British Journal of Middle Eastern Studies*, Vol. 19, No. 2 (1992): 144-158.

Black, C. E. *The Dynamics of Modernization: A Study in Comparative History* (New York: Harper & Row, Publishers, 1967).

Chu-Sheng Lin, G. "Changing Theoretical Perspectives on Urbanisation in Asian Developing Countries," *Town Planning Review*, Vol. 16, No. 1(1994), 1-23.

Doumato, E. A. "Gender, Monarchy, and National Identity in Saudi Arabia," *British Journal of Middle Eastern Studies* Vol. 19, No. 1(1992),31-47.

Doxiadis Associates, *Northern Region: Urban and Regional Plan and Program for Improving Existing Conditions, Final (Revised) Report No. 1, Volume I* (Athens, Greece: Doxiadis Associates, 1974),146.

Frisbie, P. "Saudi Arabian Urban Development in Comparative Perspective," *in Urban Development in Saudi Arabia: Challenges and Opportunities*, edited by S. Al Hathloul and Narayanan Edadan (Riyadh, Dar Al Sahan, 1996), 17-48.

Hetcher, M. and W. Brustein, "Regional Modes of Production in Patterns of State Formation in Western Europe," *American Journal of Sociology*, Vol . 85 (19~0), 1061- 1094.

Holden, D. and R. Johns, *The House of Saud* (London, Sidgwick and Jackson, 1981).

Huntington, S. *Political Order in Changing Societies* (New Haven: Yale University Press, 1968).

Huntington, Samuel P., "Political Development and Political Decay," *World Politics* (April 1965), 386-430.

Lebkicher, R. G. Rentz, and M. Steineke. *The Arabia of Ibn Saud* (New York: Russel F. Moore Company Inc., 1952).

Mark Heller and Nadav Safran. *The New Middle Class and Regime Stability in Saudi Arabia*. Harvard Middle East Papers, Modern Series: Number 3 (Cambridge, Mass.: Center For Middle Lastern Studies, Harvard University, 1984).

Mubarak, Faisal. *Urbanization, Urban Policy and City Form: Urban Development in Saudi Arabia*, Ph.D. Dissertation, University of Washington, 1992.

_____ "Oil, Urban Development and Planning in the Eastern Province of Saudi Arabia The Case of Aramco, 1930- 1970s," Paper presented at the Sixth National Conference, American History Association, Knoxville, TN, Oct. 1995.

Organisation of the Petroleum Exporting Countries (OPEC). *Selected Documents of the International Petroleum Industry, Saudi Arabia- Pre-1966* (Buffalo, New York William S. Hein & Co., Inc., 1983), 55.

Roberts, B. *Cities of Peasants: The Political Economy of Urbanisation in the Third World.* Sage Publications, Beverly Hills, Ca (1984).

Roeller, G. *The Birth of Saudi Arabia, Britain and the Rise of the House of Sa'ud.* (London: Frank Cass, 1976).

Safa, H. I. (ed.). *Towards A Polihcal Economy of Urbanisation in Third World Countries* (India, Delhi: Oxford University Press, 1982).

Saleh, H. A. "Tawteen al-Badou fee Aa'ahd al-Malak Abdalaziz" (Settling of Bedouins in the Reign of King Abdul-Aziz), In the proceddings of tthe Second Conference for Municipalities and Rural Communities, First Edition. (Riyadh: National Guards Press, 1985), 1-44 (Arabic).

PART III
THE URBAN LAND MARKET

12 Urban Renewal in Shanghai's Context

Cheng Yun

Editors' note:
Like other cities in the developing world, Shanghai has faced the dlimemma of either bulldoz-ing exisiting housing with the provision of new units in the same place or elsewhere, or at-tempting to upgrade the stock with new water, sewage and power connections. What happens depends greatly on the system of land tenure. In China the dismantling of central state control has effectively given land rights to many public enterprises, who lease the land for high profits in the new market system. The result puts pressure on communities to move elsewhere.

12.1 Introduction

One of the most difficult problems faced by "transition economics" is how to deal with massive housing and urban redevelopment problems. Since China launched its economic reform in 1978 to change to a market-oriented and more productive regime, Chinese cities have been undergoing rapid de-velopment to facilitate economic growth. Urban renewal consequently is be-ing regarded as a prominent issue and has become the essential part of urban planning and development. Associated with the economic transition, the most significant change that has taken place in Chinese cities, especially the large cities, is the break down of the state-owned land system, the development of a booming land and property market.

As the largest city in China, Shanghai is targetted to become one of China's international economic and financial centres in the next century, and urban renewal is becoming a top priority in Shanghai's urban policy. Since the 1950s, a number of different scale urban redevelopment schemes have been tried by the government to redevelop and improve the living environ-ment in old dilapidated urban areas in Shanghai, but with limited impact and slow progress. Before the adoption of new market approaches, the govern-ment played the dominant role in the urban redevelopment schemes. Now, in order to raise the initiatives of the developers (government and private) and investment (domestic and foreign) in the process of urban development and redevelopment, land leasing plays an important role in the urban renewal process of Shanghai.

This chapter examines the urban renewal process in Shanghai since the 1950s and the characteristics of different stages; then it analyzes present problems and conflicts of urban renewal in the land leasing programmes. Next it examines the two-level administrative system and its impact on both land leasing and urban renewal; and then finally it will discuss the alternative strategies for the future urban renewal of Shanghai.

12.2 Brief Review of the Urban Renewal Process in Shanghai

Shanghai was founded as a fishing village but has undergone rapid urban development in the early decades of this century. Development before 1949 was piecemeal in nature and haphazard in form. The old city proper of Shanghai is primarily the city proper formed before 1949. It covers 82.4 sq.km., with a population of 4.19 million and various types of residential buildings of 23.59 million sq.m. developed in a semi-feudal and semi-colonial society. The whole old city proper lacks a unified plan. Its layout is in a state of disorder, and the qualities of housing differ greatly from one area to another. Thus a lot of problems wait for solutions, and the redevelopment of the old city proper has been, since 1949, one of the key goals for the municipal government to achieve in the course of the city construction. For more than 40 years, urban renewal scale has escalated, and the policies have become more open. There are several stages of urban renewal in Shanghai:

Urban Renewal in 1950s

The 50s was the first stage of urban renewal in Shanghai. Because of the lack of the funds, the main objective of this period was the rapid improvement of poor living condition on the surroundings such as opening up fire fighting paths, laying underground water supply and sewerage networks, and filling dirty pools in the slum areas.

Urban Renewal in 1960s

The second stage is the 60s, which can be described as an initial stage of development. In this stage the national economy enjoyed growwth, and the emphasis of renewal was shifted from the outside to the inside of houses. The worst slums were demolished, and new projects of redevelopment were carried out. The refurbishment of Fan Gualong shantytown was well known at that time. After refurbishment it became the first large-scale workers' new residential district in Shanghai. The refurbishment was financed by the people's government, and the construction of worker's new residential districts in Shanghai started from this period.

Urban Renewal in 60s-70s

The third stage is from the late 60s to the late 70s, which is a stagnant period. The Cultural Revolution took place in this stage and the reconstruction of the old city proper stagnated. With the government's policies out of control, parts of the city, especially the western part of the city where residential houses of high quality were located and which previously had a large area of trees and grass, were destroyed.

Urban Renewal in 70s-90s

The fourth stage started from the 70s to early 90s. After a long period of unplanned and uncoordinated growth with little attention being paid towards its physical planning and environment, in 1978 the municipal government rehabilitated the Urban Planning and Management Bureau and Urban Planning and Designing Institute that are charged with the duties of urban planning and development for Shanghai. Since then, there have been three comprehensive plans for Shanghai, which were done related to urban renewal in Shanghai.

Master Plan of 1984

The development target in this plan focused on the central city development, which aims to renew and reinforce the urban centre with parallel development of southern area to the northern part of the city. Although Pudong was mentioned in the plan, it is by no means the sole focus of the Plan.

Since there was no substantial change in the planned economy system or the proportion of total revenue allocated to the central government, and the complicated web of decision making circuits within the bureaucracy also remained unchanged, urban redevelopment and urban renewal had stagnated. Urban renewal was restricted to piecemeal improvement of slums in the old urban area.

Revised Master Plan of 1991

In 1986, a new development strategy was ushered in to develop Shanghai's economy which consists of three sections: the guiding principal for Shanghai's economic development, the policies and tasks for Shanghai's Economic

Development Strategies and measures taken for the reform of the urban economic system.

In 1988, the municipal government made the development of Shanghai's tertiary sector a priority, and adopted a series of policies on land reform. In 1988, the first parcel of land was leased for the redevelopment project in the city core.

In 1990, central government announced the opening of Pudong, helping to attract large numbers of foreign investors and stimulating new urban economic activities. A series of land policies to promote urban land development were proposed and formalized.

As a result, the master plan of 1984 was revised to facilitate the rapid growth. Urban renewal at this phase paid more attention to the adjustment of industrial structure and the promotion of large-scale redevelopment projects.

Master Plan of 1995

A more favourable environment was promoted with the 1992 ' Tour of the South' by Deng Xiaoping. The land-use and land leasing system were further refined. In 1992 and 1993, more than 450 land plots had been leased. Private developers and foreign investment promoted new developments, such as the ring road construction, the subway system and the ordinary residential estate development.

In 1994, the Municipal Government of Shanghai sponsored an International Conference on the Development Strategy of Shanghai for the 21st Century. Shanghai, with the development plan of Pudong, will become the focal point of development both in China and the region as a whole due to its strategic location. Hence, the new master plan of 1995 shows not only on the redevelopment in the city proper, but also of the whole urban system. Since more attention has been paid to the roles of international economics, finance and trade, the function of Shanghai will be transformed from domestic economic centre to international metropolis. The CBD development and transportation construction is highlighted in this plan.

Compared with the past, urban renewal in this stage had the following features:

Firstly, the construction work was more concentrated on certain areas than before, and the scale became larger. According to the special plan formulated in 1984, the construction concentrated on 23 sites, which covered 415.7 hectares, involving the demolition of 124,700 housing units and the relocation of 1,387 units of enterprises and institutions.

The second feature of this stage was that the urban redevelopment projects were more comprehensive than before. The target for redevelopment was no longer the adaptation of only the surroundings. It became a combination of the renewal of the surroundings, the improvement of the city infrastructure, and the adjustment of industry.

The third feature was the integration of long term city planning goals with the construction of new development areas. In those areas densely populated or with factories and residential buildings mixed up, some measures were adopted to lessen the density of population and move out some pollutant factories, so as to alleviate the crowding of the old city proper. In this stage, the redevelopment of old areas and construction of new areas were gradually brought into balance, and 65% of the residential construction was on the nearby suburban area, 15% on the far suburban area and 20% in the city proper.

In spite of the great achievements in this stage, there was still a series of major problems waiting for solutions. Firstly, the redevelopment of the old city proper was short of funds. Although the sources of money might be multifarious, the amount was not large enough and it was hard to collect. In this stage, with the funds for the redevelopment of the city infrastructure supplied by the municipal government, the redevelopment funds for ordinary housing were collected through the following channels such as: 1) Cooperative construction through collecting money; 2) Cooperative construction between enterprises and individuals; 3) Redevelopment mainly financed by residents and aided financially by their enterprises and institutions; 4) Housing cooperative; and 5) Sales of commercial housing.

12.3 Land Leasing and its Impact on Urban Renewal

Land Leasing Since 1992

In order to accelerate the pace of urban renewal, land-leasing was adopted as a method to stimulate the land and property market of Shanghai. Under central planning the value of land had been ignored, and it had been freely used. Now, by transferring land use rights to foreign investment for periods by defined by contract, land became a source of revenue for urban redevelopment. Through land leasing programmes, municipal government could gain the revenue for the improvement of urban infrastructure, urban redevelopment and housing construction.

In June 1991, the Shanghai Urban Construction Committee investigated the slums of of ten central urban districts. This survey showed that 3.65 million sq.m. of slum housing needed to be redeveloped immediately. If it was combined with secondary terrace housing of 11.3 million sq.m. that needed to be replaced, the total amount was 15 million sq.m.. This is a major problem facing Shanghai municipal government.

In 1992, out of about hundred pieces of leased land, over 60% were in the former slum areas, which demolished slum housing of 71.54 million sq.m.. Huangpu, Nanshi, Xuhui, Yangpu, Jingan and Hongkou districts have leased series pieces of land with potential high value, most if it previously occupied by slum housing. There were almost several million sq.m. of slums housing to be demolished on each piece of leased land.

Case of "Two Steps Jump"

This type is called " two steps jump"(liangji tiao).

The first experiment was Xiesan site in Luwan district. Luwan district is one of the areas of highest population density in Shanghai. The Xiesan site occupied 1.98 million sq.m. of the district. The land use of this site was quite mixed - 1060 households and 16 factories as well. It was one of those 23 old sites, which were to be redeveloped according to the city plan, but which were held up by lack of funds. In 1992, it was leased on US$ 23 million to Haihua Property Ltd. Company (Hong Kong) to build four 31-story housing towers, to be sold on the foreign property market. Meanwhile the local government of Luwan district used the income from the land leasing for residents resettlement and factory relocation to suburban areas, improving the living and working conditions for both the residents and industries.

Through the endeavour of these years, 1.79 million sq.m. of slum housing, nearly half of the total slum area, was demolished by the end of 1995, and 1.10 million sq.m of this (62%) was tied to land leasing by foreign investment for redevelopment projects.

12.4 Problems with Land Markets and Land Leasing

Dual Land Market

The clause "the right to use of land may be transferred in accordance with law" was added to the Constitution of China in 1988. This led to a market for the paid transfer of land use rights, although an administrative allocative meachanism also survived alongside. So land allocation involves both market and non-market mechanisms. In the market-based mechanism, land is allocated to users through negotiations, tenders and auctions. Land allocated through the market mechanism is mainly for residential, commercial, and

Table 12.1   Land leasing situation in Shanghai (1992-1996)

	1992	1993	1994	1995	1996
Leasing Land (parcels)	201	251	445	258	207
Repaying land price #			35	23	34
Urban district	135	92	160	49	69
(Redeveloped area)	95	52	100	21	21
Pudong		65	24	12	1
Suburb district	66	31	40	32	33
County		63	221	165	104
Leasing Land (m²)		49677304	13046900	6403002	3786551
<u>By land use type</u>					
Residential		3136539.1	2219339	457232	717390
		6.30%	17.0%	7.14%	18.95%
Comprehensive		653280.4	2570699	303897	326329
		1.31%	19.70%	4.75%	8.62%
Commercial		13482.0	4561	334	0
		0.03%	0.03%	0.01%	-
Industrial		1330129.0	5662243	3998412	2722485
		2.68%	43.40%	62.44%	71.90%
Tourism			666778	5087	20347
			5.11%	0.08%	0.54%
Large scale development		44543874.0	1900772	1638040	0
		89.67%	14.56%	25.58%	0
Others		0	0	0	0
		0%	0	0	0
<u>By region</u>					
Urban district		740761.4	1106425	477476	700788
		1.49%	8%	7.46%	18.51%
(Redeveloped area)		413545.4	661476	202547	159373
		55.83%	59.78%	42.4%	22.74%
Pudong		46099857.0	3990250	1903046	78192
		92.80%	31%	29.72%	2.06%
Suburb district		1193900.0	1781262	533699	831100
		2.40%	14%	8.34%	21.95%
County		1642786.1	6168963	3488781	2176471
		3.31%	47%	54.48%	57.48%

Note: Repaying land price, referring to the six types of former administratively allocated land should pay on market price for transferring land use rights.
Source: Summarized by Yun, CHENG from the statistics of "Shanghai Land Market".

industrial development. Land for government and military institutions and other non-profit making institutions is allocated through the non-market administrative allocation mechanism such as public health, sports facilities, science, education and public utilities.

Underground Deal through Administrative Allocated Land

Land leasing belongs to the market-based land allocation. According to the related laws and regulations, after land is leased, it is must be developed and utilized in conformity with the leasing documents of contract.

However, in the case of transfer of administratively allocated land, the matter is more complicated, since the deals are made in various ways, such as in joint construction, joint management, hiring, exchange of land by building after construction, using land as a share, merger of enterprises, etc. What is more, most of the ways of transfers are unusual and sometimes even covert, which causes the government to lose the control over land-use and tax from land, giving rise to difficulties in the work of planning administration. This is a negative side of the land leasing process.

12.5 Urban Administrative System and its Impact on Land Leasing and Urban Renewal Process

Decentralize the Power of City Planning and Management

The two-level administration is a major reform in the urban administration and management of Shanghai, and is different from other cities. Specifically, on 2 June 1986 the Administrative Office of the Shanghai People's Government enacted the Municipal Construction commission's "Opinions Concerning Separating Powers and Defining Responsibilities in Urban Construction at the District level" (Document: Hufuban No. 69-1988), which stipulated that the city level and district level should each perform its own functions and holds its own responsibilities, so as to bring into play the enthusiasm of all districts, each in its own urban construction. Detailed plans are to be drawn up by the district authority taking charge of the organisational work and reporting to the Municipal Planning Bureau for examination and approval; while detailed plans for the Shanghai city district's major areas and sections of roads are to be drawn up by the Municipal Planning Bureau together with various district authorities.

Decentralize the Power of Land Leasing

After the decentralisation of planning control, the power for land leasing was also decentralized to the local level, which had a major impact on the process of the urban development, especially on the pace of urban renewal. On the positive side, this has raised the initiatives of the local governments and promoted the pace and the scale of redevelopment across the whole city. On the other hand, since the decentralisation of the power for land leasing, it has raised the competition for capital investment in the redevelopment projects among all districts. In order to attract more attention, especially the investment on redevelopment, in the detailed plans done by the local governments, almost every districts wants to be the sub-CBD of the city, so the FAR (Floor-Space / Area Ratio) in those plans is very high. Decentralisation of power for planning control and land leasing has weakened the macro-control of the whole city development by the municipal government, so that there is an over-supply of the land for redevelopment projects and there is an implicit waste of resources of land and funds.

12.6 Conclusion-Conflicts and Alternatives

The government should make a priority a full examination of the land supply and the price of leased land in Shanghai.

Firstly, it should have a plan for land use at both the whole city and district levels to control the total supply of land, and the distribution of land uses, to regulate the land leasing programmes by urban planning.

Secondly, since the dual land market has led to inefficiencies of land leasing system in promoting urban renewal, it should change the dual land market system to a single land market system, that aims to eliminate the proportion of land now allocated administratively.

Thirdly, it should give more priorities to both foreign investment and domestic private developers on leased land for urban redevelopment such as providing urban infrastructure and civic facilities, and reducing various taxes not relevant to urban land development. Since 30 August 1994 the publication of "the management methods of ordinary domestic commodity housing built by foreign investment", a series of policies upon promoting the development of ordinary housing by foreign investment have been published. It will have a significant influence to the housing development and urban redevelopment in future.

References

Chen, Yewei (1991) New strategy for old residential redevelopment in Shanghai, *City Planning Review*, Vol.15, No.6, pp.26-30 (in Chinese).

Dowall, D.E. (1993) Establishing urban land markets in the People's Republic of China, *Journal of the American Planning Association*, 59(2), pp.182-192.

Dowall, D.E. (1994) Urban residential redevelopment in the People's Republic of China, *Urban Studies*, Vol.31, No.9, pp.1497-1516.

Fung, K.I., Yan, Z.M. and Ning, Y.M. (1992) Shanghai: China's World City, in *China's Coastal Cities*, Yeung, Y.M. and Hu, X.W. (eds), p.124-152, University of Hawaii Press, Honolulu.

Healey, Patsy and Nabarro, Rupert (1990) *Land and Property Development in a Changing Context*, Hants, England: Gower.

Kivell, Philip, (1993) *Land and the City, Patterns and Processes of Urban Changes*, Routledge.

Liu, W. and Yang, D. (1990) China's land use policy under change, *Land Use Policy*, Vol.7 pp.198- 201.

Tang, Y. (1989) Urban land uses in China: policy issues and options, *Land Use Policy*, Vol.6 pp.53-63.

Whipple, R.T.M. (1971) *Urban Renewal and the Private Investor*, West Publishing Corporation Pty Ltd.

Wang, Tongdan (1995) Raise Funds by the land to deal with urban reform, *Urban Research*, No.1, pp.30 (in Chinese).

Zhu, Jieming (1994) Changing land policy and its impact on local growth: the experience of the Shenzhen Special Economic Zone, China, in the 1980s, *Urban Studies*, Vol.31, No.10, pp.1611-1623.

13 The Politics of Urban Development in New Bombay

Alain Jacquemin

Editors' note
After Ebenezer Howard published Garden Cities of Tomorrow *in 1898 at the end of Britain's first century of industrialisation, movements started in Europe and the USA to provide healthier, greener New Towns for people to live in. Abercrombie's Greater London Plan developed during World War II was implemented on these principles when the war was over. In India planners influenced by the idea developed Faridabad as a New Town outside Delhi – but the urban sprwal of Delhi, unchecked by a green belt, has overtaken it. New Bombay is a more recent example – which is separated from Bombay itself by water. But the method of financing the new town has effectively skewed its class composition to the middle and upper earners.*

13.1 Introduction

In the 1950s and 1960s, Bombay, a typical colonial (island) city on the west coast of the Indian subcontinent, was experiencing a tremendous growth in both population and industrial production. Because of this increasing congestion and a deterioration in urban services and transportation, and a rapidly worsening overall quality of living, a debate on the city's future was started at different political levels.[1] After more than a decade of discussion, the Government of Maharashtra, itself based in Bombay, ultimately (in 1973) accepted a grandiose plan to create a new and independent twin city on the mainland, opposite the existing island city. The new city, which was to be of metropolitan size and was to be a self-containing counter-magnet to Greater Bombay, was called New Bombay (or *Navi Mumbai*).

The creation and development of this grand new city, which was to cure most of Bombay's urban ills, was to take place on a large piece of relatively sparsely populated land. An area of 344 km^2 was designated, covering two older municipal towns and 95 villages from Thana, Panvel and Uran tehsils, spread over the two districts of Thana and Raigad.[2] Apart from

---

1 *Between 1941 and 1971, the population of Greater Bombay increased from about 1.5 million to nearly 6 million (5.97). In the decades thereafter, the population further increased to 8.24 million in 1981, and to nearly 10 million (9.93) in 1991. The population of the Greater Bombay UA was 12,596,000 in 1991, and is today estimated to be over 15 million.*
2 *The state of Maharashtra is organised upon several administrative layers. It is devided into six 'divisions', each having a specific number of 'districts' or 'zillahs' (30 in total), with every district comprising a number of 'tehsils', which in turn group a number of 'villages'. Of the 95 villages that were designated, 29 belonged to the Thana Tehsil of Thana District, 38 villages to the Panvel Tehsil of Raigad District, and the remaining 28 villages to the Uran Tehsil of Raigad District. Raigad District was formerly known as Kolaba District.*

231

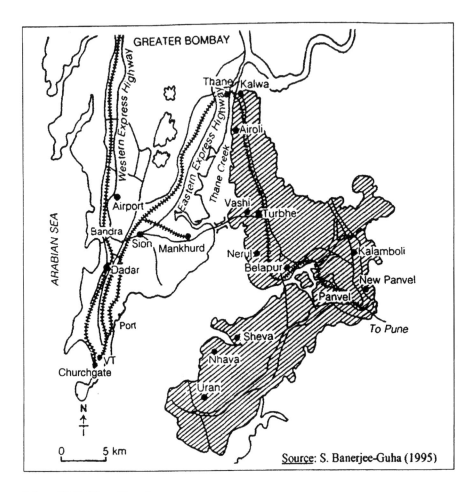

Map 13.1 New Bombay

these two towns and villages, a number of other areas also fell within the project area, but under a separate authority (MIDC, MSEB and the Defence Department), and were therefore not included in the actual development project. Of the remaining net project area of around 294 km^2, 56.3 % was private land, 34.5 % was government land, and 9.2 % was salt pan land. All privately owned land was notified for acquisition under the Land Acquisition Act 1894.

Abbreviations

BMRDA	Bombay Metropolitan Region Development Authority
BUDP	Bombay Urban Development Project
CIDCO	City and Industrial Development Corporation of Maharashtra
EWS	Economic Weaker Sections
FSI	Floor Space Index
HIG	High Income Group
HUDCO	Housing and Urban Development Corporation
JNPT	Jawaharlal Nehru Port Trust
LIG	Low Income Group
MIDC	Maharashtra Industrial Development Corporation
MIG	Middle Income Group
MSEB	Maharashtra State Electricity Board
NMMC	Navi Mumbai Municipal Corporation
NRIs	Non-Residential Indians
NTDA	New Town Development Authority
RP	Reserve Price
SES	Socio-Economic Survey
SICOM	State Industrial and Investment Corporation of Maharashtra
TBIA	Thana-Belapur Industrial Area

In March 1970, the City and Industrial Development Corporation of Maharashtra (CIDCO) was formed. Initially established as a subsidiary of the State Industrial and Investment Corporation of Maharashtra (SICOM), CIDCO is a limited company which is wholly controlled by the Government of Maharashtra. A year later CIDCO was designated the New Town Development Authority (NTDA) for the New Bombay project, and became fully responsible for planning and developing the area.[3] As the NTDA, CIDCO's functions have been planning, implementation, maintenance and administration of New Bombay. This meant that, as long as no local municipal body would be created for the New Bombay area, CIDCO would also be fully responsible for providing civic services to the residents of the new town. Only in December 1991, the state government has constituted such a municipal corporation for New Bombay, viz. the Navi Mumbai Municipal Corporation (NMMC), which was to become operative with effect from 1 January

---

3 *The setting up of New Town Development Authorities was envisaged in the amended Maharashtra Regional and Town Planning Act 1966, for implementing proposals of setting up new towns within a region; a concept which was modelled after the British New Town Development Corporation.*

1992. Due to legal and practical problems, the actual shift of administrative responsibilities from CIDCO to the first democratically elected NMMC has only been executed very recently and only partially, and the overall situation of administrative jurisdiction has been, and still is, very confusing for all parties involved.[4]

At the time of notification, about 156,000 people (24,878 households) were living in the area, of which about 117,000 (19,400 households) in the villages and about 39,000 in the two municipal towns of Panvel and Uran. The large majority of people were involved in agriculture. Most families were owning a plot of land of their own, and the number of big landlords was very limited. Other major occupations were fishing and salt making; activities which were done by many in combination with agriculture.[5] The produce of the land was usually in excess of the local demand. Apart from these primary activities, some secondary activity had already started taking place in the 2105 ha. large Thana-Belapur Industrial Area (TBIA), which in those days was the only (slowly developing) industrial area in the New Bombay region. The TBIA, which is administered by the Maharashtra Industrial Development ment Corporation (MIDC), was established after the nation-wide, post-colonial industrial location policies which prohibited a further increase in, and concentration of large-scale industries in metropolitan cities. Consequently, industries started to establish in this newly created industrial area, just beyond the borders of the Greater Bombay Municipal Corporation. Whereas the number of industrial companies in the TBIA remained very limited between 1951 (when the first company in the area started operating) and 1965, in the six years between 1965 and 1971 their number rose quickly to 44 units, in which about 15,700 workers were employed. Wages in these new industries were relatively high, as was the capital investment per person, pointing towards the presence of capital-intensive industries such as chemical and petrochemical plants, alongside a number of engineering units.[6]

Although economically necessary facilities such as water, power and roads were available (provided for by the MIDC), other essential facilities such as housing, drainage, public transport, telephone/telegraph connections, recreation, banking and medical facilities were chronically lacking. Therefore, since less than one in four (23 %) of the industrial workers in the TBIA were local village people, the majority of workers were to be brought into the

---

4 *The first elections for the NMMC took place in January 1995 and were won by the Shiv Sena-BJP combine.*
5 *For many, labour on the land was a temporary activity, limited to the four monsoon months between June and September. In the dry season, many would go fishing.*
6 *Capital investment was in 1971 Rs. 109,000 per person employed, compared to an average of about Rs. 23,000 in Greater Bombnay and Rs. 20,000 in the state of Maharashtra (CIDCO, 1973:90).*

area daily by bus from adjacent districts (mainly from Greater Bombay and Thana).

In October 1973 CIDCO published its New Bombay Draft Development Plan, as was required by the Maharashtra Town Planning Act 1966. Preparation of the plan took more than two years, as the planning officials were continuously to focus on the five broad objectives of the project put forward by the state government, viz.

(1) to reduce population growth in Greater Bombay by creating an attractive urban centre on the mainland across the harbour, which will (a) absorb immigrants who would otherwise go to Bombay, and (b) attract some of Bombay's present population, in order to keep the overall population of Greater Bombay within manageable limits;

(2) to support state-wide industrial location policies which will lead eventually to an efficient and rational distribution of industries over the state and to a balanced development of urban centres in the hinterland;

(3) to provide social and physical services, which raise living standards and reduce disparities in the amenities available to different sections of the population;

(4) to provide an environment which permits the citizens of the new town to live fuller and richer lives, free - in so far as this is possible in a large urban entity - of the physical and social tensions which are commonly associated with urban living; and

(5) to provide training and all possible facilities to the existing local population in the project area, in order to enable them to adapt to the new urban setting and to participate fully and actively in the economic and social life of the new city.

Given the objectives put forward by the state government, CIDCO defined its future principal activities as :

(1) to develop land and provide the required physical infrastructure such as roads, drainage, water supply, sewerage, street lighting, landscaping, etc.;

(2) to build as many houses and community centres, shopping centres, parks, playgrounds, bus stations, and so on, to meet the day to day needs of the population as well as for a faster take-off of new growth areas, and to make available developed plots at affordable prices to people to construct houses;

(3) to promote growth of commercial, wholesale market activities, warehousing, transport, office and other activities in order to evolve expeditiously a sound economic base for a self-sustained growth, achiev-

ing at the same time a process of relieving congestion in the old city of Bombay; and

(4) to involve agencies in the development of public transport, both road and rail, and in telecommunication.

Given that the full rehabilitation and integration of the local people into the urban fabric, or the provision of as much housing as needed, were central in the proposals of both the state government and CIDCO, the New Bombay project was thus not only to be just one of those prestigious government projects of immense scope but, more importantly, was to be a project in which all social groups, rich or poor, would find their rightful place. Housing was to be made available "for the entire range of income groups expected in the city", with the intention that "there should be no *rich* nodes and no *poor* nodes" (CIDCO, 1973:17). It was expected that "this accommodation of residents from various social and income groups... will not only make for a healthier urban environment, but will also ensure a uniform standard of maintenance of social and urban infrastructure services in all neighbourhoods so that no one class of residents is better served than another" (Ibid.:18). It was desired that "every family living in New Bombay shall have a dwelling of its own, however small" (Ibid.). Knowing the income distribution of households in Greater Bombay, as well as the average cost of a small and basic house, it was realised that at least 16 % of the prospected total population of New Bombay could not be given anything more than a plot of land of about 30 m≤. For the lowest income group, another 20 % of the population, it would be impossible to provide even this (CIDCO, 1973:56).

Therefore, well aware of what this situation had led to in Greater Bombay, with a large segment of the population living in sub-human conditions of extreme degradation, CIDCO proposed to implement a system of cross subsidies in housing, and to develop 'Sites and Services' schemes in specially outlined areas. Consequently, the first chairman of CIDCO proudly and confidently announced in his foreword of the development plan, that "The effort has been to avoid the spectacular, to provide minimally for the affluent few and to promote the convenience of the greatest numbers." And he concluded that "New Bombay, then, will not be another grand city, it will be a city where the common man would like to live".

Whereas the development plan was thus having a strong social undertone, aiming at protecting the lower income groups and providing housing, infrastructure and services to all levels of society, the state government, on the other hand, had made it clear that the entire New Bombay project was to become a self-supporting project. Apart from government loans in the initial phase, CIDCO was to finance the development of the area with other sources than government funds. Therefore, land became the main resource for development. The state government would first acquire all the private lands, after which CIDCO would start developing the land with infrastructure and services, and subsequently sell it off. With the returns of the sales, further development would be financed. The price of the land was determined on the basis of the so-called Reserve Price (RP), which is acquired by dividing the total cost of development in a given area through the total net saleable area. Financially healthy organisations, institutions, companies or individual households would pay a price above the RP, financially weaker and/or socially important organisations, institutions or households would pay a price below the RP.

Thus, the state government's decision to accept the New Bombay project on the one hand, but not to put any money in it on the other, has caused CIDCO, although a limited public sector company, to think, plan and act as a private firm. Matching the social goals of the project with the aim of financial self-sufficiency was an ambitious programme, and promised to be a difficult task.

In this chapter we will discuss to what extent New Bombay, after 25 years of project implementation, has been successful in meeting several of these initial goals. First, we will look at the twin city's overall growth and development over the last two and a half decades. Second, we will draw a socio-economic profile of the population currently living in the different newly developed townships of New Bombay (or *nodes* as they are locally called), and we will discuss to what extent New Bombay can, indeed, be called a city for all classes and all people. Finally, we will try to put forward some explanations for the socio-economic dynamics which have been at work in the area, and which have given shape to what New Bombay is today.[7]

13.2 Extent of Growth and Development of New Bombay Today

Due to the scope of this paper on the one hand, and to the limited space on the other, we will limit ourselves in this section to a number of key indicators which give a clear view on New Bombay's growth and development over time.

Whereas in 1971 the total population in the New Bombay area was about 156,000, 75 % of which was living in the villages and 25 % in the two municipal towns of Panvel and Uran, by 1995 the population was estimated to be about 700,000. Of this, about 53 % (373,001) was living in the villages and the two municipal towns, and about 47 % (326,526) in the newly devel-

7 *During my stay in New Bombay, CIDCO and its officers have been extremely helpful in providing accommodation and office facilities, as well as in helping with the practical research. I hereby wish to thank them.*

oped nodes.

Table 13.1 Population growth New Bombay 1961-1995

	Rural	Urban (a)	Total
1961	82,312	28,359	110,671
1971	116,789	39,218	156,007
1981	183,593	66,938	250,531
1991	243,901	291,308	535,209
1995 (b)	373,001	326,526	699,527

Figures from CIDCO departments.

Notes: (a) 1961 and 1971 *urban* figures are exclusively the population of the municipalities of Panvel and Uran; 1981 and 1991 figures include population of Panvel and Uran, but 1995 figure excludes this segment of the population; (b) estimate; population of Panvel and Uran municipalities (around 60,000) now included in *rural* figure; *urban* figure now called *nodal*.

The population distribution over the nodes is a reflection of the planning concept, which stipulated that the nodes were to be developed more or less one after the other. Consequently, today about 33 % (107,919) of the total population in the nodes is living in Vashi, which was the first node to be developed in the 1970s, and is located near the Thana Creek bridge which links mainland New Bombay with the island of Greater Bombay, both by road and rail. The table below gives the number of people in the other nodes which are by now fully developed or near completion. It will be noted that the total population in the nodes has increased by about 250 % within the eight-year period from 1987 to 1995.

Table 13.2 Population growth per node 1987-1995

	1987	1991	1995
Vashi	72,000	83,178	107,919
Nerul	16,000	36,768	57,597
Airoli	7,000	29,661	40,681
CBD Belapur	18,000	24,302	28,919
New Panvel	10,000	16,864	26,101
Kalamboli	6,000	15,096	29,019
Koparkhairane	-	3,489	30,168
Sanpada	-	381	6,122
Total	**129,000**	**209,739**	**326,526**

Figures from CIDCO departments.

So far as industrial growth is concerned, in 1971 the total number of workers employed in industry in New Bombay was 15,732. By 1990, nearly twenty years later, that number had only slowly increased to 55,286. In the following four years, however, the 1990 figure would almost double to 103,768, over 78 % of which was employed in the TBIA. As per 1995, the estimated total employment in New Bombay was just below 150,000, with an informal economy of around 50-60 %.[8] The labour distribution per sector and per node was as follows:

Table 13.3   Estimated total employment per sector and per node in New Bombay in 1995

	Vashi	*Nerul*	*CBD*	*Kalamb.*	*New Panvel*	*Kopar- Khairan*	*Airoli*	*Khargar*	*Total*
shops/ offices	14,889	2,412	8,391 (a)	1,047	1,362	199	1,218	-	**29,518**
wholesale	5,937	-	-	1,779	-	-	-	-	**7,716**
Mafco	3,229	-	-	-	-	-	-	-	**3,229**
truck terminal	328	-	-	165	-	-	-	-	**493**
schools/ colleges	1,221	741	295	202	380	223	497	114	**3,673**
hospitals/ nursing	822	64	73	-	22	25	21	-	**1,027**
social facilities	301	77	74	12	42	-	41	-	**547**
Thana- Bel. IA	-	-	-	-	-	-	-	-	**81,296**
Taloja	-	-	-	-	-	-	-	-	**19,315**
Jawahar. IA	-	-	-	-	-	-	-	-	**1,742**
Panvel IA	-	-	-	-	-	-	-	-	**1,415**
Total	**26,727**	**3,294**	**8,833 (a)**	**3,205**	**1,806**	**447**	**1,777**	**114**	**149,971**

Source: CIDCO departments.
Notes: IA = industrial area; (a) CIDCO's own employees included.

New Bombay's employment base today thus consists mainly of capi-

---

8 *Estimate of CIDCO's Chief Economist. This figure must be compared to the 60-70 % informal economy in Greater Bombay.*

tal-intensive industrial activities (mainly chemical and petro-chemical) in the TBIA and the Taloja industrial estate, of port-related industries in and around the newly developed high-tech port of the JNPT, of private shops and private and public offices, and of wholesale (and wholesale-related) activities in the newly developed agricultural produce markets at Turbhe and in the newly developed iron and steel markets at Kalamboli.

The JN Port Trust, a giant project which was commissioned in May 1989 and was constructed directly opposite the old port of Bombay, has showed a very rapid increase in traffic flow over the last five or six years. Total cargo handled has increased almost tenfold, from nearly 0.7 million tons in 1989-90 to almost 7 million tons in 1995-96; an increase mainly based on exports. The New Bombay port is projected to handle traffic volumes in excess of 11 million tons by the turn of the century, and a large expansion scheme is already in progress.[9]

An important aspect of the original development plan for New Bombay, and one of the proposed major catalysts for initial growth in the area, was the proposal to shift a major part of the state government's political activities out of Bombay Island, and into New Bombay's Central Business District (CBD) in and around Belapur node in central New Bombay. In total, the area outlined for this newly planned CBD was 61 ha. of land, at FSI of 1.5. This 91.61 ha. of office space is sufficient to accommodate 100,000 commercial and administrative jobs. By early 1995, 55.24 ha. of office space had been sold, of which around 27 ha. has been built up. In practice, however, business and/or administrative activity in the CBD of Belapur is very limited. At present, not more than about 8,000 people are employed in these buildings, the large majority in newly created government offices such as CIDCO Bhavan itself. The number of government offices that have been shifted from Nariman Point in South Bombay, as was planned, is thus very small. Moreover, although the majority of buildings that have been constructed have been taken up by large building corporations and other large private companies, private business in the newly created CBD of New Bombay is almost non-existent. Real private business activity in New Bombay is going on in sector 17 of Vashi node, which was not planned for this purpose, but in practice is functioning as the major commercial centre of the new city.[10]

Briefly looking at housing construction in New Bombay also shows the rapid development in the area in recent times. Whereas in 1983, only

---

9 *Figures from the JNPT. The 1995-96 figure is an estimate of the technical adviser to the Chairman. It has to be noted, that since port activities are an integral part of the central government's responsibilities, the state government nor CIDCO have had anything to do with developments in the JNPT.*

10 *Recently, this sector 17 was simply renamed as District Business Centre (or DBC).*

10,428 houses had been constructed, almost all of which were built by CIDCO, that number had gone up to 84,824 by 1991. By that time, over 19,000 houses had been built by the private sector. In 1995, the total number of houses had further increased to around 135,000, about 70 % (95,000) of which were built by CIDCO and about 30 % by the private sector.[11]

Besides population growth and economic activity, land prices and property rates are probably the most significant indicators of growth and development in a given area. If we first look at the land sales pattern in New Bombay's most developed node, the sharp increase in receipts is remarkable. Whereas up to 1980, CIDCO received for its developed land in Vashi nearly Rs. 855,000 per hectare, the table below shows that the amount has continuously and sharply increased to over Rs. 26 million per hectare by 1995.

Table 13.4   Land sales pattern in Vashi node (different sales systems), 1980-1995

	Area (in ha.)	Amount (in Rs.)	Rs. received / ha.
up to 1980	79.78	68,202,000	854,876
1980-85	108.48	294,042,000	2,710,564
1985-90	90.11	803,339,000	8,915,093
1990-95	46.46	1,227,601,000	26,422,740

Source: internal CIDCO document (20 Dec 1995).
Note: includes sales of land for residential and commercial purposes, warehousing, schools, and other social facilities.

Since land was to be the major resource for financing the development of the entire New Bombay area, it is obvious that the spurt in land prices has given CIDCO the opportunity to develop the area at a much faster pace than before. The figure below gives an idea of the fast increase in receipts and expenditure of the development corporation from the mid-1970s to 1995-96. Note, however, that the figures between 1976 and 1991 give the evolution over 5-year periods, whereas from 1991 they give the annual increase.

Looking at the evolution in residential and commercial property rates in the private market in and outside of New Bombay, not only indicates a pattern of sharply increasing property rates, but also that this pattern is not unique for New Bombay alone. The pattern of residential and commercial property rates in four nodes of New Bombay in the eleven-year period between 1986 and 1997 is given in the table below.

---

11 Figures 1975-1991 from CIDCO departments, 1995 figures from Bhattacharya (1995), which are estimates and include both houses constructed and under construction.

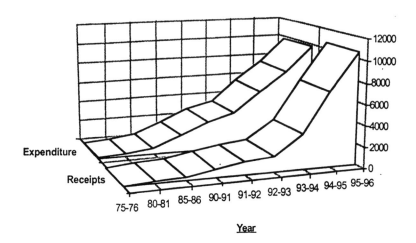

Figure 13.1 Expenditure and receipts of CIDCO 1976-1996 (in million Rs)

Source: based on figures from CIDCO departments and Gill, *et al.* (1995).

Selecting three of the four nodes, clearly reveals the sharp increase in both residential and commercial property rates between 1991 and 1995, with a serious downward correction between summer 1995 and spring 1996, and a very slow recovery of the market since then.

The same exercise has been done for two control areas outside of New Bombay, i.e. in Greater Bombay and in a number of adjacent areas to the north and north-east of Bombay and New Bombay.[12] Although the rates are obviously much higher in Greater Bombay than in the newly urbanising areas of New Bombay and other regions in the city's periphery, the same overall pattern of a sharp increase between 1991 and 1995, with a subsequent correction has emerged (see graphs at end of paper).[13]

Therefore, it can be concluded that, overall, the rapid increase in property rates has not been a distinct characteristic of New Bombay alone, but

---

12 In Greater Bombay, three functionally distinct areas (Nariman Point, Pedder Road, and Chembur) were selected; the areas to the north and north-east of Bombay and New Bombay consisted of three administratively distinct areas (Thana, Kalyan, and Virar).

13 Note, however, that in the case of residential property rates in the three northern areas the correction has been much less significant, that the fall in commercial rates has been even sharper than in New Bombay, and that no recovery can as yet be observed, not in Greater Bombay, nor in the northern areas.

Table 13.5 Property rates (built up) in four New Bombay nodes, 1986-1997 (in Rs. per ft²)

	Vashi		CBD		Nerul		Airoli	
m/year	Resid.	Comm.	Resid.	Comm.	Resid.	Comm.	Resid.	Comm.
07/86	300-600	600-800	250-300	450-800	275-300	450-700	200-250	-
10/90	500-700	1000-1500	500-600	650-900	400-500	600-800	350-450	500-900
06/91	600-900	900-1700	500-600	700-900	450-500	700-900	350-450	700-900
08/93	900-1800	2000-4000	525-750	1000-2500	550-750	1000-2000	375-525	1000-1500
07/94	1300-3000	2000-5000	750-900	1100-3000	700-1100	1100-2500	500-600	1300-1600
07/95	1400-3900	4500-8000	1800-3200	3000-7500	900-1600	2000-4000	900-1800	1300-4000
03/96	1800-2500	3500-8000	900-2000	3000-6000	900-1500	2000-4000	850-1600	1300-4000
04/97	1800-2800	2000-6000	1500-2200	3200-6000	1400-1600	2000-5000	1000-1400	2000-3000

Source: The Accommodation Times, 1986-1997.
Note: commercial includes shopping; rates may be more by 50 to 200 % depending on location, amenities, etc.

rather of the entire Greater Bombay Urban Agglomeration.[14]   The reasons behind this sharp rise must mainly be sought in the liberalisation of the Indian economy since 1991, bringing with it a greater flexibility of investment regulations, with Bombay, being the economic and financial centre of the country, acting as the major focus for incoming business and private capital from multinational corporations and non-residential Indians (NRIs). The sharp drop in prices since 1995 must be seen as a major correction to the artificially high rates of the mid-1990s, mainly due to large-scale speculation.

Thus, in terms of growth and development in general, it must be concluded that the New Bombay project is very successful, notwithstanding the fact that real progress is only a recent phenomenon, dating from the later 1980s and early 1990s, and has been very sluggish indeed during the first one and a half decades of implementation. This recent rapid growth has in part been the result of external factors, such as the opening of the Indian economy to foreign investments, which had very little to do with New Bombay itself. Within New Bombay, the construction of a first major railway line, connecting New Bombay with the main job centres of Bombay, no doubt has been the single most important factor in this sudden growth pattern. Other factors, beside these two major ones, that have played a catalysing role and have significantly added to New Bombay's potential and attractiveness, were the recent construction of a new modern port in southern New Bombay, the com-

---

14 The Greater Bombay Urban Agglomeration (GBUA) is an area delimitation which has for the first time been included in the census in 1991. The population of the GBUA was then put at 12,596,000.

missioning of a second bridge over the Thana Creek at Airoli in northern New Bombay, the streamlining of New Bombay's telecommunication system with that of Bombay, and the gradual shift of the wholesale agricultural produce markets and iron and steel markets from their traditional locations in South Bombay to newly developed land in New Bombay.

13.3 New Bombay, a City for all Classes and all People ?

In this second section we will discuss to what extent the rapid growth and development of New Bombay has been followed by the planned objectives of a social city, in which all segments of the population, rich or poor, would be able to find a place to live. For this purpose, we will draw a picture of the socio-economic profile of the population living in the nodes of New Bombay, based on the data of the Socio-Economic Survey 1995 (SES-95). This data set is the result of a fairly large questionnaire survey, conducted by outsiders under the supervision of CIDCO, which had a random sample of 25 % of all households living in the nine most developed nodes. In total, 19,727 households were included, numbering 79,075 persons.

Firstly, we will be looking at the data for all the nodes in general. Secondly, we will be looking at the data for a number of selected low-income sectors within the nodes. Finally, we will be looking at the data for a number of specially reserved areas in Airoli node, in which 'Sites and Services' schemes have been implemented.

13.3.1 The Nodes in General

Of the total number of 79,075 people included in the sample, 53.95 % were males and 46.05 % were females. The largest group fell within the category of 25 to 44 years, followed by the 10 to 15 year olds. Over 53 % of the nearly 20,000 households included in the sample count 4 to 5 persons, over 31 % 2 to 3 persons, and another 12 % 6 to 7 persons. The large majority of the population (85.5 %) consider themselves as Hindu, and the mother tongue is predominantly Marathi.(48.7 %).[15]

In 1995, over 81 % of the total housing stock in New Bombay was built by CIDCO, of which the major part (59 %) is between 16 and 35 m≤ in

---

[15] *Of the total population, 4.8 % consider themselves as Muslim, and as many as Christian, and 2.8 % as Sikh. The mother tongue of more than 14 % of the total population (14.2) is Hindi, for 8.0 % it is Malayalam, for 5.0 % Punjabi, for 4.5 % Gujarati, for 3.7 % Tamil, and for 3.1 % Kannada.*

size. Of all the houses that have been built by the private sector (nearly 19 % of the housing stock), almost 68 % is between 51 and 100 m≤. Nearly 32 % of the total housing stock (CIDCO and private) was rented from the private market, nearly 31 % was purchased outright from CIDCO and 9 % from the private market, nearly 15 % was purchased under the hire-purchase formula, and 7.5 % had been resold.[16] The average amount of money that was originally paid for a house purchased from CIDCO was Rs. 137,000, and from the private market Rs. 231,000. Of the 2,875 households that have taken a loan to purchase the house, the average monthly income per household was Rs. 4,779. Of the 9,272 households that have bought a house under a different scheme (outright purchase from CIDCO, from the private market, resale), a large part (nearly 40 %) was financed by personal savings, and another 14 % by savings in combination with a loan from the employer.

Of the total sample of 79,075 persons, 26,233 persons (88 % of which are males) are employed and are having an income. The main labour categories among males are the manufacturing industries (29 %) and personal shops or businesses (13 %). Among the females, the main labour categories are education (20 %) and the manufacturing industries (11 %).

Over 74.3 % of all households (14,427) have a single earner, another 19.5 % have two earners (3,803), 4.5 % have three earners (882), and 1.5 % have four or more earners. Every household living in the nodes is thus having at least one earner in the family.[17] Of the 53 % of households that have 4 to 5 persons, the average combined income per household is Rs. 3,507. Of the 31 % of households that have 2 to 3 members, the average combined income per household is Rs. 2,133. Of the 12 % of households that have 6 to 7 persons, the combined household income is Rs. 5,584 on average. The overall average income per household per month for the entire population in the nodes is Rs. 4,932.[18] Moreover, given the fact that these income figures have been mentioned to complete strangers in the context of a questionnaire, and given the fact that the informal economy in New Bombay is estimated to be at least 50-60 %, it may be safely argued that the actual household incomes are significantly higher then what the figures indicate.

As education is a significant socio-economic indicator, it is quite remarkable to find that 37.4 % of the total population in the nodes of New Bombay has finished (or is studying up to) at least 10th standard, and that

16 *This resale figure, however, is not worth much, as it does not reflect the resale activity as it is in practice. We will return to this at a later stage in this paper.*

17 *It must be noted, however, that 1.54 % of households did 'Not Mention', which means that given the fact that a separate category 'Not Applicable' was not included, part of this 1.54 % may have been unemployed, as they could not be put into any other category.*

18 *The percentage of households that did not mention their income was 2.64 %.*

nearly 10 % is having a graduate or postgraduate diploma or equivalent, or is studying for one.

To understand the impact of these figures, we first have to look at the definition of different income categories which are used by HUDCO, and therefore also by CIDCO. The table below shows what has been understood by Economic Weaker Sections (EWS), Low Income Group (LIG), Middle Income Group (MIG), and High Income Group (HIG) over the years.

Table 13.6 The definitions of poor groups in society

	EWS	LIG	MIG	HIG
since Jan 1986	up to Rs 700	Rs 701-1500	Rs 1501-2500	Rs 2501 & above
since Apr 1991	up to Rs 1050	Rs 1051-2200	Rs 2201-3700	Rs 3701 & above
since Apr 1993	up to Rs 1250	Rs 1251-2650	Rs 2651-4450	Rs 4451 & above

Thus, according to the latest revision, a household in New Bombay which is currently having a monthly income of say Rs. 3,200 (equivalent in mid 1998 to UK £50 or US $ 80) belongs to the MIG. Now, if the figures on household incomes in the nodes are brought into these categories, it will be observed that almost no households (all together just 359 out of 19,207 households) belong to the EWS category.[19] The number of households belonging to the LIG category is higher (all together 3,727 out of 19,207 households), but still less than one in five (19.4 %) of all households, and the number of households belonging to the MIG category is with 33.4 % still higher (6,423 out of 19,207). Most significant, however, is that the largest group of households (45.3 %), belong to the HIG category (8,698 out of 19,207). Or in other words, the distribution of households in New Bombay is such, that almost four in five households (78.7 %) belong to the higher income groups (MIG+HIG), and only 21.3 % to the lower income groups (EWS+LIG).

In 1987, when the first SES was conducted, the percentage of households in New Bombay belonging to the higher income groups was with 54 % already relatively high, but far less than what it is today. The average monthly household income in those days was Rs. 2,112, or less than half of what it is today (Rs. 4,932).

Comparing these data with similar figures in South Bombay, which is generally considered the richest part of Greater Bombay, and indeed one of the most affluent areas in the whole of India, is quite revealing. These figures, which give the household income distribution data for South Bombay in 1989

---

19 *The 19,207 figure is the total number of households in the sample (19,727) minus the 520 households who did 'Not Mention'.*

(more recent figures were unfortunately not available), have therefore been linked to the income categories of EWS, LIG, MIG and HIG as they were between 1986 and 1991 (see table above). In this way, it can be observed that in those days, 9.4 % of households in South Bombay had a monthly income of less than Rs. 750, 33.7 % had an income of Rs. 751-1,500, 26.1 % had an income of 1,501-2,500, and 30.8 % had an income of Rs. 2,501 or more. Thus, in other words, in 1989, 43.1 % of all households in South Bombay were belonging to the lower income groups (EWS/LIG), and 56.9 % of households were belonging to the higher income groups (MIG/HIG).[20] For the entire area of Greater Bombay (a separate figure for South Bombay was not given), the average household income in 1989 was Rs. 2,468.

Consequently, household income distribution figures in New Bombay in 1987 were fairly similar to the same figures in affluent South Bombay in 1989. Yet, since those days, the percentage of households in New Bombay belonging to the higher income groups has done nothing but increase, to over 78 %. What the pattern has been in South Bombay is, due to a lack of data, not known.

The overall average household income figure for New Bombay in 1987 (Rs. 2,112) was, given the two-year difference in data collection, quite comparable to that of Greater Bombay in 1989 (Rs. 2,468). Quite remarkably, this 1989 figure for Greater Bombay is almost exactly half of the average household income in New Bombay in 1995 (Rs. 4,932). [21]

13.3.2 Some Sectors within the Nodes

Apart from the nodes in general, we have been looking at the data for a number of selected low-income sectors within the nodes.[22] In Vashi node, which is the node firstly developed, we have focused on sector 2, in Nerul node on sector 10, and in CBD Belapur node on sector 2. All three sectors were planned predominantly to accommodate households belonging to the

---

20 *Figures from the Socio-Economic Review of Greater Bombay, 1993-94. The figures are based on a sample of 5,470 households, and were originally compiled by the Operation Research Group for the BMRDA.*

21 *The only figures on household income distribution that have been found for all urban households in India at large, are based on completely different categories and can, therefore, not be compared. It is worth noting, however, that based on 1988 and 1989 figures, 71.9 % of all urban households in the country had a monthly income of maximum Rs. 2,083 (Statistical Outline of India 1994-95). Thus, only the upper section of this 71.9 % (having an income between Rs. 1,501 and Rs. 2,083) would not belong to the lower income groups (EWS/LIG).*

22 *For this purpose, the raw survey data were processed specifically for these sectors.*

EWS and LIG categories.[23]

Of the 637 houses in Vashi's sector two, 285 were planned for EWS households, and the other 352 for LIG households. In reality, however, only 8.6 % of households living in this sector fall within the income categories for whom these houses were planned, and over 91 % of households in this sector actually belong to the higher income groups (MIG/HIG).[24] Education-wise, 60.3 % of the people in this sector went to school minimally up to 10th standard, and the number of people that have (or are currently studying for) a diploma of higher education is 15.5 %.[25]

In Nerul, about 81 % of the 706 houses that were included in the sample in sector 10, were meant for EWS or LIG households. In terms of household incomes, however, the figures here are not radically different from what they are in Vashi's sector 2. Less than 37 % of households living in sector 10 belong to these lower income groups (EWS/LIG), and over 63 % actually belongs to the higher income groups (MIG/HIG). Although the plan had reserved less than 2 % of the houses for the HIG category, in reality not less than one in four houses (25.2 %) are being occupied by an HIG household. The graph below shows the discrepancy between plan and real situation in Nerul's sector 10.

The same phenomenon can be observed in sector 2 of Belapur node, which was exclusively planned for LIG households, having a monthly income of not more than Rs. 2,650. In reality, just one in four of the households living here (25.7 %) belong to this category, and no less than 40 % actually belongs to the HIG category.

13.3.3 'Sites and Services' in Airoli Node

In three sectors of Airoli node (sectors 2, 3, and 4), which were part of the Bombay Urban Development Project (BUDP), 'Sites and Services' schemes have been implemented with financial support of the World Bank, to provide accommodation to the lowest income categories.[26] In these three

---

23 *Since sector 2 of Vashi might be considered as an extreme example due to its location right opposite New Bombay's main business district, and therefore as a location where the forces of the land and housing markets have been exceptionally strong, the same exercise has been conducted for sector 10 of Nerul and sector 2 of Belapur.*

24 *Over 67 % of all households living in this low-income sector had a monthly income of more than Rs. 4,451 and thus fully belonged to the HIG category.*

25 *Included in this figure are graduates and graduate students, nurses, graduates from technical institutes, graduates from law schools, engineers, architects, and post graduates.*

26 *Ultimately, under this programme over 22,000 'Sites and Services' are to be implemented, spread over six nodes of New Bombay. The three sectors in Airoli were selected because they were part of the first phase of the BUDP programme, which started off in 1986.*

sectors, nearly 97 % of houses were built to accommodate the lowest income groups. Of the 906 families included in the sample, 44.6 % admitted that they are not the first owners of the house, and another 30.6 % is just renting the house. At present, just 3 % of households occupying these houses belong to the lowest income group (EWS), and another 27.2 % to the LIG category. Thus, in total, less than one in three houses in these sectors (30.2 %) are currently occupied by the lower income households for whom they have been constructed. Just below 70 % of the houses (69.8 %) are occupied by a higher income group (MIG 41.4 % and HIG 28.4 %). These data more or less correspond with the results of an interview survey that was conducted in these areas, and has revealed that at least 70 % of the first owners of these houses, which originally, indeed, were belonging to such lower income groups, have vacated their houses and have left the area.[27]

Overall, it can be concluded that New Bombay is far from being *the city for the common man* that it was planned to be. At present, New Bombay is predominantly catering to the middle and high income groups, and over time, lower income families, even in those areas that were specially reserved for such households, have increasingly been removed by higher income families.

13.4 Factors Responsible for the Bias towards Middle and Higher Income Groups

In an attempt to explain why New Bombay has become a city predominantly catering to the higher income groups, and increasingly pushing out lower income groups, three major factors can be distinguished: (1) CIDCO's general development policies in combination with the working of the land and housing market; (2) the overall cost of living in New Bombay; and (3) the fairly homogeneous economic base of the new city. Given the limits of this paper, we will only briefly look into these three factors.

13.4.1 CIDCO's General Development Policies and the Working of the Land and Housing Market

CIDCO's policies and general attitude have played an utmost important role in this process, passively as well as actively, and mainly on two levels: in housing on the one hand, and in the provision, management and

---

27 *Several of these households can now be found in the slums in the MIDC area, opposite Airoli node on the other side of the main Thana-Belapur road.*

maintenance of infrastructure and services on the other.

In housing, CIDCO's role has been twofold. First, whereas in earlier times, plots and houses were sold at fixed price, from the moment the market picked up, plots and houses for commercial, but also for residential use, have increasingly been sold to the public by tender (at a certain minimum percentage of the RP), as well as by auction, resulting in a steep rise in land and housing prices. In the same time, plots and houses have been released in small bits in order to keep the prices high. Officially, CIDCO has always argued that an increase in land and housing prices, and the sale of commercial space by auction, helps them in collecting the necessary funds to subsidise low-income housing. Although this is, of course, a valid argument from a commercial point of view, in practice these policies have added to the steep rise in land and property rates in the area, whereby even the cheapest houses have become so expensive in the private market, that they are far beyond reach of the lower income groups; a fact which was already observed in the late 1980s by Swapna Banerjee-Guha.[28]

Second, CIDCO has also played a more passive role on the housing front. Although since the early days of the New Bombay project, CIDCO had introduced certain rules and regulations to prevent a rapid selling-off process of CIDCO housing, in practice very little has been done to prevent this. Whereas officially, selling of CIDCO housing could only be done under certain circumstances and via the development authority itself, in practice, many houses have been sold unofficially without involvement of CIDCO, or have been sold via the official way but with manipulation of the actual price that was paid.[29]   Instead of preventing such illegal practices in some way or another (which obviously push up the market rates of property), since recently the sales regulations and limitations have simply been revised and have been

---

28 *Some examples : the 30 m² plots in sector 2 of Vashi (meant for low-income households) were in the mid-1980s sold by CIDCO for Rs. 8,000. In February 1996, these houses were sold in the private market for Rs. 700,000 for houses on the inside of the sector, and for Rs. 1.3 million for houses at the commercially interesting road side. In Nerul's sector 10, an A-type house (meant for EWS) which was initially sold for Rs. 16,000, was costing Rs. 180,000 in the private market. In Belapur's sector 2, the A1 type houses (measuring 18 m² and meant for LIG) were initially sold for Rs. 8,000. When in 1992 CIDCO allowed an increase in FSI, the price of these houses in the market immediately shot up to Rs. 190,000, and in February 1996 the price for such a house with the real estate agents in the area was Rs. 450,000. (from a large number of interviews with real estate agents)*

29 *The main reasons for an unofficial sale are the rules and regulations themselves, which simply did not allow selling off in certain circumstances (hire-purchase for example), and the relatively high transfer costs involved in selling via the official way. The main reason for an official sale with manipulation of the figures, are also these transfer charges (which are based on the amount of the sale), as well as to simply reduce the high property taxes to be paid afterwards (which are based on the value of the house).*

made more flexible. Consequently, at present, even in those areas which were exclusively planned and reserved for low income households, selling of property has become fairly easy and is formally accepted.

Besides housing, CIDCO's policies in the provision, management and maintenance of infrastructure and services have been equally important. Earlier, CIDCO was planning, developing, and maintaining New Bombay's infrastructure and services. In later years, however, its role has increasingly been limited to planning and co-ordination, whereas development and maintenance have increasingly been privatised. The areas which have been prone to 'contracting out' are numerous, and include general environmental sanitation jobs such as road sweeping, cleaning of shopping complexes and markets, maintenance of the railway stations, garbage collection and waste management, and the maintenance of sewage treatment plants; further also the collection of CIDCO's service charges (for water and electricity for example), the maintenance of the water supply system, the development and maintenance of parks and gardens, the maintenance of street lighting, the management of educational and public health facilities and, indirectly, part of the public transportation sector.

Although from a commercial/financial and practical point of view, these policies have been very beneficial to CIDCO, their overall effect for the public is not unilaterally positive. No reasonable arguments can be put forward against the privatisation of urban infrastructure and services in countries or regions with limited public resources, if these mechanisms are beneficial to all parties involved. In New Bombay, for example, garbage collection, street cleaning, or service charge collection are services which have been privatised and have had positive effects for all.[30] However, privatisation is not a soul-saving policy option that can be applied at any level, at any time and at any place. Its impact has to be thoroughly researched in advance, its consequences clearly understood, and its price effect kept under strict control before it can be implemented. If, like in New Bombay, such key urban and social services like education, public health and transportation are partly or fully privatised, generating major negative cost effects to the general public, privatisation becomes unacceptable.

In the initial stages of development of New Bombay, CIDCO built and managed schools and hospitals by itself. Later, its role was limited to construction of the building or simply providing a plot of land, and management

---

30 *For CIDCO, privatisation has mainly given positive financial effects, as no staff had to be employed on a permanent basis, whereas for the local population it has given positive effects in terms of small work contracts for local businesses (with guaranteed minimum income levels for the workers), efficiency of service delivery, and community participation, without an increase in the cost of these services.*

of the school or hospital was left to a private trust. However, as there are no municipal schools left in the nodes, all privatised schools are now (unofficially) charging high donations (sometimes called *'money for the building fund'*) of Rs. 2,000 to Rs. 10,000, mostly on top of the annual tuition fees. In contrast, Greater Bombay counts numerous good municipal schools which are basically free of cost.

The same evolution could be observed for hospitals and dispensaries, most of which are now being privatised. Although New Bombay has a few municipal hospitals, they are so poorly equipped that if people cannot afford a private hospital, they have to go to one of the municipal hospitals in Bombay for treatment. The cost of medical treatment in New Bombay in such privatised hospitals or dispensaries, which are sometimes called *charitable* hospitals and where the quality of treatment is overall quite good, has over the years seriously increased, and has become unaffordable to many.[31]

Finally, CIDCO's claimed exciting privatisation experience has also caused major expenses for the public in transportation. First, train stations in New Bombay have been planned and constructed at the edge of the nodes, at a considerable distance from the node centres. For those people who do not have a bicycle, a motorcycle or a car waiting for them at the station, it is rather difficult or at least time-consuming to walk from the station to their homes. The only possible alternative is to take a privately run autorickshaw, which charges a minimum of Rs. 5.50 + metre charge. For those who do have a personal vehicle waiting for them at the station, transportation is still not free, since the parking lots at the train stations, which have been nicely planned indeed, are charging the customers. A person working in Bombay and leaving his scooter at the station on a daily basis will spend some Rs. 60 per month just for this parking service.[32]

Second, when CIDCO built the railway link over the Thana Creek, connecting New Bombay with Greater Bombay by rail, a major part of the investment was to be recovered from the operation of the bridge. Consequently, since 1992-93 CIDCO contracted private organisations to collect a toll-tax. Every vehicle, public or private, that crosses the bridge in any direction has to pay a certain amount of money. Two-wheelers for example pay Re. 1 and cars Rs. 5 per passage, an amount which is, although fairly high, affordable for someone owning such a vehicle. However, public buses have to pay Rs 15 per passage, an amount which is fully recovered from a surcharge on bus

---

31 *A friend's experience with one of these 'charitable' hospitals was, that when his wife had to give birth to a twin and was in hospital for a month, the household received a bill of around Rs. 10,000 (i.e. not much less than the Indian annual per capita GNP).*

32 *In this example, this extra cost increases the cost of a monthly train card ticket by about 40%.*

tickets, which is thus an extra cost for the public. The same system is in operation for trains, and a surcharge is to be paid on every train ticket for a Bombay-New Bombay travel, ranging from 17 % (on a single ticket) to 33 % (on a monthly card ticket) on top of the normal price.[33]

13.4.2 The Overall Cost of Living in New Bombay

Apart from the spurt in land and housing prices, and the high costs resulting from the privatisation of certain urban and social services, living in New Bombay is also relatively expensive in many other ways.

First, since in the initial phase, CIDCO has neglected to arrange its own water resources, it now has to buy water for domestic and commercial use at a high cost from the MIDC. Consequently, at present, households in New Bombay pay the highest rates for water consumption in the region.[34]

Second, since in the absence of a municipal corporation for New Bombay, it has always been CIDCO that has provided the usual municipal services to the public, households have always paid 'service charges' to CIDCO instead of the usual municipal taxes. However, since the recent establishment of the NMMC, households in New Bombay have been charged a high property tax (of 23 % for residential use and 35 % for commercial use), notwithstanding the fact that at the same time they are still paying service charges to CIDCO, and that the NMMC has not provided any services yet.

Third, the recent general rapid development in the area quite naturally has also brought in its wake a serious rise in prices of all sorts of commodities, and food prices have dramatically increased over the last decade. Earlier, all the food that was locally consumed (mainly rice, fish, milk, vegetables and fruits) was produced in the area. At present, however, with the acquisition of the lands, and the dramatic chemical pollution of the waters in and around New Bombay, very little food is still being produced locally, and the major food items have to be brought in from outside.

13.4.3 New Bombay's Fairly Homogeneous Economic Base

Apart from CIDCO's general development policies and the working of

---

33 *The amount of surcharges was proposed by CIDCO and accepted by the Railway Board of the central government.*

34 *Domestic use of water was in New Bombay in 1994 charged at Rs. 2.80 per cubic metre. In Kalyan/Dombivali this was Rs. 2.00, in Thana Rs. 1.30, and in Greater Bombay only Rs. 0.60. Since that time, the (metred) water rate in New Bombay has increased further to Rs. 3.65/m².* *(figures from CIDCO and MWSSB)*

the land and housing market on the one hand, and the overall cost of living in New Bombay on the other, the fairly homogeneous economic base of the new city is a third factor that plays an important role in the explanation of New Bombay's upper middle class profile.

Of the total employment that has been generated in New Bombay since the 1960s, the increase in new jobs for unskilled or semi-skilled workers has been lagging far behind the increase in the potential low-wage work force. None of the major job providers in New Bombay have been very effective in absorbing a significant number of such workers. The major industries in the TBIA and Taloja are capital-intensive industries which mainly need skilled workers and engineers, and the same can be said of the government and semi-government companies operating in and around the JNPT. The number of unskilled and semi-skilled workers that can be engaged in service jobs in private or government offices is obviously also very small. Consequently, most of such households, if still in the area, are engaged in some petty business, or are continuously trying to get some temporary contract work.

Therefore, it is only natural that an increasing number of such households leave the area in search of a low-wage job and a steady income, the more since the cost of living in New Bombay is relatively high and is ever increasing. For many of these families, the slums of Bombay are the main refuge, whereas others have disappeared into one of the numerous New Bombay slums, which officially do not exist.

13.5 Conclusion

Whereas in terms of urban planning and development, New Bombay had to become a counter-magnet, and whereas in financial terms the project had to become self-containing, in social terms the twin city was planned to become a model city, in which rich and poor citizens alike would have access to an affordable house, and to the excellent urban infrastructure and services that would be provided.

During the first ten to fifteen years of project implementation, the lack of certain crucial infrastructural provisions such as good transportation links and telecommunication lines, prevented the necessary interest of private developers, business and individual households. Consequently, the land and housing market did not pick up, and as the authorities were to finance the development of the area with the sales of the land, overall development in the area was hampered and proceeded at a slow pace, and New Bombay's growth remained very sluggish.

Therefore, in a number of major shifts, CIDCO drastically changed its development policies. Many urban and social services were partly or fully privatised, the sale of land and housing was increasingly done by tender or by auction, and plots were released in small bits so as to keep the prices high. In the meantime, a number of events such as the construction of the railway link with Greater Bombay and the new JNPT port, the streamlining of New Bombay's telecommunication system, and the planned shift of the wholesale markets from South Bombay to New Bombay, greatly added to New Bombay's attractiveness and potential. Both factors, CIDCO's changed policies and the provision of several crucial infrastructural facilities, caused a major spurt in land prices, partly as a result of large-scale speculation by private developers. Later, in the first half of the 1990s, a second, and unprecedented increase in land prices occured, this time mainly as a result of the liberalisation of the Indian economy. Again, a major speculation drive, of which the effect was felt all over urban India but mainly in and around Bombay, was the direct result.

Consequently, since the late 1980s, the financial returns for CIDCO sharply increased, and allowed for a much faster pace of planning and development activities. However, parallel to these changed policies and developments, New Bombay's social outlook also rapidly changed, with a population increasingly becoming predominantly middle and upper middle class, and becoming comparable to affluent South Bombay's population. Although strict regulations had to prevent a selling off process of CIDCO housing within a rapidly urbanising environment with increasing land and housing prices, in reality CIDCO's laissez-faire attitude fostered a large-scale selling-off process, in which even the areas which were exclusively planned and reserved for low-income households are currently inhabited mainly by higher income households.

The initial land use plan for New Bombay and the overall development policies were frequently changed, and were basically adjusted to the real situation. That sector 17 of Vashi was recently renamed District Business Centre, that in several sectors which were to accommodate low income households the FSI was simply increased (immediately resulting in a spurt in housing prices), and that the resale rules and regulations for CIDCO housing were drastically revised and basically abandoned, are just a few examples. In such a way, it may be questionned what exactly the use of a plan is.

Developments in recent years have proved that not much remains of the original intentions of the entire New Bombay project.[35] The twin city

---

35 *Within the limits of this paper it was, for example, not possible to discuss the development authority's performance in rehabilitating the local village population, which has dramatically been affected by developments in the area. Suffice it here to say that this performance has, overall, been quite poor.*

has become no different from numerous earlier experiences in urban planning and new town development, and has basically developed into a large real estate development project. In this sense, some other crucial questions must be put forward. What, for example, is the role of the state government in this process, which ultimately is the political authority under which CIDCO operates. To what extent have the forces, which have been at work in Bombay for quite some time, been working in New Bombay also? Although the development concept, based on financial self-sufficiency, has itself obviously been an important factor in this process (as it limited the policy choices), some other government policies may have had an even larger effect on developments in New Bombay. In contrast to what is frequently argued, in the old city of Bombay the major problem is not the limited availability of land, but rather the use of that land. Whereas the demand for private and commercial accommodation is ever increasing, partly as a result of speculation, on the supply side the availability of land, housing and office space is restricted in numerous ways. That prices increase so rapidly is, therefore, only natural. In New Bombay, where the FSI for example is being kept very low, and land is being released in small bits, similar forces may be at work. The question, who the main beneficiaries of high land prices are, in Bombay as well as in New Bombay, becomes crucial in this discussion.[36]

Whatever the driving forces working under the surface, what the ordinary citizen feels in his daily life, is that he can no longer afford to live in New Bombay, and that the financial and other pressures are increasingly building up. In his foreword of the development plan, the first chairman of CIDCO expressed his belief in the New Bombay project, which was to become "a city in which the common man would like to live". Yet, about 25 years after date, we can say that New Bombay, indeed, has developed into a city in which the common man would *like* to live... very much so. Unfortunately, for most families, it will never materialise.

References

Accommodation Times, 1986-1997

Banerjee-Guha, Swapna (1989), *Growth of a Twin City: Planned Urban Dispersal in India*. In: Costa, Frank; Dutt, Ashok; Ma, Laurence; and Noble, Allen (eds.), *Urbanisation in Asia. Spatial Dimensions and Policy Issues* (University of Hawaii Press, Honolulu), pp. 169-188.

Banerjee-Guha, Swapna (1995), *Urban Development Process in Bombay: Planning for Whom?* In: Thorner, A. & Patel, S. (Eds.), *Bombay, A Metaphor for Modern India*. (Oxford University Press Bombay, 1995)

---

36 *At the time of writing this paper, more work needed to be done on this topic.*

Bhattacharya, A. (1971), *A Study of Industries in Trans-Thana Belt* (CIDCO, June 1971)

Bhattacharya, Ardhendu (1995), *An Integrated Approach to Urban Development. A CIDCO Experience in New Bombay* (paper written to be published in All India Housing Development Association, AIHDA)

CIDCO (Oct. 1973), *New Bombay Draft Development Plan.*

CIDCO (Nov. 1986), *New Bombay, An Outline of Progress.*

CIDCO (July 1989), *New Bombay, An Outline of Progress.*

CIDCO (May 1992), *Two Decades of Planning & Development.*

Gill, G.S.; Bhattacharya, A. and Adusumilli, U. (Nov. 1995), *Sustainable Urban Development. Case Study of New Bombay, India* (CIDCO report, final draft)

MCGB (1995), *Socio-Economic Review of Greater Bombay (1993-94).* (Centre for Research & Development, Municipal Corporation of Greater Bombay)

Statistical Outline of India 1994-95. Tata Services Ltd. Department of Economics and Statistics.(Tata Press Ltd. Bombay)

14 Changing Dynamics of the Urban Land Market in Lucknow City

Amitabh

Editors' note:
UN projections estimate the population of Lucknow will grow to 4 million by 2015 – a growth of 160% on 1990. The pattern of its growth will depend on the process of land allocation. Although India's economy is more mixed than China's and has been since independence, and there are public, cooperative, and private sectors involved in Lucknow's landmarket, the public sector turns out to have a quasi-monopoly, and is as important as in Shanghai, considered in the previous chapter.
The Editors wish to record their deep regret at the untimely death of Amitabh shortly after the Fifth Urban Urbanisation Conference.

14.1 Introduction

Scholarly studies on the dynamics of urban land markets, measuring land price changes, and identifying determinants of land prices changes across Third World cities, are few in number. One of the main reasons has been the unavailability of reliable data on the variables associated with different aspects of urban land markets in Third World cities. Although this limitation still continues, a renewed interest in urban land market studies on Third World began by the early 1980s (UNCHS, 1982; TCPO, 1983). This resulted in a growing belief that urban land markets operate almost in similar ways everywhere, and that real land prices have been increasing constantly over time (Kirwan, 1987; Batley, 1989; Wadhwa, 1983b; TCPO, 1983). But the Fitzwilliam Workshop of 1991 challenged this growing orthodoxy. The memorandum published after the Workshop argued that there is a need to understand the modus-operandi of urban land markets across various Third World cities and that the so-called upward trend in land prices in many Third World cities was exaggerated (Fitzwilliam Workshop, 1991: 625). It was further argued by some scholars that many of the conclusions about urban land markets developed in the past were based upon confused methodologies, inappropriate comparisons, too short a time horizon, and the selective choice of examples (Jones, 1991; Ward et al., 1993; Jones and Ward, 1994; Amitabh, 1994 and 1996). It is against this background that this paper takes up the opportunity to focus upon how Lucknow City's urban land markets have operated at different times.

Lucknow City is one of the fastest growing metropolitan cities in India. It is the 12th largest city in the country with a population of 1.66 million in 1991 (Table 14.1). In the decade 1981-1991 Lucknow increased in szie by 66% - one of the three highest city grrowth rates. Further, In 1981, about 39% of its total population lived in 'slums' which was the second highest percentage slum population among the 1980s' twelve million-cities in India (after Kanpur). But, as the capital city of the most populated state in India - Uttar Pradesh - Lucknow is also a white-collar city. The percentage of workers in the service sectors has increased, while employment in manufacturing has decreased (see Fig. 14.1).

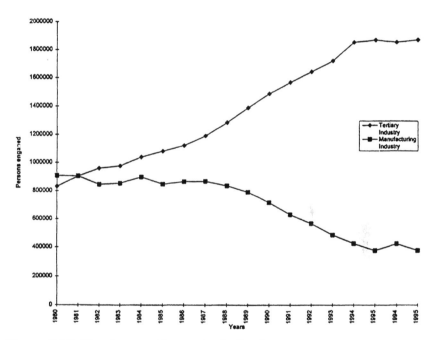

Figure 14.1 Employment by sector in Lucknow

The expansion of the built-up areas in the city mainly occurred along five major axes (roads) that connect the city with neighbouring district headquarters. This meant that land in between the radials was left in agriculture. Thus although Lucknow City's Municipal Corporation limits have been extended three times so far - during 1959, 1972 and 1987 - an analysis of Landsat 5 Thematic Mapper (TM) data of February 1987 reveals that only 36% of the total land area can be categorised as 'built-up area', while 41% land remains as agricultural land. (Subudhi et al., 1990). These estimates, though variable, indicate that physical availability of land for horizontal expansion of the city is not a major problem.

Table 14.1   Demographic, spatial and housing characteristics of Lucknow City (U.A.)

Years	Total population (in millions)	Total number of households*	Total area (in Km²)	Census Houses used as residences*[1]
1961	0.66	117, 714	NA	104, 629
1971	0.81	118, 230	96	108, 180
1981	1.01	182, 535	114	171, 625
1991@	1.66	295, 770	338	270, 571

* *Source:* Census of India, 1961, 1971, and 1981 Uttar Pradesh: 1961, Part IV-A,   Report on Housing and Establishments (Table E-l, p 318); 1971, Part IV, Housing Report and Tables (Table H-l, p 13); and 1981, Part VII, Tables on Houses and Disabled (Table H-l, p 42). @ Census of India, 1991, Primary Census Abstract: General Population. Part li-B(i), Vol 11, p 572.

The land available in Lucknow for residential (or even for other) purposes is mainly appropriated by three land delivery agents, namely, (i) the public sector (state institutions such as Lucknow Development Authority (LDA) and Uttar Pradesh Housing and Development Board (UPHDB); (ii) the private sector (individual households, property agents etc.); and (iii) the co-operative sector (co-operative housing societies). These land delivery agents appropriate land with different motives. For instance, the public sector agencies operate in the city's land market in the name of 'public purposes' that is; to safeguard the interests of 'poor people' and to check misappropriation of land by other agents of land delivery. It is expected to work to the guidelines provided by the State government. In the private sector land is sub-divided by individuals for their own needs or for the requirements of other individuals. Unlike the public sector, there is no organisation in this sector and it does not have State patronage. The co-operative sector operates through various co-operative housing societies which function differently from the other two sectors. Co-operative housing societies are supposed to be functioning on the principle of 'no profit, no loss'. However, there are not many co-operative housing societies which would meet the requirements of this principle. In most cases the officials of a co-operative housing society have vested inter-

---

1   "A 'Census House' is a building or part of a building having a separate main entrance from the road, common courtyard or staircase, etc. used or recognised as a separate unit... If a building has a number of flats or blocks which are independent of one another having separate entrances of their own from the road, common courtyard or staircase, etc. leading to a main gate, they will be considered as separate census house" (Census of India. 1991).

ests (such as escaping from the clutches of the Urban Land Ceiling Act, 1976) in forming the co-operative housing society and then sub-dividing the land. In so doing, these officials are successful in sub-dividing the land which should have gone to the government under the Ceiling Act. Another interesting feature of some of these housing societies is that estate developers buy up land around city peripheral areas in the name of a housing society but at the time of sub-division and/or building houses they function against co-operative societies principles (for details see Amitabh, 1994 and 1997).

Table 14.2 Land appropriation in Lucknow City

Land type I Built-up area within Lucknow Municipal Corporation Urban:
Land type II Remaining area within Lucknow Municipal Corporation Urban:
Land type III Area beyond LMC within Lucknow development Authority: Rural:

Buyers	*Sellers*	*Land type*	*Means of Transaction*
State Sector	Farmers	III	Compulsory
(LDA, UPHDB)	Urban Residents	I/II	Transaction*
	State Institutions	I/II/III	
Private Sector	Farmers	III	Open Market
(Individuals &	Urban Residents	I/II	
Developers)	State Institutions	I/II/III	Purchase
Co-operative Sector	Farmers	I/II	Open Market
	Urban Residents	I/II	Purchase
	State Institutions	I/II/III	

Source: Author's fieldsurvey, 1990/91
Note: *Compulsory acquisition implies that compensation is paid to the sellers, but not always at market prices.

14.2 Access to Land for Housing in Lucknow City

Demand for housing in Lucknow City is responded to in two main ways - first, by supplying only land for housing and, second, by supplying built houses. A third option can be rental accommodation; however, without increasing the stock of housing the rental market stock will not increase rapidly. In practice, then, there are only two ways adequately to meet a rising demand for housing in the long term. The first response to housing demand comes in the form of land supply, particularly for those who demand or intend to buy land in order to build houses on the purchased plot. This mode of housing supply is generally reserved not only for only those who are financially privileged, but also for those who are aware of the networks and linkages that operate in the city's urban land market.

As Table 14.3, makes clear, since land is available especially to high and middle income groups (HIGs and MIGs), it is extremely difficult for lower income groups and the economic weaker sections (LIGs and EWS) to buy land and then to construct their houses.[2] Under the State's sites and services scheme some land with certain basic urban services is sold to a few LIG and EWS people. This is an exceptional situation by means of which EWS and LIG people can be counted as clients of the city's urban land market. Even so, the scheme is not liked by most EWS and LIG people in the city. There are examples in the city, such as on Hardoi Road and in Vikas Nagar, where such a scheme has been implemented, but because of the spatial location of the scheme the take-up rate of beneficiaries is low. There are, indeed, mixed responses to the possible success of sites and services schemes in different parts of the country. Families of the Dadu Majra site in Chandigarh reported that the infrastructure facilities were adequate (Datta, 1987: 39). The Dakshinpuri Resettlement Scheme in Delhi has also been judged a success, while the Arumbakkam project in Madras yielded mixed results (Bhattacharya,

---

2   *The possibility of squatting in this case may exist. However, I do not believe that in Lucknow people belonging to the LIG and EWS categories prefer squatting to renting. My contention is (after talking to more than 25 squatters) that squatting is an action resorted to by those who do not have a steady daily source of income. This absence of a steady income restricts the social reputations and connections of many squatters. Thus squatters are not usually people who belong to the LIG and EWS categories. Such people only enjoy this Government categorisation upon the presentation of a certificate of steady work. The LDA, in reports submitted to the World Bank in 1983, mentioned that in Lucknow there are about 5,000 families squatting, mainly along the G.H. Canal and railway tracks. If one talks to these people it becomes obvious that they are entirely different from people in the EWS and LIG categories, in the sense that the squatters do not have a constant income. Until now, none of the State policies have dared to house these people, for the State understands that by doing so it will not solve the problem but will increase the problem. It is because of this type of inherent demand by the squatters, and also the economic constraints of the State policy (not to be able to give them houses free of cost), that the squatter settlements perpetuate themselves and gradually grow over time. Once they attain a reasonable size they attract the attention of political bosses who in return for favours expect their votes at the time of elections. The patronage of political bosses can also result in violations of urban development regulations.*

On the other hand, LIG and EWS people, even if they have to rent in the most unhealthy and crowded conditions, prefer renting to squatting. It is mainly because of their steady incomes and social status. These people are afraid to lose the social status to which they are entitled. Squatter settlements thus arise in response to needs which are very different from those that lead to the growth of the new colonies. Squatters do not usually discriminate between what is perceived as a less disturbed physical environment and an environment that is constantly invaded by the presence of heavy transport and industrial pollutants, for example. Since private and other developers do discriminate between these environments, they go in for areas that can provide locations that are seen as prime value residential areas.

Table 14.3 **Modes of land/house acquisition and plot sizes according to different income groups in Lucknow City**

Income Groups	Sources of Plot Acquisiton	Sources of Newly-Built Houses	*Approximate Plot Sizes in different sectors (m²)* *(Pub=public sector; Co-op = co-operative sector)*					
			Pub	Co-op	Private			
					(1)	(2)	(3)	
HIG Rs > 24000 p.a	I, NL, IA, PD, SHS	(i) SHS (L.S.), (ii) D & C (L.S.)	150-250	<500	273 to 280 Type – A	417	<500	
MIG Rs 12000-24000 p.a.	I, NL, IA, PD, SHS	(i) SHS (L.S.) & long 'Q', (ii) D&C (L.S.)	115-150	< 500	168 Type-B	292	<500	
LIG Rs 7500-12000 p.a.	Extremely difficult*	SHS (L.S. & long 'Q'), (ii) D&C (L.S.)	75-100	-	128 Type-C	188	-	
EWS < Rs 7500 p.a.	Extremely difficult*	(i)SHS (L.S. & long 'Q'), (ii) D&C (L.S.)	35-55	-	92 Type-D	-	-	

* Exception - sites and services scheme,
** squatting mav be an alternative (see the text),
*** plot sizes vary over time and according to schemes.

Notes: (I) In the Private sector column Cols. I and 2 show plot sizes of private developers (such as Udyan - ELDECO - (and South City - Unitech), while Col. 3 refers to the size of individual appropriatlon.

(2) Income limits according to government criteria.

(3)I = Inheritance; NL = Nazul Land (basically State land); IA = Individual acquaintance; PD = Property and estate dealers; SHS = State housing schemes; D & C = Developers and colonisers, L.S. = Limited supply

Table 14.4 **Plot size and income of the sampled households according to the sectors of land delivery**

Sectors	Mean Plot Size (m²)	Mean Respondent's annual income Rs/-	Mean Family's Total annual income Rs/-
Private	223	39600	52380
Public	269	49692	77292
Co-operative	249	39456	58296

Source: Author's questionnaire survey, 1990/91.

Note: The total number of sampled households in the private sector is 220, while 164 and 137 households respectively belong to the public and co-operative sectors.

1981: 94).

In order to meet a rising demand for houses one alternative to land supply is the supply of housing units directly; that is, land coupled with construction. This option is available to each income group, but there are only two major suppliers in this regard - the State agencies and those private developers/colonisers who have been entrusted with this job by the State agencies (Table 14.3). The main thrust of State housing policy is for the public sector itself to engage in constructing as many houses as it can, especially for the EWS and LIG people. State housing policy also aims to provide, in theory, housing units to socially depressed and unprivileged classes, including those who are engaged in menial jobs scavengers (sweepers), weavers, rug-pickers, workers in raw leather jobs etc.[3] The State definition of High Income starts at quite a low middle-class level. True affluence starts higher up the earnings scale.

Table 14. 4 shows that income of more than Rs 72,000 per year is required to buy a plot from the public sector and more than Rs 52,000 per month to get a plot from the private and co-operative sectors. Since the average incomes of the people who bought land in Lucknow City (as covered by my field research) is 3 to 4 times greater than the defining income bands of HIG and MIG peoples as decided by the State, it is evident that the land market is open only to the 'affluent income group', while the housing market is available to both the affluent and middle income groups.

---

3 See *U.P. State Housing Policy, Government of Uttar Pradesh, 1989.* See also the policy statement as stated in the Seventh Five Year Plan. In practice, people engaged in such menial occupations find it next to impossible to be classified as belonging to the EWS category, and thus to qualify for State housing schemes. A de facto income qualification seems to be imposed by the State, at least in Uttar Pradesh. Compare this to the rhetoric of the Seventh Five Year Plan (1985-1990) which committed the government to the 'provision of houses to the poorer sections of the society in particular and through developing the necessary delivery system in the form of a housing finance and taking steps to make developed land available at right places and at reasonable prices in general by the efforts of Public sector. Not only this, all sectors of the economy - the Government sector, the public enterprise sector, the private corporate sector, the co-operative sector and the household sector - would have to participate in housing activities in a coordinated manner. And the major responsibility for house construction would have to be left to the private sector, in particular the household sector. It has been emphasised in the Seventh Plan that now the time has come for the Government to set before itself a clear goal in the field of housing and launch a major housing effort: not so much to build but to promote housing activity through the supply of fiscal and financial infrastructure such that every family will be provided with adequate shelter with definite time horizon'(Seventh Five Year Plan, 1985-90, Government of India).

14.3 Major Shifts in the Urban Land Markets in Lucknow City

The type and nature of urban land markets in Lucknow City have changed significantly over the past twenty years. The balance of land appropriations has shifted from an individual level to institutional and group-based levels. Prior to the development of State institutions exclusively responsible for housing and urban developments in the city, land appropriations meeting a local demand for housing were carried out mainly by individuals and by groups of individuals intent on sub-division.[4] With the enactment of the Uttar Pradesh Housing and Urban Development Act of 1965 and the Uttar Pradesh Planning and Development Act of 1973, a new institutionalisation of urban land market processes started in the major cities of U.P. During the mid-1970s urban development institutions such as the LDA and UPHDB began a drive for 'controlled' and 'balanced' housing and urban development in the city. The full-fledged development of State, private and co-operative sector urban land activities happened at the instigation, on the one hand, of the pressing need for prime urban housing land, and, on the other, of the interest in capital returns on the part of these investors. One way to describe this situation is to see it as a conflict of interests which has shifted the emphasis of land appropriation processes from the individual to the institutional level. This sometimes represents a disadvantageous situation for the city residents at large, especially in terms of the prices that city dwellers pay. This is not to discount the value of the enormous improvements in urban infrastructure that this shift has brought about: the important question though is, 'at what cost'?'

14.4 Land Delivery Agents in Lucknow City

Private Sector

The private sector presents itself as the voice of enterprise, competitiveness, speculation, vigorous economic activity and aspiration. The State per se recognises the private sector as representing the creative potential of

---

4 *Individual planning should not be assumed to be a process generating illegal subdivisions. Sub-divisions of individually planned areas are largely legal. Illegalities occur where there are violations of building bye-laws and other urban development rules. This point was a matter of serious concern in the 10th National Convention on Architecture and Legislation that was held at Lucknow on 8-9 February 1991. During this convention it was emphasised that the complicated building bye-laws and regulations require urgent modification in a way that might provide a comprehensive and clearly spelt out set of building bye-laws (Akhtar, S.M., 1991). To find a summarised copy of building bye-laws that sets out regulations for subdivisions in new areas in Lucknow City, see Amitabh, 1994a (pp 259).*

the populace and it wishes to enlarge the domestic market by the incorporation of private sector investment. The State reiterates this desire in its own housing and land market policies. But private sector enterprise is motivated and regulated by what can only be described as the profit motive. Unlike the State it has no constituency to defend and benefit but itself. Its natural tendency is towards greater and greater monopolisation where an imperfect market situation exists, and towards greater and greater acquisition of powers (even political power). However, this tendency is baulked by the anti-monopolisation, pro-competition polices of the State. It is to curb the proliferation of the monopolising aspect of individual and corporate private enterprise that the State enacts legislation that prescribes land ceilings. In extreme cases the State even takes over sectors of the market that it considers to be under threat from private monopolisation (as with the nationalisation of India's banks).

Public Sector

According to the Indian Constitution, the state has a mandate from the people and its powers are legitimised by the people. The state has immense resources at its disposal and it is a repository of great power. The Indian state aims to be a facilitator of progress and development. Power comes to the state from the people and this power is supposedly used for the sake of the people. But the path from policy declaration and legislation to actual performance is mediated by several layers of bureaucracy. Even though the State is vested with great powers to effect change, after the perilous passage through the heavy and ponderous bureaucracy the main thrust of its policies often gets dampened down or even perverted. At almost every stage of the process by which policy gets translated into practice there is a danger of sabotage. Thus, even though the State is expected to safeguard the interests of the depressed classes and unprivileged classes, the bureaucrat in charge of dispensing aid may manipulate the distribution of land in such a way that government policies are seemingly satisfied on paper, while in practice they are not. Such officers then become 'Trojan horses' working on behalf of vested interests. The allocation of State resources functions in the manner of a market bargain where different interest groups solicit and demand benefits and favours in competition with each other. Since the majority of the Indian population belongs to one or other of the various categories of the depressed and unprivileged classes, they can (in theory) represent a decisive and powerful voting block in elections. No government with any ambition to be elected back into power can afford to be seen to renege upon promises made to these classes:

opposition parties are quick to make political capital from the slightest in-fringement of the rights of this quiescent but crucial majority. However, since the socio-economic background of the politicians in charge of executing these promises is almost always at odds with that of those whose interests they are expected to guard, the political will necessary to carry out the objectives of important legislation is often lacking. An already lethargic bureaucracy is thus assigned responsibilities that cannot be met without energetic and empa-thetic determination. The public sector's role as the defender of the rights of the depressed classes is too often high on hypocritical rhetoric, and low on performance. It is my contention that the power and resources that are at the disposal of the public sector are grossly disproportionate to the benefits they bring to the constituencies in whose name these resources are used.

Cooperative Sector

The co-operative sector, like the private sector, represents private ini-tiative, but in the form of collective groupings with specific common inter-ests. Since co-operative groups come into being in response to specific local-ised needs they do not represent a threat in terms of appropriating or monopo-lising urban land markets. This explains the high level of patronage enjoyed by this sector from the State. Since it is committed to a 'no profit no loss' policy in its market activities, the co-operative sector is able to deliver land to its members at rates considerably below those demanded by the private and public sectors. Since a co-operative is unified by its commitments to a com-monly agreed objective, and is unhindered by too many bureaucratic proc-esses, it is able to operate at levels of efficiency in the process of sub-division which are bettered only in the private sector. Above all, the co-operative sec-tor is able to offer to individuals a sense of being in control of their own choices and options, since the members of a co-operative decide for them-selves all the details of the land acquired (e.g. proximity to industrial areas, to parklands, to open areas, to work places, etc. and to gaining loans from the financial institutions such as banks and U.P. Co-operative Housing Federa-tion).

It would seem that the co-operative sector partakes of the positive as-pects of both the private and the public sectors; but it is also troubled by some of the problems that beset these two sectors, besides having problems of its own. Like the private sector, it too fares very badly in comparison with the power and resources of the public sector. It also caters to the needs and inter-ests of a small and closed community. But while individuals in the private sector are profit-seekers who are sometimes able to attract a large amount of capital, individuals in the co-operative sector lack similar powers of attraction and their ventures are often under-funded. As with the public sector, the co-operative sector also perceives its role as that of providing urban land for housing at a minimum cost and it is able to do this much more efficiently than both the other sectors. The problem is that, while it is able to provide land at very low prices, it is unable to follow up with a quick delivery of urban services and infrastructure, as per the public sector, because of a lack of power and resources. In theory, a co-operative venture is supposed to admit a common sense of purpose, but being by definition a collection of individuals committed to democratic decision-making, it usually produces a minority of aggrieved members every time a major decision is made. This leads to internal conflict and disputes that can sometimes be self-destructive or lead to a blockage of important proceedings. Once initial targets are met the ties that bind the members of the co-operative often weaken, and this results in the growth of communities unable to voice their grievances in a concerted fashion. Urban services then take a long time to reach them, and even then come usually through the mediation of the LMC. For all of these reasons, one finds co-operative sector colonies ranging from the successful like Ravindra Palli and Vigyanpuri; the middling, like Triveni Nagar and Garhi Peer Khan, to the quite unsuccessful, like Khurram Nagar.

14.5 Land Supply Trends in Lucknow City

Land Supply and Regulation in the 1970s

The sanctioning of the Lucknow Master Plan in January 1970 was the first major step in regulating urban planning and development in Lucknow City. The introduction of zoning regulations and building bye-laws, apart from other restrictions, was a characteristic of the 1970 Master Plan. These also feature in the Master Plan-2001 (of 1987), which is a slightly revised version of the 1970 document. The Master Plan prescribes various conditions (in bye-laws) for the sub-division of plots in new and built-up areas. These bye-laws restrict the sub-division of plots in new areas according to the width of roads. Thus, the width of a service road is to be 7.5 to 9.0 metres: this road is to serve a length of 200 to 300 metres with 100 plots in its length; collector roads are to have a width of 18 or 24 metres, and are to serve an area of 200 plots (which is less than 4 hectares); a service lane is to be permitted in areas where there is no sewerage system, and the width of the lane is to be not less than 3.5 metres and its length is not to exceed 300 metres (for further details, see Amitabh, 1994a Appendix 2, pp 259-268). Another major condition would relate to the minimum size of plots used for the construction of buildings:

> the construction of a building for residential purposes shall not be permitted on any plot which has an area of less than 90 m^2, a width of less than 6 m, or an average depth of less than 12 m, provided that in the case of reconstruction on the site of a building which has fallen down, or has been demolished, or in the case of a site lying in such main commercial and business centres and thickly populated areas in the city as may be specified by the Authority from time to tie, a building shall be allowed to be constructed where plot area is not less than 55 m^2. The Chairman of the LDA shall be empowered to exempt the restrictions of the plot size. For housing schemes put up by public agencies under special housing schemes of EWS, LIG Housing, slum clearance scheme, industrial labour housing, etc., the size of plots shall be relaxed as may be decided by the Authority' (Lucknow Master Plan-200 1 : 3 1).

In the early-seventies, the zoning and building bye-law restrictions of the Master Plan did not seem to affect the private and co-operative sector subdivision. Private and co-operative sub-divisions in this period increased along the four major axes of urban expansion in the city. But the new colonies were soon declared illegal by the urban development authorities; for example, S.G. Puram (north east) (see zone 14 on Map 14.1), Shakti Nagar (north east), Triveni Nagar (north), and many small colonies situated in Singar Nagar, Bhilawan, Geeta Palli (south west) and Thakur Ganj (north west). By the mid-1970s, the UPHDB's growing land acquisition and development activities along the North East (R. S. Misra Nagar) and South West (Talkatora Yojana) axes raised hopes in the private and co-operative sectors of more sub-divisions around the newly developing public sector colonies.

The enactment of the Urban Land Ceiling Act in 1976 was another tool used by the State to control and check the unequal possession of land in the urban land market in Lucknow, as well as in other cities throughout India. The Act stimulated the formation of co-operative housing societies, to prevent the requisition of land above the Ceiling. During the mid and late-1970s the LDA concentrated on the development of the Kanpur Road housing scheme (south west axis) and extended the Aliganj housing scheme further northwards. By all accounts, the Kanpur Road scheme of the LDA was not as well received by the people as was the case with the UPHDB's subdivisions for Indira Nagar colony (along the north east axis). This encouraged the LDA to concentrate its subdivisions along the north and north east axis - the axes which were well received by the people .

By the end of the 1970s it was obvious to the three main land delivery agents in the city that there were three major axes of urban expansion in the

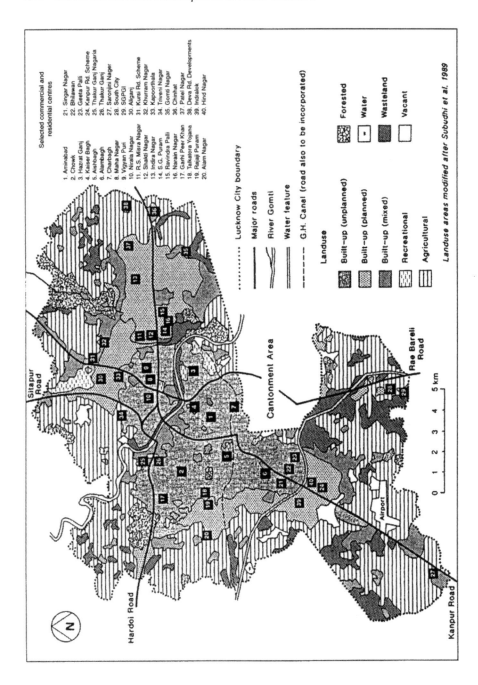

Map 14.1   Lucknow City

city. The north and north east axes urban expansions had hosted successful large scale public sector sub-divisions which encouraged a local congregation of small scale private and co-operative sector sub-divisions. The Rajaji Puram scheme of the UPHDB (located between the south west and north west axes) was also well thought of because of its proximity to the railway station and the railway colonies. The scheme was mainly taken up by railway employees who were renting railway quarters or who lived in the neighbouring old private colonies/mohallas, such as Alambagh, Aishbagh, Charbagh.

Land Supply and Regulation in the 1980s

Having learned some lessons from the mid-1970s housing and land sub-division projects, the LDA put its full weight behind launching a massive housing development project in 1982-83 to the south of the north-east axis road. This was a scheme which is now in the form of a township and is known as Gomti Nagar. For the purposes of Gomti Nagar, the LDA acquired about 1,115 acres land in 1983, which is far more than the LDA's previous land acquisitions. By 1991 the total land acquired for the Gomti Nagar colony was 2,875 acres, which accounts for more than one fourth of the total land (11,041 acres) acquired by the LDA in the city.

Since both the Indira Nagar and Rajaji Puram schemes of the UPHDB were successful in meeting local housing demands, the UPHDB did not design a massive housing project similar to that of Gomti Nagar. However, since the early 1980s the UPHDB has extended Indira Nagar from 339 acres to 1054 acres. The UPHDB also launched another housing scheme in the early 1980s, the Kursi Road Scheme of 339 acres. Hence, the UPHDB, in the early 1980s, acquired more than 1,054 acres land for its two housing schemes, which is equivalent to about 50% of the total land acquired by the LDA until January 1990.

The acquisition of 1,250 acres land by the LDA and UPHDB for three peripheral housing schemes (Gomti Nagar, Indira Nagar Extension and Kursi Road which are located along the north and north east axes) alarmed many private land owners around the periphery of Lucknow City. Many such land owners were farmers who wanted to sell to the private rather than public sector - for more money. because of the land ceiling act and because they needed to buy a single plot for their own housing purposes, unless individuals were willing to establish a corporate body and buy plots under the names of kin relations and friends, it was often not feasible for them to purchase legally a big chunk of land. On the other hand, many farmers wanted to sell large parts of their land-holding, or indeed the whole lot. Zoning laws also blocked some small scale transactions, because the potential to challenge zoning laws is very weak in the case of small scale transactions, as compared to transactions by groups of people. It appeared inevitable that many individuals would withdraw themselves from the market and let the public sector surge ahead.

The large scale land acquisition by the LDA and UPHDB was useful for the co-operative sector because the representatives of the co-operative housing societies could use the State's land acquisition as an effective 'bogeyman'. There was growing resentment among farmers about the non-payment of compensation for lands acquired from them by the LDA. On 6 April 1991 the farmers organized a protest and 'gheraoed' the LDA's premises for four hours. The protest was led by the farmers' leaders, city councillors, and opposition party Members of the Legislative Assembly. Not only did they demand the immediate payment of compensation, but they also made it clear that the compensation must be paid at the market rate. The compensation is supposed to be given at the rate of Rs 0.14 per square foot, while even the Revenue Department's minimum landprice rate for the villages which are within the LMC's limits is set between Rs 13 and Rs 19 per square foot

Thus the LDA and UPHDB found it harder acquire land because many farmers were vigilant and had already sold their land to a co-operative housing society. Once a CHS buys land in an area, the State institutions (LDA/ UPHDB) cannot acquire that land (unless a new enactment is carried out). The CHSs gained from this because farmers would rarely refuse to underreport the prices at the land registration office. Their concern was to earn money from the land and not to worry about signing papers that underreported prices. This understanding between the CHSs and the farmers was profitable to the CHSs in the sense that the CHSs' accounts would not show the unaccounted money paid to the farmers

Besides these changing relationships between the farmers, the State and the CHSs, another kind of land agent - the property agent/dealer -merged in the 1980's, to act as commissioned brokers between vendors and buyers, and as brokers to buy up land in an area in the name of kin and friends. Finally some took steps to form a CHS. Some of the property dealers/agents became so powerful that they developed direct personal links with bureaucrats and urban development ministry officials, and some of them even stood in city council elections.

The LMC's boundary adjustment became an important stimulus of a variety of changes in the city's urban land market between 1987 and 1991. Another event that took place in 1987 was the LDA's invitation to three well-known developers and colonisers in North India to work in Lucknow City. The developers, Ansals, Eldeco and Unitech, were expected to work with the LDA. Between 1987 and 1991 these developers bought a total of 949 acres of land from the LDA. Although these developers were also given opportunities to build commercial complexes (such as the Kapoorthala complex), their major assignment was to develop colonies along the south east and south west axes of urban expansion. In so doing, the LDA aimed to generate both competition in the housing market, and processes of land valorisation in those areas which were seemingly least profitable at the time of appropriation. The Dubey Lands, a private enterprise which was headed by a retired chief engineer of the LDA, started its own sub-divisions under the name of 'Saraswati Puram'.

Along the south west axis (Kanpur Road), the LDA sold land to another developer, Ansals Limited, to build a colony which is known as 'Aashiana'. According to the LDA's guide-lines to these developers, they should devote 40% of their construction for EWS and LIG people, but in reality the developers have shown no sign as yet of fulfilling these guide-lines. In the course of my research I spoke to one of these developers and he told me, off the record, that his first intention was to build houses at a profit and that any housing for poor people would be considered at a later stage, if at all.

Thus, one finds that along the south east and south west axes the public sector (LDA and UPHDB) is keeping away from sub-divisions and building activities, and has given an opportunity instead to a few private and co-operative sector agencies. Basically, what the public sector is doing is sitting back and watching until the market is ready for its intervention. The day the market becomes active because of the catalytic role played by the private and co-operative sectors, the public sector (LDA and UPHDB) will start their antimonopolisation campaigns and chop off the hands of these catalysts. During the Mughal Empire the hands of the builders of the Taj Mahal were similarly (but literally) chopped off in order to prevent another Taj Mahal from being erected. By allowing small scale private developers to flourish (e.g., Ansals, Eldeco and Unitech private companies), the State injects into the land market the stimulus of competition to which they themselves respond by setting up further colonies in close proximity to such successful ventures. While on the one hand valorising the land and making it less affordable to most of the middle and lower income group families, this kind of activity by the State on the other hand stimulates the development of successful urban expansion that places a greater and greater premium on urban services.

14.6 Conclusion

This chapter has focussed upon access to land for housing purposes by different socio-economic groups, and the changing nature of land delivery mechanisms which operate within a system of three supply-agents/interme-diaries - the private, the public and the co-operative sectors. It is evident from the foregoing discussion that the quasi-monopolistic activities of the public sector are predominant in shaping the city's urban land market. The power of the State lies with elected politicians – so politics becomes the pursuit of 'non-market' supply-side economics by other means. Although the putative beneficiaries can sometimes be aided – houses and sites *are* built for the economically weak and poor – this never happens on the scale either prom-ised or legislated for, and farmers at the periphery more often lose than gain from the demand for their land. The wealthier classes are the demand side of the equation: they can make the new colonies pay their way, and their atti-tudes do count. Although the State clearly has some control over the zoning of development, and this has resulted in better development of the urban in-frastructure, the axes which it chooses to develop show sensitivity to middle class demand. On the surface, the State controls development in the name of social justice: but economic and political power by-passes and subverts the legislation to build a city for middle and upper India.

References

Alonso, W. (1964) *Location and Land Use: Toward a General Theory of Land Rent.* Harvard University Press: Cambridge MA.

Amitabh (1994) Residential Land Price Changes in Selected Peripheral Colonies in Lucknow City, India, 1970-1990. *Unpublished DPhil Dissertation*, University of Cambridge, UK.

Amitabh (1994a) Urban Land Price Research and the Utility of Land Registry Data Sources. In Jones, G. and Ward, P. (ed) (1994) *Methodology for Land and Housing Market Analysis.* UCL Press: London.

Amitabh and Rajan Irudaya, S. (1995) Urban Growth and Infrastructure Development in the Indian Context. *Productivity* (Focus: Infrastructure), Vol. 36, No. 2, July September 1995, p 215-227.

Amitabh (1996) Estimation of the Affordability of Land for Housing Purposes in Lucknow City (India): 1970-1990. *Working Paper No. 265*, Centre for Development Studies, Trivandrum - 695 011, India.

Amitabh (1997) *Urban Land markets and Land Price Changes.* Avebury: Aldershot.

Angel, S. (et al.) (1987) The Land and Housing Markets of Bangkok: Strategies for Public Sector Participation. *A Report submitted to PADCO (World Bank)*, 1987.

Angel, S. and Hyman, G. (1972) Urban Transport Expenditures. *Papers and Proceedings of the Regional Science Association*, Vol 29, pp 105-123.

Angel, S.; Archer, R.; Tanphiphat,S.; and Wegelin, E. (1983) *Land for Housing the Poor.* Select Books: Singapore.

Ansal, S. (undated) Non Conventional Role of Private Corporate Sector in Housing & Infrastructural Development. Unpublished Paper.

Baken, R.J. and Van der Linden, J. (1992) *Land Delivery for Low Income Groups in Third World Cities.* Avebury: Aldershot.

Batley, R. (1989) The Management of Urban Land Development in Bangalore. *Papers in the Administration of Development (mimeo)*, No. 32, Development Administration Group,

School of Public Policy, University of Birmingham: Birmingham.

Bertaud, A. (1988) Efficiency in Land Use and Infrastructure Design: An Application of the Bertaud Model. *Discussion Paper*. Infrastructure and Urban Development Report, INU 17 - Report. Policy Planning and Research Staff. World Bank Publication: World Bank.

Bertaud, A. (1992) The Impact of Land-use Regulations and Land Supply and Consumption and Price. *Regional Development Dialogue*, Vol 13, No. 1, 1992, pp 35-43.

Bosque, M. (et al.) (1986) The Dynamics of Land Prices and Spatial Development in Madrid, 1940-1980. *Iberian Studies*, Vol 15, No. 1-2, 1986, pp 49-59.

Brigham, E.F. (1965) The Determinants of Residential Land Values. *Land Economics*, Vol XLI, No. 4, November 1965, pp 325-334.

Census of India, 1951: Vol II, Uttar Pradesh, Part II-A, *General Population Tables*.

Census of India, 1951: Vol II, Uttar Pradesh, Part II-B, *Economic Tables*.

Census of India, 1961, *Uttar Pradesh, District Census Handbook*, 38 - Lucknow District.

Census of India, 1961, *Uttar Pradesh,* Vol XV, Part II-A, General Population Tables.

Census of India, 1971, Series - 21. Uttar Pradesh, *District Census Handbook*, Part X-B, Primary Census Handbook, Lucknow District.

Census of India, 1971, Series - 21, *Uttar Pradesh, District Census Handbook*, Part X-C (Analytical Report and Administrative Statistics & Census Tables), District Lucknow .

Census of India, 1981, Series - 22, Uttar Pradesh, Part II-A, *General Population Tables*.

Census of India, 1981, Series 2", U.P., *Household Tables*, Part - VIII - A & B (iii), Table No. HH-6, Part A (pp 226-7).

Census of India, 1991, *Primary Census Abstract: General Population*. Part II-B(i), Vol II.

Chaudhuri, I. (1984) Land and International Year of Shelter for the Homeless. *Nagarlok*, (IIPA publication, New Delhi), Vol XVI, No. 4 October - December 1984.

Diamond, D.B. Jr. (1980) The Relationship Between Amenities and Urban Land Prices. *Land Economics*, Vol 56, No. 1, February 1980, pp 21-32.

Dowall, D. (1992) The Benefits of Minimal Land Development Regulation. *Habitat International*. Vol 16, No. 4, 1992, pp 15-26.

Dowall, D. and Leaf, M. (1991) The Price of Land for Housing in Jakarta. *Urban Studies*, Vol 28, No. 5, 1991, pp 707-722.

Drewett, R. (1973) The Developers: Decision Processes. In Hall, Peter (et al) (1973) *The Containment of Urban En~land (Volume Two - The Planning System: Objectives. Operations. Impacts)*. George Allen & Unwin Ltd: London.

Evans, A. W. (1985) *Urban Economics: An Introduction*. Basil Blackwell: Oxford.

Evans, A.W. (1983) The Determination of the Price of Land. *Urban Studies*, Vol, 10, No. 2,May 1983,pp 119-129.

Ferguson, B.W. and Hoffman, M.L. (1993) Land Markets and the Effect of Regulation on Formal-sector Development in Urban Indonesia. *Review of Urban & Regional Development Studies*, vol 5, No. 1, pp 51-73.

Fitzwilliam Memorandum (1991) on Land Markets and Land Price Research. *International Journal of Urban and Regional Research*, Vol 15, No.4, December 1991, pp 623- 628.

Gnaneshwar, V. (1986) Land Value Management: The Experience of Andhra Pradesh. *Nagarlok* (IIPA publication, New Delhi), Vol XVIII, No. 2, April - June 1986, pp 71 - 81.

Goodall, B. (1970) Some Effects of Legislation on Land values. *Regional Studies*, Vol 4, No. 1, 1970, pp 11-23.

Haddad, E. (1982) Report on Urban Land Market Research in Sao Paulo, Brazil. In Cullen, M. and Woolery, S. (ed) (1982) *World Congress on Land Policy*. Lexington Books: MA.

Haig, R.M. (1926) Towards an Understanding of the Metropolis. *Quarterly Journal of Economics*, Vol 40, pp 179-'~08.

Harris, C.D. and Ullman, E.L. (1945) The Nature of Cities. Annals, American Academy of Pol. and Sci., Vol 242, pp 7-17. Reprinted in Mayer, H.M. and Kohn. F. (eds(1959) *Reading in Urban Geography*. Chicago University Press: Chicago.

India Today (1993) Plot Sale Advertisement by Unitech Ltd. *India Today* (A Monthly News Magazine). October 31, 1993, pp 63 (International Edition).

Jones, G. (1991) The Impact of Government Intervention upon Land Prices in Latin American Cities: The Case of Puebla, Mexico. Unpublished PhD Dissertation, Cambridge University, Cambridge.

Jones, G. and Ward, P. (ed) (1994) *Methodology for Land Market and Housing Analysis.* UCL Press: London.

Kirwan, R.M. (1987) Fiscal Policy and the Price of Land and Housing in Japan. *Urban Studies*, Vol 24, No. 4, October 1987, pp 345-360.

Kumar, S. (1989) How Poorer Groups Find Accommodation in Third World Cities: A Guide to Literature. *Environment and Urbanisation*, Vol. 1, No. 2, October 1989, pp 71-85.

Lowder, S. (1993) The Limitation of Planned Land Development for Low-income Housing in Third World Cities. *Urban Studies*, Vol 30, No. 7, August 1993, pp 1241-55.

Lucknow Master Plan (1969) Town and Country Planning Department, Uttar Pradesh.

Lucknow Master Plan-2001, (Revised Draft), Town and Country Planning Department, Uttar Pradesh.

Mitra, B. (1990) Land Supply for Low-Income Housing in Delhi. In Baross, P. and Van der Linden, J. (ed) (1990) *The Transformation of Land Supply Systems in Third World Cities.* Avebury: Aldershot.

Muth, R.F. (1969) *Cities and Housing.* Chicago University Press: Chicago.

Ravindra, A. (1996) *Urban Land Policy: Study of Metropolitan City.* Concept: New Delhi.

Richardson, H.W. (1977) *The Urban Economics and Alternatives.* Pion Ltd.: London.

Sah, J.K. (1967) Land Policies for Urban and Regional Development in the Countries of ECAFE Region. *Urban and Rural Planning Thought*, Vol X, No. 1-2, January-June 1967, pp 3-37.

Singh, B.N. (1983) Research Model for Urban and Infrastructure, Pricing, Costing and Design: A Case Study of U.P., India. Cited as an Appendix (pp 69-94) in, Bertaud. A. (1988) *Efficiency in Land Use and Infrastructure Design: An Application of the Bertaud Model.* Discussion Paper, Infrastructure and Urban Development Report, INU 17 - Report. Policy Planning and Research Staff. World Bank Publication: World Bank.

Subudhi, A.P.; Sharma, N.D. and Mishra, D. (1989) Use of Landsat Thematic Mapper for Urban Land Use/Land Cover Mapping (A Case Study of Lucknow and its Environs). *Photonirvachak (Journal of the Indian Society of Remote Sensing)*, Vol 17, No. 3, 1989, pp 85-99.

Task Force on Planning and Urban Development (1983) Planning Commission, Government of India, 25 January 1983. *Journal of Indian School of Political Economy*, Vol III, No. 1, January - March 1991, pp 145-162.

TCPO (1984) (Town and Country Planning) A Study of Urban Land Prices in India. *(mimeo)* Ministry Of Urban Development, Government of India, New Delhi.

The LDA (undated) Synthesis. A Publication of the Lucknow Development Authority (LDA), Lucknow.

The Uttar Pradesh Urban Planning and Development Act, 1973, (with Short notes). Eastern Book Company: Lucknow.

U.P. State Housing Policy (1989) Uttar Pradesh Government, 1989.

UNCHS (1982) United Nations Centre for Human Settlements (Habitat). Survey of Slum and Squatter Settlement. Tycooly International: Dublin.

Wadhwa, K. (1983b) *Urban Fringe Land Market.* Concept: New Delhi.

Ward, P. (1989) Land Values and Valorisation Process in Latin American Cities: A Research Agenda. *Bulletin of Latin American Research*, Vol 8, No. 1, 1989, pp 47 - 66.

Ward, P. Jimenez, E. and Jones, G. (1993) Residential Land Price Changes in Mexican Cities and the Affordability of Land for Low-income Groups. *Urban Studies*, Vol 30, No. 9, November 1993, pp 1521-42.

15 Private Residential Developers and the Spatial Structure of Jabotabek

Haryo Winarso

Editors' note:
Ja- (Jakarta) bo (Bogor) ta (Tangerang) and bek (Bekasi) is a conurbation which the UN projects will have a population of 21 million by 2015. Only three other Asian cities – Tokyo (28 million), Mumbai (Bombay) (27 million), and Shanghai (23 million) will be bigger. The policies for land allocation here have been different from those in Shanghai and Lucknow – they favour the private sector and they have had different results – during sequences of booms and busts.

15.1 Introduction

The word Jabotabek has been coined to signify the urban agglomeration centred on Jakarta which has also engulfed nearby Bogor, Tangerang and Bekasi. In this Indonesian megaloplois, private developers have sold twenty five thousand units of houses annually[1] for the last twenty years and transformed 16.6 thousand hectares of land[2] into residential areas.

This chapter discusses the role of private developer in land development process in the Jabotabek area. First, it explains the process of urbanisation and reconstruction in the region. Second, it describes the involvement of private sector land developers in the development of Jabotabek, whose activities have significantly affected the recent restructuring process. Lastly, it assesses the recent real estate boom arguing that there are significant problems that might arise from the activities of the private developers in the region.

15.2 Jakarta and Jabotabek: Urbanisation and Spatial Reconstruction

It was not until the 1970s that Jabotabek earned its name, underlining the cycle of restructuring by means of which the tiny city of Jayakarta experienced one of the most astonishing periods of city growth in the Third World. This cycle was supported by foreign investments in the area as spurred

---

1 Calculated from Central Bureau of Statistics data presented in the Indonesian Team (1995)
2 Calculated from BPN data 1997

by government policy.[3] This was followed by an oil boom[4] which encouraged the development of infrastructure and laid down the basic shape of the Jabotabek of today.

Jakarta, the core city of Jabotabek region and originally was called Sunda Kelapa, was founded in the 1300s as a small trading port at the sanctuary of Ciliwung river. After the triumph over the Portuguese in 1527 by a Sundanese ruler, Sunda Kelapa was named Jayakarta, which means 'the victory'. Jayakarta was renamed Batavia by the Dutch in 1618, who expanded the city as one of the VOC - *Vereenigde Oost Indische Compagnie* (The [Dutch] United East Indies Company) base ports in Indonesia.

Over the next century Batavia grew into a city with a population of around 500,000. Inter-city rail lines connected Batavia with Tangerang in the West, Serpong and the Sunda straits in Southwest, Bogor and Bandung in the South and Bekasi and Cirebon in the East. By then, the expansion of Batavia had started and the demand for land for residential purposes was apparent as Karsten[5] noted:

> ... The rapid expansion of the Town naturally led the Europeans to buy more and more land, preferably along the existing highways. This was to large extent not farm and nature land....... but land where there already were Kampungs (in Wertheim, 1958).

During the Japanese Occupation (1942-1945) a plan for new satellite town in Kebayoran at the Southwest of Batavia was completed by the last remaining Dutch Administration. The new satellite town was designed to accommodate 100,000 inhabitants in an area of 730 hectares. The development, however, was abandoned because of the Second World War.

In 1950 when the Republic of Indonesia was fully recognised internationally, Batavia was renamed Jakarta and was maintained as the capital of the new nation. Between 1945 and 1965, Indonesia experienced a transformation from a society under colonialism to that of a free nation. This transformation was not an easy process. The economy of the country experienced

---

3 *In 1973, the government of Indonesia experiencing a massive windfall of revenue from oil. The Real GDP increased at an annual average rate of 7.7 percent. An Investment Co-ordinating Board was set up in this year (see Hill, 1996)*

4 *The oil boom (for oil-exporting countries) was triggered by the hostilities that broke out between Israel, Egypt and Syria in the end of 1973. This event had pushed the oil price from $3 per barrel to over $5 per barrel within one month, and by three months the crude oil price had reached $12 per barrel. This oil boom eventually had enormous impact on total government revenue (Winters, 1991).*

5 *Ir. Thomas Karsten was the architect of the first Indonesian Town Planning Act and the principal planner for several Indonesian cities including Batavia*

only modest progress in the early years of independence, but it became a nightmare at the beginning of 1960s, and so did the social conditions. Foreign and domestic investment had decreased because of the anti-capitalist policy of the Sukarno administration. Finally 1965 saw the collapse of the Sukarno Government in a very violent upheaval.

In the 1950s the population had already reached 1.5 million, more than double that of 1945. By the year 1961, Jakarta's population had reached 2.9 million, which made Jakarta one of the largest cities in the world. Most of its population were living in dense kampung (from which the English word 'compound' is derived) areas with bad infrastructure conditions. Public transportation systems were largely neglected. The first Master Plan for Jakarta prepared in 1952, which envisaged an urban area with a ring road as the limit and surrounded by a green belt following the principles of Ebenezer Howard's Garden city, had never been realised. However, the next 35 years witnessed rapid growth and structural change in the city. As the economy grew, Jakarta emerged as the centre for the development of the country. From 1961 to 1971, Jakarta's urban population had almost doubled, from 2.9 million to 4.6 million with an annual growth rate of 5.8 percent. This was the fastest urban population growth in the country and the fastest in Jakarta's history (see Table 15.1). Furthermore, it became obvious that, combined with the industrialisation process in the region, urbanisation was expanding beyond the administrative boundaries to include the adjoining *Kabupatens.* Thus in 1967 the second Master Plan, this one for the period of 1965 - 1985, was introduced to deal with the massive new development in the area.

During this period Kebayoran Baru had arisen as an area for middle and high income residence in Jakarta. This was followed by several other

Table 15.1 Population in Jabotabek ('000's)

Region	Population				Annual growth %		
	1961	*1971*	*1980*	*1990*	*61-71*	*71-80*	*80-90*
Jakarta	2905	4579	6503	8254	5.8	4.7	3.0
Bogor	1303	1662	2494	3736	2.8	5.6	5.5
Bogor Municipality	146	196	247	272	3.4	2.9	1.1
Tangerang	848	1067	1529	2765	2.6	4.8	9.0
Bekasi	690	831	1144	2104	2.0	4.2	9.3
Botabek	2987	3756	5414	8877	2.6	4.9	7.1
JABOTABEK	5892	8335	11917	17131	4.1	4.8	4.9

Note: Botabek is the summation of Bogor, Tangerang and Bekasi.
Source: Calculated from the Indonesian Team (1995).

residential developments, but the most prominent was in Slipi which marked the first housing area in Indonesia developed by a real estate company. It was created by the Government owned company PT (*Perseroan Terbatas* means Limited Company) Pembangunan Jaya under the chairmanship of Ir. Ciputra[6] (Properti Indonesia, February 1994: 10). The first residential premises built by a foreign company in this period was perhaps Pertamina's[7] housing complex which was developed by *Tosho Sangyo*, a Japanese company in the late 1960s (Properti Indonesia, February 1994: 16).

15.3 The Involvement of Private Developers in the Development of Jabotabek, 1970 and 1980s

In early 1970s a new cycle in the restructuring process was started by the institutionalisation of housing production and the involvement of the formal private sector in residential development to accommodate the ever increasing demand for housing. The economic condition, which was pictured as a "sustained economic growth" by Hill (1996) brought a new era in the physical development of Jakarta and later in the Jabotabek area. The average economic growth rate of 7.7 per annum had an impact on the development of the formal private sector. The growing need for offices, combined with a government policy to develop commercial areas encouraged the participation of the private sector. The first private sector involvement in commercial property development was perhaps PT. Metropolitan Development, founded by Ir. Ciputra in 1971 (Properti Indonesia, February 1994:10), which constructed several office buildings in Jakarta.

Almost at the same time, Ciputra, together with the wealthy Indonesian Lim Sioe Liong,[8] established a company named PT Metropolitan Kencana and transformed 720 hectares of rubber plantation at the south of Kebayoran Baru into a residential area, named Pondok Indah. This area became the first large planned residential area for middle and high income groups

---

6 *Ciputra became the most prominent person in land development in Indonesia. He was the president of the International Real Estate Federation (FIABCI) and owned several large real estate companies in Indonesia*

7 *Pertamina is the only state Oil Company which had an important role during the first decade of the Suharto's administration. According to Winters (1991) Ibnu Sutowo, the president of the company, acted as the president's political financier. In interview with Winters, Sutowo said that : "You can't find a single road or school or hospital that wasn't at least partly funded by the money I borrowed through Pertamina". Winters dissertation on the political economy in Indonesia gives good picture of the role of Pertamina at that time.*

8 *For the discussion on the emergence of large and dynamic private sector in Indonesia , see for example Hill (1996). Hill's book also provides a list of 25 major business conglomerates in Indonesia in which he put Lim Sioe Liong at the first rank.*

in Indonesia to be developed privately. Its success caused land prices in the area to soar and made Pondok Indah one of the exclusive areas of Jakarta. Following Ciputra, several other small and large real estate developers began to emerge. In 1972, to further strengthen their lobbying power, private developers established an association called Real Estate Indonesia (REI), the industry's official trade organisation with Ir. Ciputra as the first President. At the beginning only 33 developers were registered as members. (REI, 1997a)

In 1973, a National Housing Development Corporation (Perumnas) was set up to provide housing, primarily, for low income people. To support the policy, a financial institution which could provide mortgage finance was founded. This was the State Saving Bank (BTN), created to provide mortgages for low-middle income groups. One of the first large projects undertaken by the National Housing Development Corporation (Perumnas) was the low income housing in Depok in an area of more than 400 hectares (Silas, 1995). Later in 1980, a Housing Finance Corporation (PT. Papan Sejahtera) was also established to serve the higher income groups.

The restructuring process occurred not only in Jakarta but also spilled over into adjoining *Kabupatens* as the demand for urban land for housing increased to match the ever expanding population of Jakarta. In the early 1970s it was realised that the Master Plan of Jakarta of 1965 was not viable anymore, and thus a new concept labelled the Jabotabek Development Plan which incorporated the development of the surrounding *Kabupaten* was introduced in 1974 following a report by the Dutch team working for the Ministry of Pubic Works.[9] By 1980, the population of Jabotabek area had reached 11.9 million which made Jakarta the largest metropolis in Southeast Asia. According to Douglass (1991), in 1979 Jakarta and its surrounding area accounted for 42 percent of the total value added and two third of total employment in medium and large-scale manufacturing in Indonesia. By that time the physical development had reached as far as 20 km from the city centre.

At the beginning of 1980s, following the construction of toll roads which connected Jakarta to Tangerang and Merak in the West, and to Bekasi and Bandung in the East, a new trend of residential land development emerged. It undoubtedly marked one important step in the space restructuring of the Jabotabek area. The trend was the development of new towns, pioneered by

---

9 *The Jabotabek Plan was initially developed with the assistance of the Dutch Government in 1970. The planning concept was inspired by the model of "Ransstad" in the Netherlands (Giebels, 1986). The concept includes self contained growth centre. Basically two main models were introduced. One was a concentric model and the other was a linear model . Since then, the plan has been reviewed three times, however there is no significant alteration to the basic concept, except for the prediction of population growth. The last review conducted in 1992 re-stressed the proposed development along an east- west and a north-south axis. The plan, however, is not legally binding..*

Ciputra and 10 other large companies. Together they launched the idea of a new town in Serpong located about 20 km south-west of Jakarta. The proposal was surprisingly unchallenged, despite the fact that it did not conform with the Jabotabek Metropolitan Development Plan of 1980 and the West Java Urban Development Project (WJUDP) finished in 1985.[10] Bumi Serpong Damai (BSD), the new town, was designed to house 600.000 people by the year 2005 in an area of 6000 hectares. This was then followed by other new towns scattered throughout *Kabupaten* Tangerang, Bekasi and Bogor. Thus, by the end of the 1980s more then 10 new towns of 500 hectares or more had been initiated in the area. (Properti Indonesia, June 1995: 28). The concept behind these new towns is more comparable to the US experience rather than British, in the sense that the proposal, site selection and implementation of the new town are undertaken by private developers.[11]

The construction of new towns has increased the production of formal housing[12] by private developers. As illustrated in Table 15.2., by 1987 private sector housing had significantly exceeded that built by Perumnas. These figures clearly show the essential role of private developers in providing formal housing in the region. Thus by the last of the 1980s, the penetration of development into former agriculture farm lands was obvious, first at the periphery of the city and then jumping deeply into the area beyond as far as 30 to 45 km from Jakarta.

The immensity of construction can be seen in the change of agricultural land into built-up areas. According to a recent study (Indonesian Team, 1995), in 1971, the built up area of Jakarta, the core city of *Jabotabek*, had reached 17,878 hectares or 31 percent of its administrative area. This was almost doubled in 1988 for at that time the built up area covered 61 percent of the city (39,734 hectares). The study also revealed that the built-up areas of Jakarta have over-spilled into the immediate *Kabupatens*. Built up areas in *Kabupaten* Tangerang have increased from 14,033 hectares (corresponded to

---

10 *The first Jabotabek Plan was silent on the development of the Serpong area, while the West Java Urban Development Project, financed by the World Bank, which had finished its final report in 1985, only predicted that at best Serpong would be developed as a large-scale dormitory area for middle to high income commuters with only limited local employment and a narrow cross-section of income groups (Douglass, 1991: 227)*

11 *Corden (1977) comparing the new towns development in the US and Britain, notes that a basic difference is the initiator. In Britain the role of the Government is strong, while in America private developers are the engines of new towns construction.*

12 *In Indonesia, as revealed by a housing study, production of urban housing is largely done by popular and professional house builders. Popular are those developed by individuals without reliance upon either government or formal private sector institutions, while the professional are those created by private or government owned companies (Struyk, Hoffman and Katsura, 1990). Formal housing development has to comply with certain building standards set up by the Government.*

Table 15.2 Housing units built by Perumnas and by private firms in Jabotabek

Year	1987*	1988	1989	1990	1991‡	1992	1993	Average/year
Perumnas								
Jakarta	103	4	0	NA	200	2230	0	362
Bogor	885	85	220	NA	212	442	57	271
Tangerang	365	7578	1218	NA	2016	2075	1872	2160
Bekasi	5221	2342	128	NA	549	2349	692	1611
Botabek	6471	10005	1566	NA	2777	4866	2621	4043
Jabotabek	6574	10009	1566	NA	2977	7096	2621	4406
Pr. Firms								
Jakarta	513	0	515	147	1	0	72	178
Bogor	4338	5432	6600	3973	2828	2797	1759	3961
Tangerang	11709	15887	18353	9656	7996	4485	1194	9897
Bekasi	15819	14756	17788	14271	13526	7022	2434	12230
Botabek	31866	36075	42741	27900	24350	14304	5387	26089
Jabotabek	32379	36075	43256	28047	24351	14304	5459	26267

*Note:** Figures for housing built by Perumnas in this year are cumulative from 1985.
  ‡ Figures for housing built by Perumnas in this year are cumulative from 1990.
  All the figures are for housing financed by KPR-BTN (State Mortgage Bank)
Source: Calculated from Indonesian Team (1995).

11 percent of the Kabupaten area) in 1980 to 44,214 hectares (corresponded to 34 percent of the Kabupaten area) in 1992. The same situation also occurred in *Kabupaten* Bekasi and *Kabupaten* Bogor. The constructed area in *Kabupaten* Bogor increased from 33,150 hectares (15 percent of the Kabupaten area) in 1974 to 64,782 hectares (34 percent of the Kabupaten Area) in 1994, while in *Kabupaten* Bekasi it increased from 14,310 hectares (10 percent of the Kabupaten area) in 1980 to 27,378 hectares (18 percent of the Kabupaten area) in 1993. Maps 15.1 and 15.2 show this growth of Jakarta and the location of the new towns.

15.4 The Real Estate Boom and Space Restructuring, 1985 - Present

Within the period 1982 to 1986, Indonesia's economy was pictured as puzzling:

> .. *The macro indicators are encouraging... Yet investment is sluggish,...*"
> (Hill, 1996).

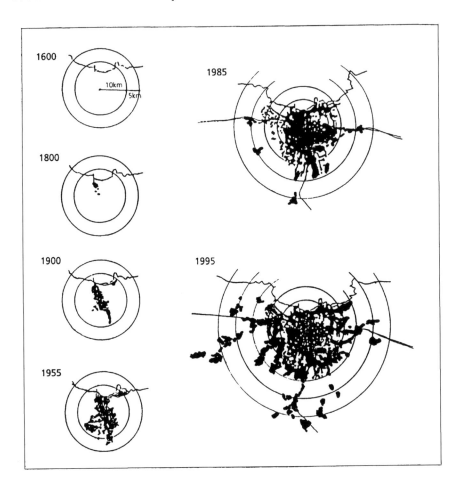

Map 15.1 The growth of Jakarta

The economy continued to grow at an average of 4 percent per year. Hill (1996) labelled this period as "Adjustment to lower oil prices". But from 1987 into the 1990s the economic conditions changed, and the economy grew at an average of 6.7 percent annually. This new cycle of change actually began with a series of deregulation policies during the years 1983 -1988.[13] The policies were aimed at improving domestic savings, improving resource allocation and developing a framework for monetary management in particu-

---

13 *Winters (1991) dissertation provides a good account in this series of deregulation. He puts it under "Jaman Deregulasi" in which he analyses the power-dynamics involved in the deregulation. More detailed analysis of the reform can be seen in Hill (1996), and Booth (1992)*

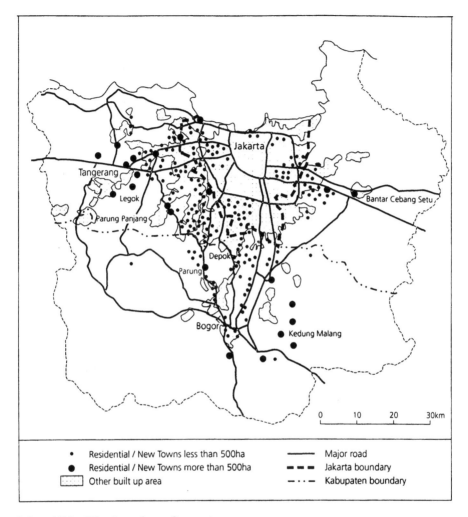

Map 15.2 The location of new towns

lar through indirect intervention rather than direct regulatory control (Hill,
1996). The most important deregulation policy was perhaps the 1988 financial, monetary, and banking reform.[14] This was particularly true for the development of real estate industries. As the Government, by then, allowed the
entry of more foreign banks in the form of joint ventures, the private banks
were enable to offer genuine competition (Hill, 1996:36). The result of the

---

14 *The package was aimed to increase economic growth, non oil export and to expand job*
opportunities. This deregulation was also aimed to encourage mobilisation of funds, efficiency
of banks and non banks institutions, and to develop capital markets (Winters, 1991)

policy was that the banking system expanded and was awash with liquidity that need to be tapped. Private banks started to offer a wide range of products including attractive term deposit rates. Hill (1996) noted that between March 1989 to June 1993, the number of private banks' branches were almost doubled, while the state bank in the same period only expanded 24 percent.

Meanwhile, the Government policy to subsidise the State Mortgage Bank (BTN) and Bank Papan Sejahtera in order to keep their housing loan interest rates low has increased the number of lenders as well as the value of the loans as can be seen in Table 15.3. According to Struyk, Hoffman and Katsura (1990:171), in April 1989 BTN's interest rates for mortgage credit were 9 to 15 percent and they averaged half that of the commercial banks which was at 24 percent:

Thus between 1987 and 1989, the Jabotabek region saw the first boom in the real estate industry, as can be seen in the Figure 15.1. Within three years, more than 111 thousand houses were produced by private developers. Adding it with the housing built by Perumnas, the Property Magazine reported that more than 160 thousand units were sold within a year (Properti Indonesia, November 1994, pp. 24). The realisation of housing loan by BTN

Table 15.3   Realisation of housing loan by Bank Papan Sejahtera ('000,000's) in Indonesia

Year	Lender/unit	Value/Rp.
80	5	4.4
81	184	181.7
82	376	423.1
83	649	917.2
84	1034	1684.6
85	535	931.8
86	1474	2442.7
87	4195	6342.7
88	4853	7445.9
89	4373	7072.1
90	4194	7267.0
91	3101	6435.5
92	1608	3853.4
93	2849	8006.9
94*	5525	9699.9

Note: * 1994 is estimated figure.

Source: Data Bidang Perumahan dan Permukiman (Housing and Settlements Data), Ministry of Public Housing, April 1994: 42.

and Bank Papan Sejahtera show the boom of housing construction in Indonesia (Figure 8). The first boom reached its peak in 1989.

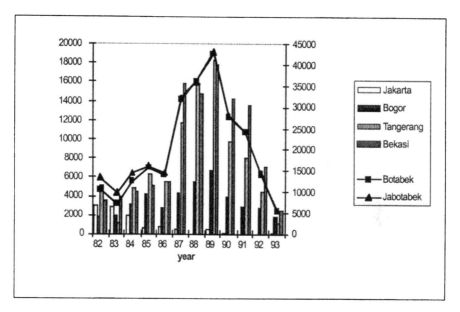

Figure 15.1 Realisation of housing construction financed by KPR-BTN in Jabotabek

Note: BTN (Bank Tabungan Negara)= State Mortgage Bank;
KPR (Kredit Pemilikan Rumah) = Housing Loan.
Source: Calculated from Housing Construction Statistic in Indonesia quoted by the Indonesian Team (1995).

The series of economic reform policies stimulated Indonesia's industrialisation progress and generated employment (Chia as quoted by Firman, 1994: 204). However this progress was spatially unbalanced, being concentrated in Jabotabek Region. Jakarta and west Java alone, in 1990, contributed one third of the nation's non-oil GDP (Hill, 1996: 226). The agglomeration of economies in the region increased the demand for urban land as shown by the land use changes during 1990 - 1994 (Table 15.4).

It is particularly interesting to note that the infertile land (unproductive) increased by 1600 percent which partly indicates the increasing abandonment of land for speculation.[15] The land for settlements (indicated as "Village") also increased significantly, from 70 thousand hectares in 1980 to 118 thousand hectares in 1994, a growth of 68 percent. Given the quality of

15 *The fact that only 14 percent of land owned by developer under location permit system (see table 5.7) already developed explains why the infertile land increased remarkably.*

Table 15.4   Land use change in the Botabek area (in hectare)

Land Use	1980	1994	increase	%	decrease	%
Village	70794	118842	48048	68	-	-
Wet Rice Field	246805	213594	-	-	-33211	13
Dry Agricultural Field	31719	33516	1797	6	-	-
Mixture Garden	125880	114314	-	-	-11566	9
Plantation	44185	28277	-	-	-15908	36
Forest	74117	69201	-	-	-4916	7
Underbrush, meadow	6485	5582	-	-	-903	14
Embankment	16225	18700	2475	15	-	-
Lake, Marsh	860	860	-	-	-	-
Infertile land	90	1600	1510	1678	-	-
Others	3530	15404	11874	336	-	-
Total*	617160	619890	65704	11	-66504	11

Note: * The figure for 1980 is different from the figure in 1994. To some degree, this shows the inaccuracy of data; nonetheless, the table provides a picture of the land use change.
Source: Calculated from BPN (National Land Agency) data as quoted by the Indonesian Team (1995: 26).

data and the unclear definition of village settlements,[16] the figure can also indicate the increasing residential land uses. On the other hand, wet rice field declined, from 246 thousands to become 213 thousands hectares, a reduction of 13 percent within 5 years.

The combination of accelerated population growth, deregulation policy and the increasing involvement of formal private sector in land development had an immediate impact on land price in Jabotabek area. Thus, by the late 1980s, the land prices in central areas had sky rocketed. As can be seen from Figure 5, the price of land doubled and even tripled within a single year, an experience which may only be compared with the escalating land prices in Tokyo in the early 1970s[17]. Concentration of land in the hands of a few large developers was immediately apparent. According to Leaf (1991) in Jakarta between 1974 and 1989, from 325 location permits[18] granted to 183 names,

---

16 *Most of new towns and real estate developments were in Kabupaten area, which in the Indonesian administration system, is considered as rural area. Accordingly, land for new towns and real estate is considered as village, regardless their quality.*
17 *In 1970, as noted by Mike Douglass (1993), Tokyo experienced the first escalation in land price caused by the a combination of population growth and the increasing command of large scale corporation over the economy.*
18 *Location permit is permit granted to formal private developers which would enable them to acquire and title land for development (see Struyk, Hoffman and Katsura, 1990; Leaf 1991; Ferguson and Hoffman for further discussion of location permit).*

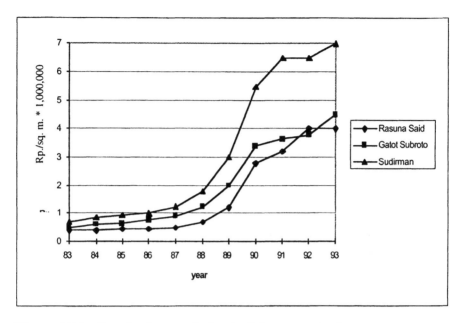

Figure 15.2   Land price escalation in Jakarta central area
Note: Unadjusted to inflation. Inflation at an average of 6.4 percent per annum
Source: Panangian Simanungkalit and Associate Properti Indonesia, May 1994

more than half of the land under permit was actually controlled by 16 (8.7 percent) developers.

Several "new" players had obviously emerged which can be seen from the REI membership. Between 1983 and 1990 its membership grew from 261 to 900 companies. By its 25th birthday in 1997, REI membership had growing remarkably to reach the figure of 2,400 developers as can be seen from Figure 15.3 below (REI, 1997b). Land speculation and concentration of ownership were apparent, for by the end 1996 it was recorded that in Botabek there were 15 companies holding land over 1, 000 hectares. (Table 15.5)

In the early 1990s, as the inflation rate approached 10 percent, the Government applied tighter monetary policy (Hill, 1996) which resulted in loan interest increasing up to 25 - 26 percent for the commercial Bank and 23 percent for BTN. The first boom ended the economy seen to be overheating. By 1993 housing production in Jabotabek area was at the lowest level since 1983, although national housing production had increased, already starting the second boom (see Figures 15.4, 15.6).

By 1993 the medium and large housing production in Indonesia had dramatically shifted into a boom, as can be seen from Figure 15.5. This second boom came partly as the result of the government's new approach to housing finance and construction permit procedures as part of deregulation policy to encourage more private participation in land development

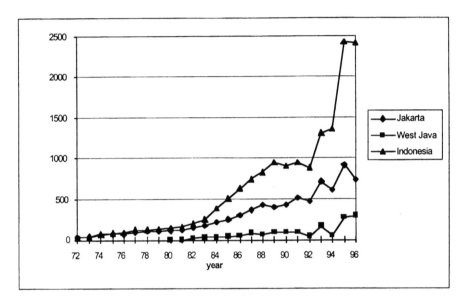

**Figure 15.3 REI Membership in Jakarta, West Java and Indonesia
      1972-1996**
Note: Kabupaten Bogor, Tangerang and Bekasi are under jurisdiction of West Java Province.
Source: REI membership list, REI 1997b.

The deregulation in 1992 allowed 100 percent foreign investment with a minimum of only two millions US dollars to invest in Indonesia, as the result of the more liberal economic policy (Hill, 1996: 76-77; Firman, 1995: 210).The new regulation also simplified the procedure for obtaining location permits, and for satisfying the requirements of environmental impact analysis and public nuisance avoidance (Firman, 1995). The economic reform policies also accelerated the domestic investment, and as Hill (1996) noted, the domestic conglomerates became a significant commercial force. Investment by domestic as well as foreign investment rose strongly, with the involvement of foreign firms in hotels, real estate and commercial services leading the way. Thus, as noted by Data Consult (1996), by early 1996, 11 companies were approved to invest in housing development projects by BKPM (Badan Koordinasi Penanaman Modal - Investment Co-ordinating Board). One of

Table 15.5   Land holdings over 1,000 hectares by companies in the Botabek area

Company	Area (hectare)	Location
1. Citra Raya (Si Pengembang)	1,000	Tangerang
2. Gading Serpong	1,000	Tangerang
3. Pantai Indah Kapuk (Si Pengembang)	1,000	Tangerang
4. Modernland (Modern Group)	1,570	Tangerang
5. Bintaro Jaya (Si Pengembang)	1,700	Tangerang
6. Puri Jaya (Si Pengembang)	1,750	Tangerang
7. Kota Legenda (Putra Alvita Pratama Group)	2,000	Bekasi
8. Royal Sentul	2,000	Bogor
9. Kapuk Naga Indah (Si Pengembang)	2,000	Tangerang
10. Kota Tiga Raksa (PWS Group)	3,000	Tangerang
11. Kota Tenjo (BHS Group)	3,000	Tangerang
12. Lippo Cikarang (Lippo Group)	3,000	Bekasi
13. Cikarang Kota Baru (Grahabuana Group)	5,400	Bekasi
14. Teluk Naga (Si Pengembang)	8,000	Tangerang
15. Bukit Jonggol Asri	30,000	Bogor

Source: Warta Ekonomi, December. 1996, Properti Indonesia, April 1997

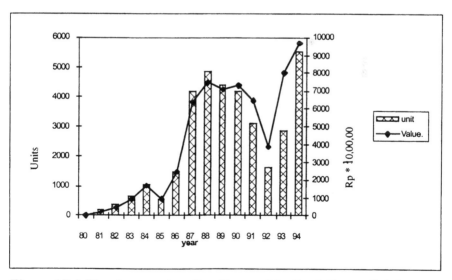

Figure 15.4 Housing construction financed by Bank Papan Sejahtera (PTPS) in Indonesia, 1980-1994

Note: PTPS = PT Papan Sejahtera (Housing Finance Corporation).
Source: Ministry of Public Housing, 1994.

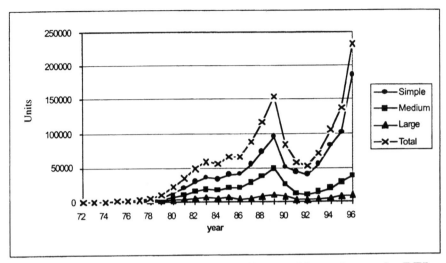

**Figure 15.5 Construction of simple, medium and large houses by REI
in Indonesia, 1972-1996**

Source: Calculated from REI, 1997c.

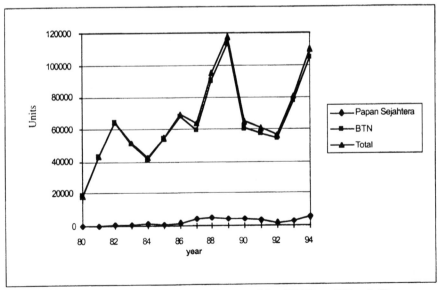

**Figure 15.6 Housing construction financed by BTN and Bank Papan
Sejahtera in Indonesia 1980-1994**

Note: BTN = State Mortgage Bank
PTPS = PT Papan Sejahtera (Housing Finance Corporation)
Source: Ministry of Public Housing, 1994

the biggest was the project by PT Pan Malayan Development, which proposed an investment of US $ 50 million for houses in Bogor.

Table 15.6   Bank lending to property market 1993-1996 (x Rp. 1,000,000,000) in Indonesia

	Dec. 93	Dec. 94	Dec. 95	Apr. 96	Jul. 96	Dec. 96*
1 Construction	10,038	13,368	15,631	17,480	19,130	21,080
Simple housing	929	1,294	1,590	1,760	2,090	2,150
Other buildings	9,109	12,074	14,041	15,720	17,040	18,930
2. Real Estate	5,513	9,715	13,455	15,380	17,210	18,680
Simple housing	805	1,101	1,670	2,060	2,360	2,540
Other buildings	4,708	8,614	11,185	13,329	14,850	16,140
3. Housing Loan	6,157	10,110	13,694	14,360	14,860	16,385
4 Total property Loan	21,708	33,193	42,180	47,229	51,200	56,145
5 Total National Loan	163,456	204,407	249,367	269,829	278,421	295,500
4/5 percent	13.3	16.2	16.9	17.5	18.4	19.0

*Note: * Estimated Figure.*
Simple Housing: low cost housing, the price of this house is fixed at Rp 12.5 million by the government. (Rp. 1.000.000 approx. US $ 360)
Source: Bank of Indonesia as analysed by Panangian Simanungkalit and Ass., Properti Indonesia, October 1996.

15.5 The Second Boom 1992-1994

The second boom which began in 1992 produced big profits in property development and encouraged banks to lend to this sector. By December 1993, the commercial banks had cut their interest rate down to between 17 and 19 percent for housing mortgages, maturing at 10 to 15 years (Properti Indonesia May, 1994:67), this further encouraged commercial lending for housing development. Thus, within 5 years the proportion of property/real estate loan has increased remarkably, despite a warning from analysts that many of these loans were potentially bad debts.[19]

---

[19] *Newspapers and magazines have reported the potential of bad debt in property sectors. (See for example, Media Indonesia Friday 20 December, 1996; Properti Indonesia , October 1996, Warta Ekonomi 2 December 1996).*

This second boom between 1992 - 1994 was fuelled by the fact that house prices were still increasing, even though the first boom had already ended in 1989. Unverified[20] data from Properti Indonesia (October, 1996:35) shows the continuation in housing prices. The magazine also reports an increase of around 40 percent within one year. This figure, however, is reasonable if we look at the following illustration. The price of type 21/60 houses (21 square meters building on 60 square meters plot) was marketed in 1990 at Rp. 8,500,000 in Depok and Rp. 5,000,000 in BSD.[21] Within six years the price rose sharply, far beyond the inflation rate of around 8 percent annually (BPS, 1995). In 1996, in Depok the price of the same type of house has reached Rp. 44,108,064 (Properti Indonesia, January 1996:108) while in BSD a slightly bigger house[22] (36/72) was marketed at Rp. 48,233,800,[23] which means price increases of 320 percent and 600 percent respectively.

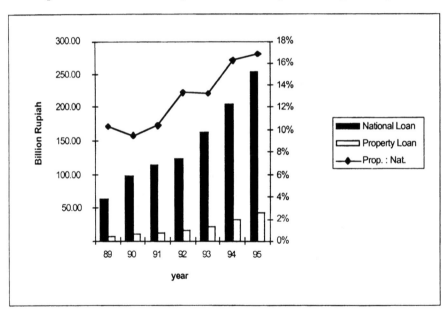

Figure 15.7 Proportion of property lending to national lending in Indonesia, 1989-1995

Source: Adapted from Panangian Simanungkalit and Associate, 1996.

---

20 *Data presented in the magazine was not adjusted to the inflation rate and was not clear on how the data was collected.*

21 *The prices are quoted from Leaf (1991: 187).*

22 *The 21/60 type of houses has not been built in BSD anymore except for the very simple houses as required by the Government, and the price of this type is fixed at Rp. 12,500,000.*

23 *This price is quoted from Price List provided by BSD, printed in 15 November 1996*

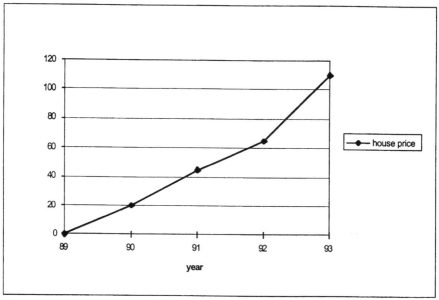

Figure 15.8 Increase in house price in Jabotabek area (percent), 1989-1993

Note: Unadjusted to the inflation. Inflation rate approximately 7 percent annually
Source: Properti Indonesia October 1996.

Mortgage finance also played an important role in the boom as it offered possibilities of magnified capital appreciation as illustrated in the following example: If a house was offered at Rp. 100 million in 1989, a typical buyer might borrow Rp 80 million and put up Rp. 20 million of his/her own money. Due to demand and inflation, by 1992 his/her house would be valued at Rp. 170 million, 70 percent more than it cost (see Figure 12). The buyer had a profit of Rp. 70 million on his/her Rp. 20 million, a capital appreciation of 350 percent, which he/she earned within three years alone. The buyer only had to offset this against relatively small interest charge and depreciation of capital due to inflation.

Essentially, the expectation for a quick return on investment as illustrated above makes the basic difference between the first and the second boom. The first boom was propelled by the Government monetary policy, particularly the banking reforms, while the second boom was largely caused by speculative buying in expectation of the rising housing prices, as experienced in the first boom, and was further pushed by the simplification of land permit granting procedures.

There was a demand for medium houses, as is shown by the realisation of housing loans[24] (See Figure 7 and 8). However, the demand was apparently a reflection of a pseudo-market, stimulated by speculative buying as illustrated in the above example. Speculators were attracted by the rising price of medium and luxurious housing which rose uncontrollably. When they began to release their stock to gain profit the market experienced over-supply. Accordingly, the medium and large housing development slowed down in 1994-1995. Correction, then, was made by large developers by shifting to simple housing production which was still in demand. Meanwhile some developers were also offering a price discount up to 20 - 30 percent for medium and large types.[25] However, by then land held by companies under the permit system in Botabek had already reached 121, 631 hectares (Table 15.7), an

Table 15.7   Location permit issued in Botabek area, October 1996

	Kab. Bogor	Kod. Bogor	Kab. Tangerang	Kod. Tangerang	Kab. Bekasi	Total	%
No of Location permit	492	122	374	207	397	1,592	-
Area (Ha)	43,842	2,462	44,094	8,359	22,874	121,631	100.0
Land Released	13,664	182	21,765	1,522	9,506	46,639	38.3
Land with SKPH	2,969	182	21,336	852	1,420	26,759	22.0
Land With Master title	2,707	133	9,869	852	1,405	14,966	12.3
Land developed	8,704	157	2,306	674	4,768	16,609	13.7

Note: Kab.= Kabupaten.
Kod = Kotamadya = Municipality.
SKPH = Surat Keputusan Pemberian Hak = Decree for granting Land Right
Master Title = Title of land granted to developer before further split for individual buyers.
Source: BPN as quoted in Properti Indonesia , April 1997.

24 *It is common in the real estate industry in Indonesia that the developer implements a "pre-project selling" strategy. In these circumstances, the realisation of a housing loan by BTN or other banks can be assumed as an ex post demand.*

25 *Compiled from the developers brochures and leaflet printed in September to November 1996, also from the Catalogue of Housing Expo held by REI in Jakarta 1 - 10 November 1996*

area more than enough for 6.08 million houses, if it is assumed 50 houses can be built on a hectare of land.[26] Yet, up to late 1996 only 16,609 hectares (14 percent) had been developed (Properti Indonesia, April 1997:23). If it is further assumed that one household consists of 4.3 persons,[27] then the land under location permit alone can accommodate 26.15 million people.

On the other hand, the real demand for medium and luxurious housing is not that high. In Botabek, as suggested by Table 15.2, the market can only absorb 26,089 units per year. This figure is not far from the estimation made by the vice president of REI, who calculated that in Botabek the market can only absorb 5000 units of luxury housing and 20,000 units of small and medium housing per year.[28] This means that the currently available supply of land for housing in Botabek would not be absorbed for 201 years. This condition is more problematic if the size of the loan for property development in Indonesia is considered. Table 15.6 above suggests that the credit expansion to housing in 1996 has reached Rp. 21,075 billion.[29] With an interest rate of 22 percent per year, the interest to be paid in one year would be Rp. 4,636 billion, which correspond to Rp. 386.375 billion per months. This sum means that, in order to pay the interest alone, 3,863 unit houses of Rp. 100 million should be sold within a month. Considering the market absorption capacity in the last two years, which is slowing down,[30] the figure is unlikely to be reached.

What is of great concern in this situation is the possibility of a real estate crash which will have major effects throughout the Indonesian economy. An illustration may be taken from the British experience with a property crash in 1973/1974. The price rise experienced during a commercial boom in the early 1970s in Britain had induced a massive expansion in credit, based on the use of the rising value of the real estate as the collateral security. But when the price began to fall because of an oversupply, the collateral also

---

26 *In a low density housing project such as BSD the gross densities would be 27 unit per hectares (Kota Mandiri Bumi Serpong Damai, 1995), while in a high density Kampung areas the density would fall between 150 to 190 unit per hectares. If we assumed that in a residential development only 50 percent of land is designated for housing (52.7 percent in the case of BSD) and that the medium houses would need 100 square meters plot, a hectare land would be then enough for 50 units of houses*

27 *Struyk., Hoffman and Katsura. (1990) reports household size in Jakarta dropped from 5.6 to 4.2 person per households. Bandung Urban Development Project estimated that household sizes in Bandung is 4.2 person in 1992*

28 *Author's interview with Herman Sudarsono, vice President of REI, on 17 November 1996*

29 *This is the sum of : Construction of simple housing + Real Estate, simple housing + Housing = (2,150+2,540+16,385)= 21,075 billion Rupiah. (see Table 2)*

30 *Panangian Simanungkalit, a prominent real estate analyst, estimates that in 1996-1997 the market can only absorbs 700 units housing per month (Properti Indonesia, October 1996)*

declined in value to the point where it could not support the outstanding credit which it had secured (Scott, 1996: 194). The land and housing prices in Jabotabek have not declined yet, but if this should happen in the Jabotabek area, a remarkably high value of credit would become bad debt, producing a crash in the financial market.

This period also witnessed the entrance of private developers into the stock exchange which was reopened in 1977. Developing a new town on a large area of land definitely requires a huge investment. As an illustration, Bintaro Jaya, a new town developed by Ciputra on 1,700 hectares required Rp. 10, 000 billion (Warta Ekonomi, December 1996:18). Herman Latief (1995), the president director of Lippo new towns, estimated that at least Rp. 500 billion is needed to develop 1.000 hectares of land. This sum of money is not readily available even from commercial banks. Therefore, securing fresh and liquid funds from the stock exchange is one alternative for private developers and, obviously, only large companies which satisfy certain requirement can be listed in the stock market. Thus, by 1996, 17 companies engaged in property development (including hotels, offices and shopping centres) were already listed on the Jakarta Stock Exchange. They include Jaya Real Property, Ciputra Development, Dharmala Intiland, Lippo Land Development, Pujiadi and son, Modern-land Reality and several others (Properti Indonesia, January 1996: 31). It should be noted here that the first two companies are under the ownership of Ciputra.[31]

Although, compared to popular housing, the contribution of formal developers on housing production is very small,[32] nevertheless the impact on the space restructuring in Jabotabek is tremendous. This is because the developers can consolidate small parcels into very large developments for low density housing, provide public facilities such as roads, piped water and, in the case of Lippo Karawaci, a regional shopping centre, regional hospital, international hotels and office buildings (Master Plan Lippo Karawaci, 1995). As the land prices in Jakarta continue to increase, only lands a considerable distance[33] from the centre are still available for this kind of development. Thus, by 1995 the space occupied by Jabotabek had expanded to a radius 30 to 40 km away from the centre. This was worsened by the fact that the, supposedly self-contained new towns are still dependent upon Jakarta. With a poor transportation system, a one-way commuting time of up to two hours is becoming

---

31 See Table 7
32 It is estimated that up to 1989 only 15 percent of annual housing production is developed by professional sector (Struyk, Hoffman and Katsura, 1990: 69)
33 Dowall and Leaf, (1991) demonstrate that the land prices decline with the distance from the centre, they also show that there is significant effect of infrastructure upon land price. Land with infrastructure would be valued up to 90 percent more than land with no infrastructure

a common experience for the commuter (the mean travel time to job places is 60 minutes while the mode is 120 minutes).[34]

This period also witnessed a new era in Jabotabek development for by then Jabotabek had been integrated into a global city system.[35] In these circumstances, the role of the private sector is particularly important especially in restructuring the urban space. The accumulation of land by a limited number of people has become very apparent. Table 15.8 illustrates the accumulation of land under one group of companies. By early the 1990s, the new town[36] were mushrooming: within ten years, 20 new towns had been created, a number which is remarkable.[37]

15.6 The Private Sector and Road Infrastructure

The involvement of private developers in the restructuring process is not only in the accumulation of land and the production of formal houses, but also in the construction of new roads. By the early 1990s as the traffic to and from Jakarta worsened, the private developers start to invest in the development of toll roads. The first proposal of this kind was to link Bintaro Jaya, BSD, Citra Raya, Puri Jaya, Teluk Naga and Pantai Kapuk Indah,[38] a road plan which had never been considered in the Jabotabek Plan or the WJUDP plan.

The road plan was then discussed in the preparation of the Spatial Plan of *Kabupaten* Tangerang which is yet to be approved.[39] Nonetheless, construction of a 14 km long toll road connecting Jakarta- Serpong trough Bintaro Jaya and BSD has already been started. The development of the first

---

34 As revealed by a households survey in Kabupaten Tangerang conducted by the author in November 1996

35 Tommy Firman in his paper delivered in the International Conference on Cities and the New Global Economy has nicely shown the transformation of Jakarta into a system of a global cities. He shows the increasing role of Jakarta as an international management centre, and the locus of advanced production centres. He also shows the increasing number and amount of both foreign and domestic investment in the Jabotabek area (Firman, 1995)

36 The definition of a new town is debatable. The new town here is loosely defined as residential development in an area of more than 500 hectares which provides minimum urban facilities such as, shop houses, small offices, post offices and schools

37 Between 1946 and 1977, 31 new towns were built in Britain, including 12 second generation new towns which were produced in 1961 to 1972 (Cresswell and Thomas, 1972; Aldrige, 1979; Corden, 1977). In France, 9 new town were developed within 25 years (Roullier, 1993). In the United States, more then hundred new towns were developed since the end of World War II (Corden, 1977)

38 This six new towns are under the management of Ciputra (see Table 7)

39 The Spatial Plan for Kabupaten Tangerang is still in draft version. The draft plan, however, has accommodated the plan to develop the toll road which already initiated by private developers.

Table 15.8   Project and area of land hold under "Si Pengembang" in Jabotabek area

Companies under Management of Ciputra		Area (hectares)
A *Bumi Serpong Damai*		6000
1. BSD (Tangerang)	6000	
B. *Ciputra Group*		1947
1. Citra Garden I, II, III (Jakarta)	430	
2. Citra Land Sentra (Jakarta)	4	
3. Citra land Kuningan (Jakarta)	13	
4. Citraland Industrial Estate (Tangerang)	500	
5. Citra Garden Grand City (Tangerang)	1000	
C. *Metropolitan Group*		2775
1 Kosambi Baru (Tangerang)	15	
2. Harapan Baru (Tangerang)	25	
3. Industrial Estate Jababeka (Tangerang)	510	
4. Kota Cikarang Baru (Tangerang)	1400	
5. Bekasi Metropolitan Mall (Bekasi)	11	
6. Kali Baru Bekasi Barat (Bekasi)	72	
7. Medan Satria Bekasi Barat (Bekasi)	41	
8. Wisma Metropolitan I, II, III (Jakarta)	700	
D. *Jaya Group*		3712
1. Bintaro Jaya (Jakarta/Tangerang)	1700	
2. Taman Impian Jaya Ancol (Jakarta)	250	
3. Garden Residence Kemang (Jakarta)	12	
4. Puri Jaya (Tangerang)	1750	
E. *Pondok Indah Group*		1690
1. Pondok Indah (Jakarta)	450	
2. Puri Indah (Jakarta)	180	
3. Bukit Cinere Indah (Jakarta)	60	
4. Pantai Kapuk Indah (Jakarta)	1000	

Total Land hold under this group in Jabotabek alone   16125

Source: Adapted from Properti Indonesia February 1994.

phase, which required investment of Rp. 350 billion (Warta Ekonomi, December 1996), is funded by a consortium of three companies namely BSD, Jaya Real Property and PT Jasa Marga[40] (Properti Indonesia, June 1996: 29).

Although the involvement of the private sector in Indonesia's economic development is not unplanned,[41] nonetheless, the magnitude of private sector development in space restructuring in Jabotabek is extraordinary, and arguably its contribution to the provision of housing in the region could also be increased significantly. Therefore, the rise of private sector as a major actor in the development of the region, undoubtedly, marks one of the important cycles in the space restructuring process.

15.7 Conclusion

Restructuring processes taking place in Jabotabek have transformed what was the tiny city of Jayakarta into one of the world's biggest. At least three major cycles can be identified. The first finished in 1945 when the city changed from a colonial city into the capital of the Republic of Indonesia. The second was largely accomplished in the 1970s, when the economic policy of the government brought a high economic growth and energised massive urbanisation in the area. The third, propelled by the flourishing private sector, particularly in land development, had been shaping Jakarta into a city which economically integrated into a global system of urban areas.

High economic growth and rapid urbanisation have been experienced in Jabotabek area. These factors, combined with the more liberal investment policy of the government have, encouraged the increasing role of the private sector in land development. The emergence of private land developers in the Jabotabek area has been particularly important because: First, private developers are now the leading institutions who can create market and provide a significant amount of formal housing. Second, private developers are now expanding their activities from merely investing in housing development to the provision of 'public goods' including roads and piped water. Third, private developers, with their ability to mobilise funds and their control over considerable amount of land, are influencing the making and implementation of statutory spatial plans for the region. Therefore, it can be argued that the most recent restructuring processes undergone in the area has been driven, to

---

40 *The first two companies are owned by Ciputra, while the last is owned by the government*
41 *Hill (1996) noted that the shift to the private sector for funding the development is revealed in both the Government's Repelita (Five-years Development Plan) and in the actual investment. In particular, the 5th Repelita (1981-1986) stipulated that private sector would provide over 50 percent of total funding.*

the greatest extent, by large-scale land speculators and developers. Under such conditions the sprawl of development and the choice of formal housing types are largely determined by the decisions of the private sectors.

References

Aldrige, Meryl., (1979), *British New Towns: A programme Without A Policy*, Routledge & Paul Kegan, London.

Angel, S., Archer, R.W., Tanphiphat, S. And Wegelin, E., eds., (1983), *Land for Housing the Poor*, Select Books, Singapore.

Baken, R.J. and van der Linden., (1993), "'Getting the incentive Right"; Banking on the formal private sector - a critique of World Bank thinking on low-income housing delivery in Third World Cities' in *Third World Planning Review*, (15) pp. 1 - 22.

Blusse, Leonard., (1981), "Batavia 1619-1740: The Rise and Fall of a Chinese Colonial Town", in *Journal of Southeast Asian Studies*. Vol. XII, No. I, March.

Booth, Anne., (1992), "Implementing Monetary Policy", in Booth, A., ed., *The Oil Boom and After: Indonesian Economic Policy and Performance in the Soeharto Era*. Oxford University Press: Singapore.

Browder, J.O., Bohland, J.R., Scarpaci, J.L., (1995), "Patterns of Development on the Metropolitan Fringe: Urban Fringe Expansion in Bangkok, Jakarta and Santiago" in *Journal of the American Planning Association*, Vol. 61, No. 3, Summer.

Bumi Serpong Damai (1995), Data dan Penjelasan Proyek Dalam Rangka Keikutsertaan Pada REI Awards (Data and Projects Briefs for REI Awards).

Castles. L. , (1965), Socialism and Private Business: the Latest Phase" in BIES, No1 pp 13- 14, quoted by Hill, Hall., (1996), The Indonesian Economy since 1966, Cambridge University Press, Hong Kong.

Chia, R. et al (1992), Globalisation of the Jakarta Stock Exchange, Prentice Hall , Singapore. Quoted by Firman , T., (1994), Urban Restructuring in Jakarta Metropolitan Region: An Integration into a System of "Global Cities", in *Cities and the New Global Economy , Conference Proceedings Volume I* , Australian Government Publishing Service.

Cresswell, P. and Thomas, R., (1972), "Employment and Population Balance" in Evans, H., ed., *New Towns: The British Experience*, Charles Knight & Co, Ltd. London.

Corden, Carol., (1977), *Planned Cities, New Towns in Britain and America*. Sage Publication, London.

Data Consult, (1996), *Real Estate Business Sluggish*, in Indonesian Commercial News Letter. No. 197, June, available from http://www.idola.net.id/dc/dc/htm.

Devas, N., and Rakodi, C., eds., (1993), *Managing Fast Growing Cities*, Longman, New York.

Douglass, M. , 1993, "The "New" Tokyo Story": in Fujita, K. And Hill, R..C., eds. (1993), *Japanese Cities in The World Economy*, Temple University Press, Philadelphia.

Dowall, D.E. and Clarke, G.A., (1991), *Framework for Reforming Urban Land Policies in developing Countries: Policy paper*, Washington; Urban Management programme.

Dowall, D.E. and Leaf, M., (1991), " The Price of Land for Housing in Jakarta" in *Urban Studies*, Vol. 28, No. 5. pp. 707-722.

Farvaque, Catherine and McAuslan, P., (1991), *Reforming Urban Land Policies and Institutions in Developing Countries*, Urban Management Programme.

Ferguson, B.W. and Hoffman, M. L., (1993), " Land Markets And The Effect Of Regulation On Formal-Sector Development In Urban Indonesia" in *Review of Urban and Regional Development Studies*, Vol. 5

Firman, Tommy., (1994), Urban Restructuring in Jakarta Metropolitan Region: An Integration into a System of "Global Cities", in *Cities and the New Global Economy*, *Conference Proceedings Volume I*, Australian Government Publishing Service.

Foo, Tuan .Seik, (1992a), "The Provision of Low-cost Housing by Private Developers in Bangkok, 1987-89: The Result of an Efficient Market?" in *Urban Studies* Vol. 29, No 7.

Foo, Tuan Seik (1992b) "Private Sector Low-cost Housing" in Yap, K.S., ed., (192), *Low - income Housing in Bangkok A Review of Some Housing Sub-markets*, HSD Monograph 25, Division of Human Settlements, Asian Institute of Technology.

Giebels, L.J., (1986), " Jabotabek: An Indonesian-Dutch Concept on Metropolitan Planning of The Jakarta-Region" in Nas, P.J.M., eds., (1986), *The Indonesian City*, Foris Publication, Dordrecht.

Hill, Hall., (1996), *The Indonesian Economy Since 1966*, Cambridge University Press, Hong Kong.

Indonesian Team, (1995), *Study on the Emerging International Urban System of Megacities in The East Asia: The Case of Jabotabek Region*, The National Development Planning Agency, Indonesia in collaboration with The Nippon Life Insurance Research Institute.

Ministry of Public Housing, (1995), Data Bidang Perumahan dan Permukiman (Data on Housing and Settlemets), Kantor Menteri Negara Perumahan Rakyat, Jakarta.

Lasserve, Alain Durand., (1990), "Articulation Between Formal and Informal Land Markets in Cities in Developing Countries: Issue and Trends" in Barros, P. And Van der Linden, J eds. *The Transformation of Land Supply System in Third World Cities*. Avenbury, Aldershot.

Latief, Herman (1995), *Kemitraan dan Sinkronisasi Investasi Dalam Pembangunan Permukiman Skala Besar/Kawasan Berkembang Pesat*, (Partnership and Synchronisation of investments in Large scale Residential Development), paper presented at URDI (Urban Development Institute) Seminar, Jakarta, 23 - 24 September 1996.

Leaf, M., (1991), *Land Regulation and Housing Development in Jakarta, Indonesia: From the "Big Village" to "Modern City"*, unpublished Ph.D Dissertation, University of California at Berkeley.

Leaf, M., (1993), "Land Rights for Residential Development in Jakarta, Indonesia: the Colonial Roots of Contemporary Urban Dualism", in *International Journal of Urban and Regional Research*, Vol. 17, No. 4, pp. 477-491.

Leaf, M., (1994), "Suburbanisation of Jakarta: A Concurrence of Economics and Ideology" in *Third World Planning Review*, (16), 4, pp 351 - 352.

Marcussen, L., (1990), *Third World Housing in Social and Spatial Development: The Case of Jakarta*, Avebury, Aldershot.

McAuslan, P., (1985), *Urban Land and Shelter for the Poor*, Earthscan, London.

Payne, G (1990), *Informal Housing and Land Subdivision in Third World Cities: A Review of The Literature*, Centre for Development and Environmental Planning (CENDEP), Oxford.

REI, (1997a), *Seperempat Abad REI (REI, a quarter of century)*, REI, Jakarta.

REI, (1997b), REI membership, 1972-December 1996, unpublished data, REI.

REI, (1997c), Housing construction by REI, 1992 - 1997, unpublished data, REI.

Roullier, J.E., (1993), 25 Years of French New Town, Translated by Alan Lee, np.

Scott, Peter., (1996), *The Property Masters: A History of British Commercial Property Sector*, E & FN Spon, London.

Silas, Johan., (1995), *Perum Perumnas dalam Tantangan Tugas* (Perumnas in Challenging Task), Perum Perumnas, Jakarta.

Sivaramakrisnan, K.C., and Green, L., eds., (1986), *Metropolitan Management : The Asian Experience*, Oxford University Press, Oxford.

Struyk, R.J., Hoffman, M.L. and Katsura, H.M. (1990), *The Market for Shelter in Indonesian Cities*, The Urban Institute Press, Washington

Surjomihardjo, A (1977), *The Growth of Jakarta*, Djambatan, Jakarta.

Winters, Jefry, A. (1991), *Structural Power and Investor Mobility: Capital Control and State Policy In Indonesia, 1965 - 1990*, Unpublished Ph.D. Dissertation, Yale University.

Sources without named authors

"Bekasi dan Tangerang Akan Jadi Incaran Utama" (Bekasi and Tangerang will be the First Choice), *Properti Indonesia*, February 1994, pp 101

"Berharap Pada Investor Asing" (Put a Hope on Foreign Investor), *Properti Indonesia*, January 1996, pp 30 - 32.

"Boom Kota Baru Di Pinggiran Jakarta" (New Towns Boom in Jakarta's Fringe Areas), *Properti Indonesia*, June 1995, pp 27 -28.

"Dua Catatan Agar Pasar Membaik" (Two Notes to Improve Market), *Properti Indonesia*, October 1996, pp 34-35.

"Hadir Dengan Gagasan-gagasan Besar" (Coming with Big Ideas), *Properti Indonesia*, February 1994, pp 8-11.

"Jor-joran Kredit" (Competition to lend), *Properti Indonesia*, November 1994, pp 24

"Memang Perlu Napas Panjang" (The Need of An Endurance) Warta Ekonomi, No. 28/VIII/2 Desember 1996, pp 22.

"Menguak Peta Bisnis Para Raja Properti" (Disclosing the Business Map of The Property Kings), *Properti Indonesia*, February 1994, pp 6-7.

"Menjual Superblok Dengan Kepercayaan" (Selling Superblock with Trust), *Properti Indonesia*, February 1994, pp 16-17.

"Menyulap Lahan Seluas 30,000 Hektar" (Transforming 30,000 Hectares Land), *Warta Ekonomi*, No. 28/VIII/2 Desember 1996, pp 18.

"REI: Older and (Hopefully) Wiser, *Properti Indonesia*, April 1997, 101

"Suku Bunga KPR Turun Moderat" (The Moderate Decline of Housing Loan Interest), *Properti Indonesia*, May 1994, pp 67.

"Tak Bebas Di Jalan Bebas Hambatan" (Traffic Jam in Freeway), *Properti Indonesia*, June 1996, pp 28 -31.

"Tanah Terlantar, Salah Siapa?" (Abandoned Land, Who Is To Be Blamed?), *Properti Indonesia*, April 1997, pp 22 -25.

PART IV
URBAN LAND USE AND
ARCHITECTURE

16 East meets West in Bangalore: The Impact of Economic Change on the Built Environment

Kalpana Kuttaiah and Gail Gordon Sommers

Editors' note:
Since indpendence the built environment of Indian cities has undergone vast change. Older traditional and colonial architectural styles have been replaced with new forms built with new materials – above all steel reinforced concrete. In Bangalore, as elsewhere, the older heritage is being erased, but with little debate and no policy on the wisdom of this course.

16.1 Introduction

Bangalore, a cosmopolitan city located in Southern India, has emerged as the high technology center of India. This was made possible primarily by the economic reforms in 1991 that lowered import barriers, created free market competition, and allowed greater foreign investment. But the economic changes did more than affect the local economy. They had a major impact on the built environment of the city. Since 1991 the physical appearance of the city has changed drastically, resulting in the loss of much of the traditional character of Bangalore.

Added to the effects of the economic changes has been the unprecedented growth in the last two decades. Bangalore recorded a growth rate of 76% for the decade of 1980-90, making it one of the fastest growing cities in the world. This trend is continuing into the 1990s, with Bangalore's population growing from approximately 3.5 million in 1980 to an estimated 7.0 million in 1995. The result is that the city is losing its reputation as the pensioner's paradise and the garden city.

16.2 Traditional Bangalore

The city was founded in 1537 AD as a mud fortress and was replaced by a stone fort one hundred years later. This formed the original nucleus of the city, and is today called Bangalore City. A second center developed when the British established a garrison, which is referred to as the Cantonment, in 1809. These two centers functioned as independent administrative units until

1949, when they were merged into one city corporation. It was from these beginnings that the modern city of Bangalore emerged.

As population increased so too did the area of the city - from 20.7 square miles in 1901 to 60 square miles in 1971. Much of this growth occurred it what were called extensions in Bangalore City and towns in the Cantonment. Examples of these planned suburbs are Chamarajpet, Malleswaram, Basavangudi, Richmond Town, Cleveland Town, and Frazer Town, which were begun in the early 1900s. While each suburban development was planned, there was no integrated planning (Rao and Tewari, 1979, pp. 175-266; Gowda, 1977, pp. 295-296). The result of continued growth is that Bangalore's urban structure today consists of the historic core, the Cantonment, the pre-1970 suburbs, and the newer suburbs built since 1970.

Historic Core

The historic core, or Bangalore City, began as the native area with its highly skilled artisans, bazaars, and wholesale and retail traders. Focusing purely on the then compact historic core's physical morphology, construction incorporated environmental considerations. Its buildings provided comfort and shade for most of the day. It effectively created a hierarchy of micro-environmental levels as in thermal, acoustical, lighting, etc. Its vernacular building techniques, with its socially and culturally responsive traditional housing/building types, had environmentally sound design values. Many of the construction types used renewable natural resources; wood, brick, bamboo, lime, and clay to name a few (fig. 16.1). Building designs responded to climate and provided environmental comfort through courtyards, buffer spaces,

Figure 16.1 Traditional materials in ceiling construction

Figure 16.2 Older roof-lines in Bangalore

Figure 16.3 A private back yard

skylights, etc. (see figs 16.2 and 16.3). Building design and construction methods were also very cost effective and efficient. The traditional houses reflected the lifestyles of the inhabitants through spaces such as verandahs, courtyards, and service courts.

The successive growth of the city occurred through the formalisation of planning and development, resulting in planned neighborhoods. Characteristics similar to the historic core can be observed in these older planned neighborhoods. While they are not as compact as the historic core, they do reflect similar values in terms of scale, spatial relationships, land-use characteristics, and their socioeconomic structure.

Cantonment

Much of the architectural tradition of Bangalore will be remembered by what it inherited from the British and this was located mostly in the cantonment area. This was a military cantonment with a surrounding civilian settlement for people providing services to the British. Development in the cantonment included barracks for the soldiers, spacious bungalows for the officers and their families, a parade ground, mall, bars and night clubs, and polo fields and race tracks, all typical of the garrison culture in India (Rao & Tewari, 1979, p. 175).

The Cantonment developed as a self-contained township. Land was plentiful and originally lots were sold in acres rather than in square yards or feet. Many houses were built with the intention of renting them out to the British officers and their families. The cantonment was noted for its large open spaces and tranquil setting. The British troops were housed in large cool colonnaded barracks lining the vast expanse of the parade grounds and riding schools. Roads in the cantonment seemed endless and were lined with shady trees. Buildings in the cantonment were of many architectural styles. The bungalow style was typical for private residential buildings (see fig. 16.4). The European-Classical or Greco-Roman style was used for public buildings. A mixture of Indo-Saracenic and Dravidian styles was used for certain other types of buildings. Numerous public gardens, parks and buildings were

Figure 16.4 An ornate colonial period bungalow

built in the Cantonment in the late 1920s, mostly commissioned to celebrate the Silver Jubilee celebration of the King. It was this that gave Bangalore its lasting reputation as a Garden City.

Development of the City Prior to 1991

The 20th century has seen Bangalore change from a small city surrounded by rural areas with major green space dividing the historic core from the cantonment into a major city. However, much of this growth has been unplanned and often detrimental. One author, writing in the 1970s, provided the following description of the changes that were occurring:

> "One who lived in Bangalore about 50 years ago, generally describes these changes ruefully. Even in 1930 the town was quite small in size with rural surroundings. The great green belt, which was purposely left from the Indian Institute of Science extending up to Mental Hospital connecting the Palace Orchards, High Grounds, Cubbon Park, Wilson Gardens, Lalbaugh, and separating the City area from the Cantonment area, could not be preserved, as the two cities have merged with each other encroaching upon most of these green areas. One could still witness the existence of this green area here and there, but it has considerably reduced in extent except in Cubbon Park and Lalbaugh areas.... Vast compounds of old bungalows are being subdivided to form plots. At the periphery of the city also the position is not better. The city is sprawling along all the diverging highways, eating away precious agricultural lands indiscriminately. Sub-standard developments and haphazard constructions are seen enveloping even the very attractive and well-planned extensions of the city. Within these extensions also, land uses are being mixed up indiscriminately. Calm residential localities are disturbed by the establishment of commercial and industrial units. Absence of control on the use of land all these years allowed individuals to utilize their properties for the best personal advantage rather than for the benefit of community at large. Vested interests have their own hand in deciding great many issues in urban development, starting from land speculation up to industrial locations." (Gowda, 1972, p. 42).

As Bangalore grew, additional infrastructure was needed to support the growth in population and economic development. This consisted of expanding the water supply and sewerage system, building schools and universities, constructing and maintaining roads, and providing housing. Unfortunately, the city was not able to cope adequately with the demand for services. Estimates in the 1970s were for a population of over two million in 1981 and three million in 1991 (Gowda, 1977, p. 303-12). Instead it was already over 3 million in 1981 and estimated at over 7 million in 1995.

16.3 Indian Economic Reforms of 1991

When India gained independence from Great Britain, it established state planning for the public sector 'commanding heights' of the economy. Also instituted at this time was legislation aimed at developing the nation's economy. This was done through import barriers, state monopolies and state licensing. These policies created a huge public sector that accounted for half of India's gross investment. Even if it did provide the foundations for a new industrial power, this system isolated India from the rest of the world, resulted in higher prices for lower quality consumer goods, and forfeited the benefits of foreign investments and modern technology (World Bank, 1996, p.xvii). A major effect of India's post-independence economy, however, was in international reputation for hostility to foreign investments and a bureaucracy that hampered rather than assisted development.

This began to change in the 1980s with the implementation of reforms designed to encourage business development and international investment. When Rajiv Gandhi became Prime Minister in 1984, he began to liberalize import policies and encourage exports. A major focus for Gandhi was technology, which was advantageous for Bangalore. This began the movement to remove or minimize the obstructionist policies of the past and to institute greater public and private sector cooperation (Vikas Deshmukh, 1993, pp. 1-2). But the changes that have had dramatic impacts on Bangalore were initiated in 1991 under the leadership of Prime Minister P.V. Narasimha Rao.

Two events - the Persian Gulf War and the breakup of the Soviet Union-led to a major crisis, with India being virtually bankrupt. The Persian Gulf War ended a steady supply of money from expatriate Indians working in Persian Gulf countries, and the breakup of the Soviet Union cut off major markets and a source of cheap oil. The Indian government faced the task of making fundamental economic changes in order to avoid bankruptcy (Public Broadcasting System, December 1,1996). It was desperation as much as conviction, however that led to deregulation. Reforms were the "tacit prerequisites for getting the International Monetary Fund, the World Bank, and foreign aid donors to grant the credits India . . ." needed to survive its financial crisis (Pura, October 21,1991, p. A-15).

The balance of payments crisis that occurred in 1990-91 led to both immediate stabilisation measures to restore confidence in the government and structural reforms that have had a major impact on the economy. Fiscal stabilisation had the highest priority because it would determine the success of economic reforms. Next in priority was a series of economic reforms. Changes in industrial policy removed some of the barriers that hampered the establishment of new industries. Industrial licensing was reduced to only those industries that have environmental or pollution impacts. The number or industries reserved for the public sector was greatly reduced, allowing much more private sector competition. Foreign investment is now encouraged rather than limited, with 51 percent foreign equity allowed in a large list of industries. Trade policy liberalisation opened India to raw materials and manufactured products and custom duties have been lowered considerably (Cassen and Joshi, 1995, pp. 1520). The results of economic reforms are evident in the competition for former state monopolies such as Indian Airlines, a growing middle class with money to spend on consumer products, and multinational corporations flocking to India. Examples of the latter are Coca Cola, Pepsi, Kellogg's, American Express, Motorola, Hewlett Packard, Digital, Lee, Levi, and Merrill Lynch. It was not just the economic liberalisation that encouraged these companies to locate in India but also India's large, educated, English-speaking workforce (PBS, December 1, 1996).

All of these changes have had a major impact on the economy of Bangalore, which in turn has drastically changed the physical appearance of the city. According to Deshmukh,

> "The emergence of Bangalore as a world-class software manufacturing center . . . represents perhaps the best example of what political and economic reforms can do to unleash the vast potential of India's entrepreneurs. Those not acquainted with this city's meteoric rise to prosperity may well ask, what turned a sleepy army cantonment town from the days of the British Raj into the cradle of high technology in India? The answer: Effective public-private sector cooperation, coupled with an abundant supply of cost-effective, well-qualified technical talent" (Deshmukh, 1993, p. 1).

The result is a booming metropolis that symbolizes the new India. This Silicon Valley of India has become a Mecca for the young middle class. But more importantly, Bangalore is second only to the United States in the export of computer software (PBS, December 1,1996).

16.4 Bangalore: Current Characteristics

As Bangalore heads into the twenty-first century, the residents remain unfazed by the rapid changes that have been occurring. But then, Bangalore has always been a very flexible city, ever willing to accept new people, new ideas, and new cultural traditions. One would expect that a city that had changed little for decades would not necessarily remain the same and would experience its share of new developments. However, few anticipated the onslaught of development that occurred after economic liberalisation. The city

Figure 16.5 Old buildings still forming a street front

has changed dramatically and no longer looks familiar to people associated with it for the past few generations. Former residents who have returned after years abroad are amazed by the differences in the city. The entire architectural character of Bangalore has changed in the last decade and there are very few old buildings left for one to project what the city might have looked like (see fig 16.5).

Since the economic liberalisation policies were adopted, a large number of new businesses and industries, including multinational corporations, have been established in Bangalore. This has spawned rapid urbanisation, growth and development, migration and globalisation, which are threatening the city's physical development. Although this liberalisation and resulting globalisation is bringing in many opportunities, it is also threatening the city's ability to cope. The city is hardly recognizable now as the one that bore the sobriquets of "Garden City" and "Pensioner's Paradise." The city has lost much of it green space and is no longer attractive for the older generation because of the city's faster pace.

In the last decade the city has grown so rapidly, that it is virtually impossible to plan much less provide the infrastructure that is needed to accompany such growth. Paucity of funds may be the main cause but the fact does remain that few could have had the foresight to anticipate the effects of economic liberalisation on the city and estimate the city's infrastructure needs. New locations, mostly residential areas, are cropping up with remarkable frequency with an equally remarkable scarcity of facilities. There are several residential areas that have no power supply, no good roads, no public transportation, and other civic amenities that are considered essential in a modern industrial city. Adequate sewerage disposal is a major concern throughout the city. What Bangalore is experiencing now regarding infrastructure is just the beginning; the next five years will see the serious effects of this development.

Building Style and Type

As the social and economic climate of Bangalore has changed, the results can be seen in its building styles and types. Instead of building religious institutions and more traditional structures, Bangalore is becoming a city that caters to business. Thus new construction focuses on business needs. New industrial headquarters, banking institutions and companies are setting shop in Bangalore and all are trying to buy office space in the heart of the city (1995 National Trade Data Bank, March 21,1995). This has unleashed a frenzy of real estate speculation, which has resulted in sky rocketing real estate prices (1995 National Trade Data Bank, March 21,1995). The land prices in many areas have quadrupled in the last two years. On an average, prices have gone up a minimum of 50 percent a year. This demand for space has triggered indiscriminate development patterns, resulting in multistory buildings in that part of the city that once boasted of single-floored offices. Today, nearly all the commercial buildings are several floors high and the old buildings are being demolished and replaced by buildings of steel and concrete. The influence of the west and transplantation of models and designs can be seen in the new wave of architecture and materials: mechanically-oriented office build-

Figure 16.6 A modern 'anywhere' department store

ings with air conditioning and other amenities; glass and steel construction; enclosed shopping arcades and malls (see fig 16.6); departmental stores (mini-Macy's), etc.

Figure 16.7 The new night-life

Social Life and Habits

Traditional Indian life styles may not be disappearing but they are certainly less apparent in Bangalore. The Western influence can be seen in the bars, pubs (fig. 16.7), fast food restaurants (fig. 16.8), and discotheques that are becoming a common feature everywhere. Indicative of the changes that have occurred is the fact that Bangalore prides itself on being the pub capital of India (Los Angeles Times, Dec. 17, 1991 & New York Times, Dec. 29, 1993). The proliferation of satellite dishes, television cable networks and western programs that are now offered to viewers have a strong influence on local social lifestyles. The younger yuppie generation, with a degree of disposable income previously unknown, has elevated consumerism to a level also previously unknown. As a result the demand for consumer goods-name brands, entertainment, luxury holidays, and time sharing holiday resorts-exceeds the scale of anything previously known in India. Most of this can be traced to economic liberalisation that has led to the development of an extensive middle class in India and especially Bangalore. This in turn created a market demand for consumer goods such as household appliances, electronics, hi-tech gadgets, name brand clothing and other quality goods. The economic liberalisation of the 1990s has also made it possible to meet these market demands.

Figure 16.8 The Kentucky Fried Chicken restaurant: scene of an anti-western siege

New Suburban Subdivisions - Gated Communities

The American model of the gated community has been imported to Bangalore and incorporated into new residential developments (see fig 16.9). In the last three to four years residential construction has moved further away from traditional Indian models. Many new residential subdivisions are on $^1/_4$ to $^1/_2$ acre plots, with no fencing, two-car garages, club houses, tennis courts, and broad boulevard streets, enclosed by secured fencing. Besides being alien to Bangalore or Indian traditions for land development, the new types of development consume large chunks of land and require much more extensive, and therefore expensive, infrastructure. The encroachment of all the new development that is occurring in Bangalore and the surrounding area on open space, agricultural lands, and natural resources is having environmental repercussions. Besides being cost inefficient and negatively impacting the environment, these adaptations of American development patterns promote sprawl and isolate the individual from the community. Equally important for Indian culture, these development patterns encourage nuclear family lifestyles rather than the traditional Indian joint family system.

Figure 16.9 The prominent entrance to a new gated community

Planned Layouts and Suburbs

Another residential development pattern that is affecting the urban landscape is the planned layouts and extensions built by public agencies to provide low- to moderate-income housing (see fig. 16.10). These are usually about four to five stories high and their floor area ratio is relatively low. The intent is to provide smaller dwellings for a larger number of families. These developments tend to be very dense and congested. In addition the quality of construction and materials used are poor. The developments are often built by government contracts and not by private developers to keep the cost of housing affordable. Added to this is poor maintenance. The result is that these

Figure 16.10 Public sector middle income flats under construction

developments tend to be a blot on the visual landscape. Although these are not as insensitive pedestrian-focused life style typical of India as the new suburban subdivision movement, it incorporates the same segregated land use and inappropriate zoning ordinances, density requirements, linkages to shared or common space, the concept of street, and the cultural realm of public spaces.

The Automobile Revolution

The automobile revolution of the late 1980s and 1990s has had the most influence on the current city structure and its growth and development pattern. Public transit or two-wheelers were the most predominant form of transportation. The urban structure supported public transportation and commuting patterns within walking distances. Growth coupled with the number of automobiles changed the commuting patterns and is thus leading to urban

Figure 16.11 Traffic - most of it diesel or two-stroke powered

sprawl. Both the two-wheeler and the four-wheeler have become affordable and within the reach of the middle and upper-middle classes of Bangalore society. In fact the current trend is to own two or three cars among upper middle class families; one car and one two wheeler among the middle class families; and one or two two-wheelers among middle/lower middle class families. The result is that traffic has grown at least 6 to 8 times in the last decade (see fig. 16.11). This has serious implications on pollution, parking, gas consumption, waste disposal, sprawl, etc. This again is the result of an emphasis on the individual consumption and isolated social living pattern that has become possible since economic liberalisation. The present street structure and its hierarchy are inappropriate and are incapable of accommodating such development patterns. The current trends are another indication of western influence in Bangalore.

While developed and western societies are trying to curb the automobile and its adverse effects on society, Bangalore seems to be embracing the automobile. In the last two years at least a dozen foreign car companies have unleashed their cars and bikes on the Indian society. The scant attention being paid to the enormous amount of toxic fumes being spewed out by the hundreds of thousands of two wheelers, auto rickshaws, buses, etc. is making the air so polluted that it is a common sight to see people with air filter masks.

16.5 Conclusion

The most notable effect of economic reform has been the influence of the west on Bangalore. In a metropolis like Bangalore, where so many cultures have come together, it is the western culture and the desire for all things western that has become the only common, cementing force (Los Angeles Times, Dec. 17, 1991). In terms of the physical appearance of the city, this means that all the growth and change being experience in Bangalore is erasing the older architectural heritage. This may be too kind a way of expressing what is happening. In reality, older structures are being demolished to make way for new buildings to meet the growing demands of business and industry and the desires of the middle class that has emerged in such numbers in recent years.

The problem is as much in the character of new buildings as in the fact that older buildings are being demolished. No architectural guidelines exist to guide construction of new buildings. If these existed, the guidelines could at least require that new buildings reflect some of the characteristics of the older built heritage. This would provide the city with some native Indian character, rather than the strictly international approach that can be seen in Bangalore today. This could be accomplished without diminishing the ability of buildings to function in ways appropriate to the needs of business and industry.

A major concern emerging from the rapid growth of Bangalore is environmental. The region's climate is becoming a hot and dry one rather than the cool temperate climate which made it such an attractive place to live and visit. There is less rain and more intense sun and solar radiation, resulting in a very uncomfortable thermal environment. This overloads the power supply as the use of air conditioners and fans increases. This just adds to the infrastructure problems that already exist.

On one hand the change can be argued as good in light of economic and technical prosperity, but the effects of sprawl, the change to the nuclear family living, pollution, the effects of new consumer consumption, and environmental problems are questionable. In these respects these models of change deviate from traditional Indian culture. But what is imperative is to recognize the effects of the irreversible adverse changes and reconcile them with maintaining vestiges of traditional Indian culture, which will determine whether Bangalore remains an Indian city in more ways than just location. Change cannot be stopped, but it can be managed.

What is especially important for Bangalore is what has always made it attractive. Besides business considerations, it was the climate, the tree-lined streets, parks, the modern shopping malls, good schools, and recreational choices. For tourists, however, the lures are Bangalore's elegant old colonial buildings, dating back to the days of the British Empire and the maharajah princes. These buildings are interspersed among high-rise office towers of steel, glass and polished granite (New York Times, August 28,1995, p. A-1). Once all of this is gone, it cannot be created. Thus the people of Bangalore have some major decisions that need to be made quickly.

References

Cassen, Robert & Josi, Vijay, India The Future of Economic Reform, *1995, Delhi: Oxford University Press.*

Deshmukh, Vikas, Bangalore: India's Hi-Tech Birthplace, *1993, Washington, D.C.; Center for International Private Enterprise.*

Gowda, K.S. Rame, 1977, "Bangalore: Planning in Practice," in Indian Urbanisation and Planning: Vehicles of Modernisation,ed. *Allen G. Noble and Ashok K. Dutt, New Delhi, Tata McGrow-Hill Publishing Company, Limited.*

Gowda, K.S. Rame, Urban and Regional Planning, *1972, Myssore; University of Mysore.*

"India-Gangalore Business Climate," 1995, national Trade Data Bank, Market Reports, March 21, 1995.

"India Booming as a Leader in Software for Computers." New York Times, *December 29. 1993.*

"An Indian City of the Future with the Lure of the Past," New York Times. *August 28, 1995.*

"India's New Middle Class Finds Home in Bangalore," Los Angeles Times, *December 19.*
    1991.
Public Broadcasting System (PBS), December 1, 1996.
Pura, Raphael, "India Pins Hope for Future on Capitalism," Wall Street Journal, *October 21,*
    1991.
Rao, V.L.S. Prakasa & Tewari, V.K., 1979. The Structure of an Indian Metropolis: A Study of
    Bangalore, *New Delhi: Allied Publishers.*
World Bank, India: Five Years of Stablisation and Reform and the Challenges Ahead, *1996,*
    Washington, D.C.: The World Bank.

17 Shanghai's Land Development: A View of the Transformation of Mainland China's Administrative Systems since 1978

*Hung-kai Wang and Chun-ju Li**

17.1 Introduction

Changes in the institutions related to land development are among the central elements of the economic reforms in mainland China that started in 1978. They also generated some of the most critical issues faced by the nation which has been under the transition from a basically orthodox socialist economy to a "socialist market economy with Chinese characteristics". The purpose of this paper is to clarify the relationship between urban land development and the institutional changes as they are carried out by the administrative systems at local levels. Because of its phenomenal growth since early 1990s, the city of Shanghai is used as an example in the analysis.

The chapter is composed of three major parts. First, we briefly review the development and changes that have taken place in the institutions related to land and its usage, emphasizing the differences that existed among institutions before/after the 1949 "liberation", and before/after the on-going economic reform's commencement in 1978. Then, based on some findings from our empirical studies on the post-1978 Shanghai, we try to show how land has been treated as a means to help achieve the all-important economic development. Finally, the interactions among institutions of land use, city planning, enterprise reforms, and delegation of administrative powers between local governments, that determine the geographic pattern of spatial development, are to be clarified. Hopefully this type of empirical research will shed light on the way to a fuller understanding of the spatial structure of China's transforming socialism.

17.2 A Brief Description of the Development of Land Policies and Related Institutions in Mainland China

There are two types of land ownership in mainland China: national ownership and collective ownership. The former applies to towns and cities, and the latter to rural areas.[1]   No individual or establishment can occupy, trade, lease, or transfer rights on land by means not allowed by law. This socialist public land ownership was of course based on communist doctrines.[2] A land revolution was deemed the only solution to old China's land problem and thus determined the land policies of the early-stage Chinese Communist Party.[3]   However, there were some technical variations among different time periods and geographic regions.

17.2.1 Before the "Liberation" (1927-48)

Communist forces were basically contained in the rural areas of the nation, land policies being therefore focused on those territories, and the central issue was how the old "feudal" land ownership system could be abolished and mass impoverished peasants could have their own land ("land to the tiller", as an element of agricultural production). Policies fluctuated between outright confiscation of land from the landowners (and the head-count-based even distribution to poor peasants), and more moderate means of reducing the land rents or interests, depending on national political situations and local conditions.

17.2.2 Between the "Liberation" and the Economic Reforms (1949-77)

With the formal establishment of the socialist government, necessities rose for policies and administrative systems to be modified to cover urban land and non-agricultural land user entities. The "Land Reform Act of P.R.C." was enabled in 1950 to liberate agricultural productivity. In 1951, the "Land Reform Regulations for Cities and Suburbs" was enacted, so local governments could confiscate or commandeer suburban land owned by individuals or organisations. By 1952-53, basic tasks of land reform were considered accomplished under the system wherein land was controlled by the state while peasants could own their land. An important decision was made by the central government in 1954 to relinquish the land-use charges and rents on state-owned enterprises as well as educational, military and other

---

1 See article 10 of the Constitution of the People's Republic of China and article 2 of the Land Management Act. But the state retains the right to commandeer collectively owned land when it is needed for urban expansion or public interests.

2 See, for example, the Communist Manifesto among numerous other communist statements and conference resolutions.

3 See a resolution adopted by the 6th plenary meeting of the Chinese Communist Party, 1928.

types of establishments.[4]   After the mid- and late-1950s when the peasant private land ownership was circumscribed and transformed by a number of regulations and policies (including the development of agricultural cooperatives and finally the formation of the communes), the co-existence of the dual national as well as collective land ownership was secured until the commencement of the current economic reforms in 1978.[5]

The notion then held by the Chinese Communists was the utopian assumption that once the private land ownership was done away with, so would be the land rent. But curiously, the usage of land, one of the most basic elements of production, was not subject to the strict central planning control as was almost any other resource in those hey days of the socialist economy. Possible reasons include the fact that: a) due to the rent-less notion of land and the vest land area of the nation, of which the Chinese have long been boasting, land as a form of resource might have failed to assert its fundamental significance in national development; b) agricultural land supplies were not subject to apparent competition pressures from the slow urbanisation process of land, because of the insufficient investment in non-agricultural infrastructure; and, probably most importantly, c) the power of land management was split among so many functional branches of the government (e.g., agriculture, forestry, water resources, transportation, energy, and urban development), there was no one unitary local administrative authority powerful enough to carry out effective planning and control (Chou,1993). A myriad of problems had thus resulted, such as lack of land-use efficiency, illegal occupancy and construction, shortage of funds for construction of urban infrastructure (partially due to the inability to recapture adequate portions of profits from urban land development), chaotic land-use patterns, and the impossibility of economic implementation of the state ownership of land (the effectual paring of the rights to benefit from and handle the land on the part of the state).[6]

17.2.3 Since the Economic Reforms (1978 to present)

There have been drastic changes in the conception of land held by the Chinese Communists since the economic reforms went under way. Among the most influential and fundamental is the switch from the assumption that

---

4 As long as the land area is commensurate to the land user agent's actual needs and future development; see Chou Cheng, 1989. P.263.

5 See Hsiao Tze-yueh,1991; Wang Shung-san,1990; and Gue Chiang-fang,1994.

6 Numerous authors have pointed out land-use problems associated with the non-market land development system, e.g., Ma Keh-wei, 1992 and Chou Cheng, 1993.

land was an unredeemed social resource to be allocated by the state according to its all-embracing planning apparatus, to the conception of land basically as an element of production (and profit-making) and as a commodity to be priced and appropriated by the market. From this change of heart, starting in 1979, came a series of new legislations, amendments, and administrative adjustments, including the decision by the central government in 1983 to levy new land-use charges, the enabling of the Land Management Act and the establishment of the National Bureau of Land Management in 1986, the all important amendments to the Constitution and Land Management Act in 1988 to allow the lawful transfer of the right to use land,[7] and the Temporary Regulations on Land Value Increment Tax in 1993. In addition to the chronic financial difficulties on the part of local governments, judging from the fact that the first cities designated by the central government to try out the new system of landuse-right transfer are all in the coastal areas along the Pacific Ocean where international investment has been thriving, the involvement of foreign capital in the real estate industry seems to be another critical catalytic factor in the design and adoption of the system of transferring landuse rights for fees and various compensatory charges related to land development (e.g., expanses induced by relocation of previous occupants and provision of urban services).[8]

For the purpose of stimulating the economic growth through development of real estate industry, numerous relevant new laws and regulations and changes to old ones have been put into effect. Besides those mentioned above, many real estate-related laws, regulations and organisational modifications have also been made, e.g., the 1989 Urban Planning Act, the Regulations for Demolishing and Relocating Urban Buildings in 1991, and a series of management regulations for urban real estate development, rental, and rights transfer in 1995. In addition, up to late 1992, the following measures, among others, have been put into effect: a) creation of profit-oriented real estate companies, mostly financed with public sector equities, with a mandate to development commercial estate or properties, and sell these at prices that include locational premiums; b) promotion of economic development zones, often far from the city center, to attract foreign investments; c) introduction of land use development fees, taxes, and other exactions; d) promotion of spot redevelopment of inner-city areas, with occasional recourse to "value added" strategies linked to a change in land use; e) reintroduction of land and building registration (World

---

7 *In 1988, article 10 of the Constitution and the Land Management Act were amended to this effect.*
8 *The six designated cities were Shenzheng, Shanghai, Xiamen, Guangzhou, Fuzhou, and Tianjin.*

Bank,1994:xii-xiii). All these are for the multiple purposes of making the whole apparatus of real estate management system compatible with the larger context of the economic reforms, rationalizing the allocation of spatial resources by the market mechanism and the logic of commodity economy, refining the spatial pattern of urban land-use, improving the efficiency of land-use, enhancing finances for the city and the nation, creating a better and fairer environment for conducting businesses, and deepening the reforms of state-owned enterprises.

As it now stands, state-owned land is allocated to various uses by two co-existing systems (the so-called dual system): uncompensated administrative allotment and paid usage. The former basically covers the assignment of land to all government units and various enterprises which are owned by the central government or by governments at the city and county levels and not involved with foreign capital;[9] the latter is the core of a wide range of measures inaugurated since the economic reforms to marketize the urban land. It is a landuse-right leasing system initially devised to recapture, for the local government, portions of profits from land developments involving foreign investors, but later extended to cover certain types of domestically funded commercial developments.[10] There are two ways of compensation in this system: landuse fee and land leasing. Landuse fees (an institution set up in 1979 during the early stage of economic reforms) apply to enterprises that are partially owned by foreign investors. In Shanghai, landuse fees are determined by landuse types and location of the land.[11] Land leasing is a mechanism to transfer landuse rights from the local government, for a certain period of time, to foreign-funded enterprises and some specified types of domestic ones. As one of the four pilot cities, Shanghai began this new system in late 1987.[12] Initial transfer of landuse rights may only be conducted between the local government and foreign investors, and can be done through negotiation, auction, or open bidding; but establishments holding land from administrative allotment can enter the land market by

---

9 The expected land-user agent gets to use the land for unlimited period of time without paying any price. But it has to pay costs of related compensations (e.g., commandeering, relocation of the present land user) if there is any, and , since November, 1988, the landuse tax.

10 Since the beginning of 1995, land used for domestically-funded businesses in commerce, travel, recreation, finances, services, and commercial housing in Shanghai has to be obtained through this system.

11 Matrix with fee standards currently ranging from RMB 0.5 to 170 per square meter per year.

12 The policy of paid landuse rights transfer was proposed by the central government in early 1987, and Tianjin, Shanghai, Guangzhou, and Shenzheng were assigned the task of trying out the new policy.

re-compensating the local government with appropriate land price (bu-di-jia).[13]

17.3   A Brief Description of Recent Reforms of Administration and Enterprises in Shanghai

In addition to these land-related institutional reforms, Shanghai has launched other also influential amendments in its administrative and enterprise fronts. In this section we will briefly touch upon two with the most far-reaching effects.[14]

One of the basic post-reform attempts of the leadership of the country has been eradicating bureaucratic inefficiency. The strategy has been clarifying the relationship between the central and local governments, between the state and the enterprises, and among the party, the state and the masses (Ju,1991:145).

Regarding the delineation of administrative powers between the city and the lower level government, one of the relevant important events is the so called "two-tiered management" in planning. It was implemented in a staged fashion and has proved highly effective in stimulating the administrative enthusiasm and resourcefulness on the part of the local government. (Zheng, 1996:35) In 1986, the upper limits of site area a district government was allowed to be autonomous in cases of landuse rights leasing were raised. In 1987, after the central government's approval of a contracted tax package with Shanghai (which was a part of central government's delegation of financial autonomy to the city), the city government in turn entrusted district and county governments[15] with financial sovereignty as well as a whole array of

---

13 *Shanghai's first case of landuse rights transfer happened in 1988. This market has been booming since 1992. Until the end of 1995, the total number of cases was 1303 with a total land area of 103,920,000 sq. meters. See issues of The Shanghai Real Estate Market, China Statistics Publishing Co., and The Land of Shanghai, The Land of Shanghai Publishing Co., 1990-1996. The system of land price re-compensation (which excludes land related to industrial relocation and housing policies) is important in that it further revitalize the market by allowing the administratively distributed land, which is still the absolute bulk of the land stock (the percentage figure varies between 95% and 99% according to sources), to enter the market. It thus also greatly enhance the financial benefit of the local government.*

14 *There have also been organisational and administrative adjustments in the areas of urban planning and land as well as real-estate management. We will only consider administrative changes deemed most obviously influential in terms of urban land development.*

15 *Currently there are fourteen "districts" (or "chu", for more urbanized areas) and six "counties" (or "hsian", for less urbanized outlying areas) under the jurisdiction of the city of Shanghai. Each chu or hsian has its own administration with an appointed head and a council of elected representatives.*

administrative powers, including planning and related management authorities.[16] Thus districts and counties were spurred to adopt development strategies centered on economic growth and infrastructure building. Planning and land-use regulating powers were further granted to the district and county governments in 1992 and 93, allowing them to draft "detailed plans" and grant permits to land development projects virtually of all types and in all areas (except major cases in important areas or along major streets, which have to be controlled by the city).[17]

Based on interviews that we conducted in November, 1996 and March, 1997,[18] and surveys by Zheng Shi-ling and Tsao Jia-ming (1996:36-38) and by urban planning students of Tong-ji University (Urban Planning Department, Tong-ji University, 1996), we summarize the problems attributed to administrative amendments as follows:-

The conflict between the city as a whole and the districts as independent entities:

This is the first and most obvious difficulty involved. It is indisputable that Shanghai needs a well conceived, relatively comprehensive "master plan" (or at least a set of widely accepted, workable guidelines) for its overall development as a metropolitan region. This overall plan or set of guidelines will hopefully lead to an orderly manner of regional development on the long run, and thus escaping a myriad of urban and regional problems, many of which have proved to be extremely costly to correct in social as well as financial terms. With districts and counties acting basically as independent (and thus localistic) decision-making authorities, the coherence of overall development inevitably suffers.[19] Liberating the local energies and resourcefulness is a necessary and plausible policy, but the accompanying question of maintaining a balance between the wholeness and oneness of overall growth,

---

16 *In addition to urban planning and related functions, the delegation included powers in the areas of financial/fiscal management, public works, labor affairs, personnel, pricing of staple commodities, foreign trade, industrial and commercial management., etc.*

17 *Worried by the omnipresent over-development induced by the relatively high autonomy of lower levels of governments, there had been an attempt to retake some of the powers in granting land leases by district and county governments, but was eventually thwarted by strong opposition from the districts.*

18 *The interviews were carried out in Shanghai. Interviewees included local planning professionals, relevant officials in various city and district governments, academics in related fields, and investors from Taiwan.*

19 *The development of hierarchy of city-level and district-level commercial centers is a case in point: e.g., the emergence of Hsu-jia-hui as a city-wide commercial attraction, Kung-jiang Street as a Yang-pu District's commercial core, and many other similar example are basically results of local efforts instead of being planned and promoted by the city; while the city-designated centers of Zhen-ru and Hua-mu have so far failed to grow.*

and the vitality of local administration has to be tackled seriously and promptly.

Developmentalist localism:

Developmentalist mentality is common among political and bureau-cratic decision-makers in developing countries, and present Shanghai has more than its share. This "the bigger the more the merrier" and "the economy is everything" attitude, which permeates through all levels of decision-making, has combined with the competition among localities and the incentives generated by differences in tax revenue sharing to make local governments greatly interested in luring high-intensity commercial and service sector growth.[20] The local interest is so high that, it is reported many times during our interviews, the district and county governments would rather grant unauthorized variances in landuse type and increase in building floor area ratio to satisfy pressures from the land developer than uphold city-imposed plans and regulations. Because planning is usually idealistic and restrictive owing to considerations on behave of city-wide reasons, disagreements and tension may evolve between planning and development forces. This is nothing new in capitalist societies all over the world. What is really academically intriguing is that, according to our interviews, strong distrust or even animosity exists between local governments and city-level planning, i.e., the division of attitude between levels of administration instead of between public and private sectors. As will be touched upon later in the paper, this has to do with the particular roles played by the local governments in China's present transitional process toward a "socialist market economy" and the awkward position of planning in such a process.

The dragging of urban planning behind urban development:

The recent speed of economic and physical growth of Shanghai has been astonishing and unprecedented. But preparing and ratification of city plans is time consuming. There has been an unavoidable time lag between

---

20 *Since 1994 a new system of taxes has been in effect, causing basic changes in tax categories, rates and sharing of tax revenues between the central and local governments. In the system, there are centrally collected taxes (e.g., excess tax, import tax), locally collected taxes (e.g., personal income tax, land value increment tax), and taxes that are shared by the central and local governments (e.g., tax on securities trading, sales tax, and value-added tax). The implementation of these taxes has resulted in various fascinating phenomena (one of them that closely relates to our subject is the fact that many localities, reasoning that land value increment tax being a local tax, lowered or waived the tax on their own in order to attract more real estate development). Because the value-added tax (25%) and business income tax (varies with the capital structure and ownership level) has become the main revenues for the city and district governments, the tertiary industries which generate these tax moneys have thus been vigorously courted by local governments.*

planning (and the ability to control and regulate) and the need of guidance in the real world of development (one conspicuous example is that the prolonged failure on the part of the central government to ratify the badly needed new Master Plan has let Shanghai's recent rampant growth run on an uncharted path and thus worsening the situation). The magnitude and the fact that there are fourteen districts and six counties under the city that are fiercely competing for growth only made the present condition more confusing and urgent.

The disparity between high-intensity development and low standards of urban infrastructure and services:

Ever since its early stages of development, Shanghai has been a high density urban settlement.[21] The almost forty years of orthodox Communist rule before the economic reforms had done little in providing basic urban infrastructure.[22] With the rapid growth in recent years, the hardly-controlled inter-district vying, and the rising expectations on the part of the residents, the shortage of urban infrastructure and services has become a critical problem to be seriously faced by the city and local governments.

On the front of reforms of enterprises, the basic strategies have been the separation of administration and enterprises (or the separation of the ownership and management of state-owned enterprises) and decentralisation of power. Under these strategies, many reform measures have been put into effect. Among the most significant are: capitalisation of properties owned by

---

21 *At the time of liberation (1949), the average population density was around 60,000 persons/km² and per capita living space was only 3.9m². As the city's jurisdiction extends over the years, especially when the Bao-shan and Pu-dong new districts were annexed in 1989 and 1992 respectively, overall density lowers. However, the densities remains very high in the old central areas, such as districts of Huang-pu, Nan-shi and Lu-wan. In some extreme areas, the figure can be as high as 160,000 persons/k, and the per capita living space has merely increased by 0.4 since the reforms started. The problem of housing quality has been very serious. The recent increase of investment in and construction of housing has caused the rise of housing standards. See the table below for density changes:*

Year	1952	1962	1975	1982	1992	1994
Density	61.38	45.14	39.54	27.32	10.00	4.63

(unit: 1,000 persons/sq. km)

Source: Shanghai Municipal Statistics Bureau, Statistical Yearbook of Shanghai, China Statistical Publishing House.

22 *Before the reforms, investments in industrial capital improvement had out paced those for urban construction and improvement. Quantities and qualities of urban public services, including housing, transportation, and utilities could not keep up with demands generated by urban growth. From 1949 to 78, the total investment in urban infrastructure was 5.38 billion RMB, averaging merely 0.185 billion per year. Investment figures for other periods are: 4.84 billion RMB for 1979-84 (0.807 billion per year, or more than four times the previous period); 15.38 billion RMB for 1985-89 (3.08 billion per year, or 3.8 times the previous period); 19.3 billion for 1990-92 (6.43 billion yearly, or more than 2 times the previous period); 68.45 billion for 1993-95 (22.82 billion yearly, or about 3.5 times the period before). Thus the total investment between 1978 and 1995 was 47 times the amount before 1978.*

the state, and delegating managerial powers, responsibilities, and interests to the enterprises per se (instead of governmental regulatory or supervisory offices which used to assert direct control of the enterprises' daily operations). Therefore governments at various levels are hoped to relate to business only through financial and fiscal policies and regulations, while enterprises are left alone with enough incentive and decision-making powers to respond to and compete in the market on behave of the state ownership. Since these policies are not yet fully implemented, it is difficult to speculate on their final outcomes. But because land, especially urban land, is the major element of state properties, these reforms, through their effects on behavior of the management, have tremendous consequences on urban land development in China. So far the management has behaved just like landowners in any capitalist society—profit-motivated, development-oriented. However, there is a crucial difference between the capitalist landowner (or developer) and the land-user agent in China: the latter lacks a consciousness of the risk associated with investment. The main reason, we speculate, must be that, unlike the real landowner or real estate investor, the manager for state-owned (or collectively owned, for that matter) properties does not have to shoulder the loss that may result in risky investment decisions—the state or the collective will absorb it. This difference makes them that much more adventurous. Because this type of mentality can be widespread, we suspect, it must have had very significant effect on the supply of urban space (both in terms of quantities and location) in Shanghai where land is highly valuable and public landuser agents abound.

There have also been problems stemming from the reforms aimed at state-owned enterprises. One type of the problems is due to the self-perpetuating nature the bureaucracy. According to Wu Guo-guang, the delegating of management power toward the lower-tier enterprises has great impact on local governments. Because of the fear of losing the benefits and influences associated with directly controlling or running government-owned enterprises, the local government may "intercept" the management power meant by the central government for individual enterprises. (Wu, 1995:27-28) As a result and also owing to the possibility of wielding the enormous government-owned properties in the emerging markets, the local government may now enjoy more influences on the local economy through policy or non-policy measures.[23] Another type of problems generated by the great trend of marketisation is the confusion and negative effects caused by the double-roles played by the

---

23 *Judging from the fact that bureaucratic decision-makers are prone to expansion of power through inflated budget, and engaging in real estate development is a good way to do just that, we believe the general tendency is for the local government to undertake more development projects rather than less.*

local government. The local authorities in China are now simultaneously playing two roles which, in capitalist eyes, are conflicting: on the one hand, it performs all the functions which are expected from a local government as provider of public services and promoter of public interests, usually by restricting the market rationality and supplementing services that are not likely to be rendered by profit-motivated investors in the market; on the other hand, it joins the profit-seeking market through the government-owned businesses either by directly running their operations by way of functional branches in the administration, or by high-ranking officials' serving in the management of the enterprises. So we see in Shanghai a remarkable phenomenon that can only be called "marketisation of the state". The problems are twofold: unfair market competition with privately-owned businesses, and the possible sacrificing of public interests which must be provided for and protected by the government against market rationales.

17.4 Effects of Administrative Reforms on Land Development in Shanghai

The first effect of all the reforms which enable the acquiring and transfer of urban landuse rights (thus urban space) as a "commodity" is of course the great increase of the marginal productivity of urban land (and space). Urban land being the major property of the local government, this increase of land productivity greatly enhanced the fiscal situation of the local government (and to a large extend, of all the units in the government or in government-controlled businesses which posses land or urban space). Thus, through the urban land market, part of profits generated in China's current economic growth is being recouped for a large portion of the masses. The possibilities of keeping the basic system of public land ownership (this is the socialist ideological part) while taking the advantages of the market mechanism (this is the Chinese pragmatism part) seem to exist, though imperfections in the newly designed reform measures abound.

In addition to the conflicts between city government and district government, as well as among different branches of government, there are other discernible difficulties resulting from the reforms. Since the economic reforms taken as a whole is a wholesale substitution of an economy based on market principles for a centrally planned one. The decline of the stature of urban planning institution which used to be an integral part of the state economic planning system is all but destined, as its wisdom is continuously called into question by land developers and even local government officials. The problem of the discord between plans and real-world landuse practice is only a reflection of this larger revolutionary contextual change. The task now is to develop a mechanism which is reasonable, flexible, and sensitive to market needs, but at the same time capable of maintaining acceptable environmental qualities by minimizing externalities of urban development and providing adequate urban services. However, there also seems to be conflicts and discordances among levels and sectors of governmental planning itself.[24] One of the most obvious effects that is directly related to our interests here is of course the oversupply of urban space[25] resulting from the unwary or even reckless investments made by the management of state-owned enterprises and other public land user entities, due to their lack of risk consciousness. Doubtless, this should be stressed as one very essential flaw in the present assortment of reforms. It not only endangers public properties, wastes the resources that go into the construction of urban space, but also de-stabilizes the real estate market and contribute to chaotic land-use patterns which in turn generate a wide range of environmental problems. The answer to this flaw lies in the combination of a fair merit-based salary system, and (in the case of capitalized establishments) an effective supervision mechanism on behave of the shareholders or (in the case of non-capitalized ones) political pressure from the public. At the present time none of these conditions exists in China. We believe that one of the goals of real reforms must be to create these elements.

17.5 Conclusions

New system of paid transfer of land-use rights and the decentralisation

---

24 *For example, there are obvious conflicts among the so-called "four plannings": National Land Development Planning, City Planning, Comprehensive Agricultural Development Planning, and Master Landuse Planning. The object of the first planning was natural, social and economic resources, and its duty focuses on the overall allocation of population and man power for national economic development. City planning tries to develop orderly town and city systems and provide the environment for economic as well as social development by planning and control of landuse in cities and towns, and by construction of various types of public works in the planned areas. The aim of comprehensive agriculture development planning is the expansion and improvement of agricultural land and the related infrastructure. The landuse master planning covers the total land body of the country. It is supposed to consider the natural characteristics of the land, and the landuse requirements for national economic and social developments in order to set goals, strategies, and patterns for national landuse. It also should establish control guidelines for important landuse sectors. In real practice, however, since the involvement of a great number of administrative units, causing serious overlaps, inconsistencies, and bureaucratic conflicts and buck passing.*

25 *From this perspective, what we now observe in Shanghai's real estate growth is not a true expression of its economic strength but an exaggerated statement of over optimism, and the slowdown since 1994 must be deemed inevitable.*

of decision-making powers (both between levels of government and within the managerial system of state-owned enterprises) are the double-power propeller of the astonishing post-reform urban development, because they provided the basic incentives which stimulate the aggressiveness and resourcefulness of the front-line decision-makers involved in the land development process.

To put simply, the total and final effect on urban land development of all the related reforms is the increase of land-use efficiency and over supply of urban space, the latter being due to the lack of risk consciousness and the almost congenital bureaucratic over supply of "services" on the part of the local government and other public land managers.

The whole assemblage of reforms has been conflict-prone. There is the conflict of roles played by the local government: the government as implementer of national policies and protector of public interests vis-a-vis the new role of aggressive, profit-oriented land developer; there is the conflict between the central and the local governments as well as between the city and district/county governments: overall interests and direction vis-a-vis partial and local autonomy; there is the conflict between public interests and commercial profits: the contradiction of planning and economic growth; and there is the conflict among different functional branches of the government. Conflicts and contradictions teem in other countries, too. What makes these types of inconsistency particularly significant is the fact that social conflicts have to be checked or supervised in order to avoid severe adversary consequences, but there is no mechanism in present mainland China to perform this function (because these contradictions involve the local government which enjoys many powers and privileges not subject to market competition, and neither is there any meaningful checks and balances in the political arena). This is one of the most serious and fundamental dilemmas in the process of economic reform which must be faced squarely, if the great reform is to be really successful.

What is now happening in mainland China is an endeavor of marketisation without (or with limited) privatisation. This is a daring (if not clever) attempt of historical proportions. How likely is this to be successfully accomplished? Going one step beyond what is pointed out by Peter Marcuse that ownership is a bundle of rights to be divided between government and individuals (Marcuse, 1996: 119-122), we can say that land ownership in present mainland China is to be divided between government, bureaucrats, and, to a far lesser degree, individuals. These rights are being re-divided among these actors during the economic reform process. How the partitioning is devised and carried out will largely determine the outcome of the reform. With luck and prudence, the possibility of success appears to be there.

Switching the basic system of a society is a complex and time consuming process. Establishment and amendment of legal and administrative systems are necessary but not sufficient conditions. Social, cultural, and bureaucratic factors always play significant roles in shaping a country (and a society). For the reform to be successful much more attention has to be paid to these fronts.

References

(All references are in Chinese except those indicated otherwise.)

Chou Cheng, (1989), *Land Economics*, Beijing: Agriculture Publishers. (1993), "Macro-Tuning the Land-use in Mainland China", A Collection of Essays for Cross-Taiwan-Strait Academic Exchange. Taipei: Dept. of Land Economic, National Chengchi University Press.

Gue Chiang-fang, (1994), *How Foreigners Can Invest in Real Estate in China*, Taipei: Yung-ran Publishers.

Hsiao Tze-yueh, (1991), *Real Estate-related Institutions in Mainland China*, Taipei: Wan Guo Wei Li Publishers.

Ju Shin-min, (1991), *A Study on Chinese Communists' Reforms in Political Institutions, 1978-1990*, Taipei: Yung Ran Publishers.

Ma Keh-wei, (1992), *A Comprehensive Book on Chinese Reforms:1978-1991*, Dai-ren: Dai-ren Publishers.

Marcuse, Peter, (1996), "Privatisation and Its Discontents: Property Rights in Land and Housing in the Transition in Eastern Europe", in Andrusz, G., Harloe, M. and Szelenyi, I. (eds.), *Cities after Socialism: Urban and Regional Change and Conflict in Post-Socialist Societies*, Oxford, UK: Blackwell. (in English)

Wang Shung-san, (1990), *A Study on Land Problems in China*, Taipei: Ching-yu Publishers.

World Bank, (1994), *China: Urban Land Management in an Emerging Market Economy*. Washington, D.C.: World Bank Publishers. (in English)

Wu Guo-guang & Jeng Yeong-nian, (1995), *On the Central/Local Relationship: the Pivotal Issue in Institutional Transformation in China*. Hong Kong: Oxford University Press.

Zheng Shi-ling and Cao Jia-ming, (1996), "Planning and Administration of Shanghai as an International Metropolis" in Zheng Shi-ling, *The Renewal and Redevelopment of Shanghai*, 31-42. Shanghai: Tong-ji University.

Urban Planning Department, Tong-ji University, (1996), *The Implementation of City Planning in Shanghai: a Report of Survey* Shanghai: Tong-ji University Press.

18 Characteristics and Problems of Asian Urbanisation from the Viewpoint of City Planning

Tomoyoshi Hori, Keiichiro Hitaka, Satoshi Hagishima

and Shinji Ikaruga

Editors' note:
The core areas of some Japanese cities are experiencing depopulation just like some western cities, and just like Mumbai, Calcutta, Seoul, Singapore. Urban land values have increased and journey to work times extended. Japan has attempted to differentiate between adminstrative and industrial cities, to provide better environments, mixed land uses and shorter journeys to work.

18.1 Introduction

The techniques of city planning have been developed since the 19th century and through the early decades of the 20th century mainly in West Europe and North America. This has given a common language to city planners, and it is a common intellectual heritage of mankind. The tools have been developed to curb the injury to the public and the damage to the lived environment that emerged during the industrial revolution.

We do need the techniques of city planning to cope with this situation. Differing from the natural sciences and engineering sciences, city planning embraces many forms of social analysis and policy making. However, compared with the history of the West, urbanisation in Asia is progressing more rapidly and in a different technological era. Additionally, because the social and cultural circumstances and the policy environment of Asian cities are different from those of the west, we should not necessarily expect to be successful in applying western techniques in Asian cities.

The purpose of this chapter is to analyze the present condition of the urbanisation of Asia, with the background and the awareness of the issues mentioned above, and then to clarify the problems of Asian city planning. Therefore, to begin with, we review the development stages of Asian economies and analyze the urbanisation of Asia from the aspects of population, industry, improvement level of urban facilities, urban life, education level, land use, the diffusion of consumer durables, and so on. Next, we will analyze the present condition and problem of the urbanisation of Japan, which is one step ahead in Asia, in order to find out the concepts that are helpful in guiding the urbanisation of Asia. We have applied the methods of field investigation, statistical collection and literature review from the standpoint of city planners

18.2   Development Stages of Economy in Asia

In Asia, the first country to achieve economic take off was Japan. After the Second World War the other economies began to develop from a colonial or semi-colonial inheritance. They most developed new industries dependent on investment and technologies from overseas, and on overseas development aid. In the latter part of the 1960s the four Tigers (Hong Kong, Taiwan, Korea, Singapore) began their acceleration with export-led growth. In 1980s these four were joined by another group of countries acclerating their economic growth - Thailand, Malaysia, Indonesia, the Philippines, and then China. In the 1990s Vietnam embraced market reform, and India adopted a policy of liberalisation.

This pattern of economic growth has sometimes been called the "flying-geese-shaped economic growth", in which Japan took off first followed by the Tigers and then ASEAN, and finally vast China. Each country and area developed export industries in turn, earned foreign money, improved the economic infrastructure of their countries and achieved economic growth.

18.2.1 Growth Rate and Scale of Asian Economy

The growth rate of the Asian economy was 8.4% in 1994 and 8.0% in 1995. In 1998 Asia faces recession, but witnesses expect this 'correction' to be a passing and not permanent affliction. The average growth rate of 8.0% of these countries was about four times that of the advanced countries - at about 2.2 %. The rate of economic growth in Asia (except Japan) was increasing in the long turn. It was 5.6 % in 1970s, 6.9% in 1980s and 7.4% from 1990 to 1994. As the result, the position of Asia in the world economy   improved. The total amount of the GDP of Asia in 1994 was $ 2.3 trillion, which is around the half of Japanese GDP of $ 4.7 trillion. It accounted for 8.9% of the world GDP (Japan 18.2%, USA 25.7%). However, as for the GDP per capita, it varied widely and in the aggregate remains fairly low. It is over $20,000 in Singapore and Hong Kong, but less than $500 in China, India and Vietnam.

18.2.3 Foreign Trade and Industrialisation

In 1944 the total export trade of Asia (1994) surpassed the total amount of the 3 member nations of North American Free Trade Agreement (NAFTA; USA, Canada, Mexico). The imports are less than those of the three countries of NAFTA. However, it nearly equals the total of the import of the three main European countries (Germany, France, UK). While the trade of Asia has grown rapidly in quantity, the change of quality is also remarkable. Formerly, the export of many Asian countries had depended on primary commodities. However, it is shifting rapidly to depend on industrial commodities with the progress of the export oriented industrialisation. Now three fourths of the Asia's exports are industrial.

The progress of the industrial power, that is the background of this trend, is also remarkable. The share of Asia in the production of basic industrial commodities in the world is 19.4% for coarse steel, 41.5% for cement, 16.1% for naphtha, 24.4 % for family refrigerators, 24.5% for family washing machines, 47.6% for TV sets, 20.6% for semiconductors, 52.2% for acoustic apparatus, 29.1% for merchant ships and so on.

18.3 Present Condition of the Urbanisation in Asia

18.3.1 Urban Population

According to the "Demographic Yearbook; UN, 1994", the population of the world amounted up to 5,700,000,000, of which one billion people live in the urban areas of Asian countries. There are 35 cities with a population of more than three million. Examples are Seoul (10,727,000), Bombay (9,926,000), Manila (8,594,000), Shanghai (8,206,000), Tokyo (7,968,000) and so on, all of which are bigger than the older metropolises - New York (7,312,000), London (6,905,000), Berlin (3,475,000), Paris (2,152,000) - of the advanced countries in the West.

Editors' note:
Different population figures based on different aggregations and definitions may be found elsewhere – e.g. the UN World Urbanization Prospects: 1994 Revision. But the general ranking and the general point remains valid.

18.3.2 Rapidity of Urbanisation

Many cities in Asia have ancient foundations, and often were continuously occuped or occupied for long periods as the seat of power e.g. Beijing. Another kind of city derives from the impact of colonial and trading powers who founded new urban settlements – sometimes where previously there had been nothing but scattered villages – e.g. Batavia (Jakarta), Singapore, Hong Kong, Tsing Tao, Harbin. They are rather new cities from the view point of history. Many cities in South East Asia were formed in an appropriate place for a trade center or a supply center of commodities. Since the 1950s either or both of these have grown rapidly, while in addition in some countries this has also been a period of yet more town foundation – sometimes associated with new national and provincial governments, sometimes with planned industrialisation.

18.3.3 Characteristics of the Urbanisation and National Development

Much of the discussion in the following paragraphs can be understood with reference to Table 18.1. The quality and quantity of data available for Asian countries varies greatly – so that on only a few indicators is it possible to conduct a proper comparative survey. This Table includes representative comparative data that we believe useful to this discussion.

18.3.4 The Primacy of the Capital Cities

Most Asian countries show quite a high degree of urban primacy. Often the capital city embraces a major fraction of the national population - Seoul and neighboring Incheon in Korea have nearly 30% of S.Korea's population: it is 13% for Taipei, 10% for Bangkok in Thailand and 13% for Manila in the Philippines. If we think in terms of their share of the non-agricultural (mostly urban) population, these fractions would double.

The size of the agricultural/rural population is still overwhelming in many ASEAN countries, China and Indochina. Table 18.1 shows that the average for these countries is 50 % of the population still supported by primary economic activity. In these areas, increases in population have been absorbed to some extent by the increase of the production capacity and the development of poorer lands, often because of new the introduction of new technology. However, the generation of new rural employment is now approaching the limit, and there are even some areas that show absolute decreases in rural incomes. Although industrialisation is rapidly progressing in many cities, there is concern that the cities will not be able to provide adequate employment or shelter for the excess population in countries where population growth rates still remain high. Again, it is the same group of countries that is picked out – the average for China, Indochina, ASEAN and India must be near 2%.

18.3.5 Development Level of Communications

The comparison between Western and Asian economies is made quite clear in Table 18.1. Apart from the small city states like Hong Kong and Singapore, road densities in Asia are lower, car onwership is lower, telephone sets per capita are lower - in the case of India hardly even 1/50th of those in the West. We do not have data for urban facilities like parks or sewage systems, but it is our belief that the standards of urban services in most countries is very low compared with the Western countries.

18.3.6 Urban Life and Education Level

Here we look over some indexes of urban life and education level given by the "Statistical Yearbook 1995" of UNESCO and the "Statistical Yearbook 1993" of UN. First, we look over the number of TV sets per 1000 persons. The diffusion level for the Asian countries, except some advanced areas of Singapore, Hong Kong and so on, is about one 10th of that for the advanced countries. In the same way, the number of telephone circuits per 100 persons is one 50th. As for the circulation of news papers, it is noticeable that the number for Hong Kong, Japan and Korea is higher than in Western advanced countries, but for other countries it is only 10th. As for the number of publications, it is noticeable that it is high for UK and low for USA, but for most other countries except Korea it hardly reaches 1/20th of the UK level.

On the other hand, if we look over the rate of attendance at schools of primary, secondary and higher education, primary education has gained in popularity for the most part, but iin secondary education it is about half Western levels. As for higher education, when face to face with the advanced countries atabout 40% level, most of the Asian countries are lagging far behind, with the exception of Korea (64%) and Japan. It is very low level for the other countries compared with the advanced countries.

18.3.7 Land Use in Urban Area

The Industrial Revolution in the West produced squalid and insanitary

Table 18.1 Asian urbanisation indices compared with advanced countries

INDEX	Growth Rate of Real Economy			GDP per capita (US $)	Population Growth Rate	Population Ratio of Primary Industry	Population Ratio of Secondary Industry	Population Ratio of Tertiary Industry
PERIOD -YEAR	1970-75	80-85	90-95	1995	1990-94	1993	1993	1993
U.S.	2.3	2.5	2.3	27,547	1.1	2.8	23.7	72.8
U.K.	2.0	2.0	1.2	18,990	0.4	2.2	25.9	71.9
Germany	2.2	1.2	1.8	22,539	0.6	3.0	34.7	62.3
France	3.5	1.5	1.1	26,447	0.5	4.7	26.5	68.8
Japan	4.4	3.7	2.6	40,819	0.3	5.8	33.4	60.3
NIEs								
Korea	8.9	8.4	7.5	10,156	0.9	15.6	33.4	48.5
Taiwan	-	-	-	-	0.9	-	-	-
Hong Kong	-	-	-	-	1.5	0.6	28.0	71.4
Singapore	9.5	6.2	8.5	28,463	2.0	0.3	33.2	63.5
ASEAN (4)								
Thailand	5.6	5.4	8.4	2,812	1.4	62.1	13.3	21.7
Malaysia	7.4	5.1	8.4	3,594	2.6	30.6	22.0	47.4
Indonesia	8.0	4.7	7.1	1,022	1.7	54.7	12.7	29.7
Philippines	6.1	-1.4	2.2	1,054	2.2	44.6	15.8	39.6
China	5.0	10.1	11.7	420	1.1	56.4	22.4	21.2
India	3.0	5.4	3.9	328	1.9	62.6	12.3	16.1
Indochina(3)								
Vietnam	-	-	-	-	2.3	-	-	-
Laos	-	-	-	-	3.1	-	-	-
Cambodia	-	-	-	-	2.8	-	-	-

Urban Population Ratio	Population Ratio of the Capital City	Ratio of Paved Roads	Roads Length per Area (E/E)	Number of Passenger Cars (per 1,000 persons)	Number of TV Sets (per 1,000 persons)	Number of Telephone Circuits (per 100 persons)	Number of News Paper Publication (per 1,000 persons)	Number of Book Publication (per 10,000 persons)	Population Ratio of High Education (per 1,000 persons)	
1995	1994	1995	1995	1995	1993	1993	1992	1995	1995	
76.2	0.2	90.0	0.64	565	816	57.4	236	1.89	65.9	U.S.
89.5	11.9	100.0	1.58	355	435	49.4	383	16.36	36.9	U.K.
86.6	4.3	99.0	1.80	482	559	45.7	323	8.26	41.1	Germany
72.8	3.7	100.0	1.47	431	412	53.6	205	7.14	43.9	France
77.6	6.4	73.0	3.01	341	618	46.8	577	2.84	35.4	Japan
81.3	24.1	77.8	0.74	116	215	37.8	412	6.94	63.6	Korea
-	12.6	-	-	180	-	-	-	-	-	Taiwan
-	-	100.0	1.54	55	286	51	822	-	-	Hong Kong
-	100.0	97.2	4.76	111	381	43.4	336	-	-	Singapore
20.1	9.9	92.8	0.10	21	113	3.7	85	1.28	17.5	Thailand
-	5.9	75.0	0.20	-	151	12.6	117	1.92	7.4	Malaysia
35.4	4.1	-	0.13	7	62	0.9	24	0.33	12.9	Indonesia
54.2	12.8	-	0.54	8	47	1.3	50	0.15	21.9	Philippines
30.3	0.6	-	-	-	38	1.5	43	0.77	5.3	China
26.8	0.03	48.2	0.61	4	40	0.9	31	0.14	10.5	India
20.8	1.5	-	-	-	-	-	8	-	-	Vietnam
-	9.6	-	-	-	-	-	-	-	-	Laos
-	9.9	-	-	4	-	-	-	-	-	Cambodia

Sources: Demographic Yearbook (1994) (United Nations), *Statistical yearbook (1993)* (United Nations), *Yearbook of Labour Statistics (1993)* (International Labour Organisation), *Statistical Yearbook (1995)* (United Nations Educational ,Scientific and Cultural Organization) and *World Road Statistics (1995)* (International Road Federation).

slums with high densities of settlement close to foundries and factories. It was partly as a response to this and the percieved unhealthiness of the cities that land-use planning evolved to rate residence, commerce and industry, and the semi-rural semi-open ideal of the lower density suburb evolved. The ideal became the efficient complementary practical combination of these land-uses, without 'improper' mixing. This was economically possible in an age which prospered on cheap energy and cheap transport. One consequence has been in part the increasing abandonment of city centres. Many people in the West no longer know how to live in a traditional urban culture.

On the other hand, most Asian cities retain the trait that residences, commerce, and industry are intermixed. Given that the advanced countries are working hard to renew the de-urbanized city centers, we can also say that the existing forms of the urban areas of Asia are anticipating the future, for the land use mix of the urban area in Asia guarantees the bustle of prosperity that western city centres have often lost.

18.3.8 Urban Structure for Saving Energy

The inhabitants of western cities rely heavily on driving cars, on sprinkling water on large gardens in the summer, and on sewage and garbage collection from sprawling suburbs so roads, water systems, sewage systems and electricity circuits are spread all over the country. Residences in the suburbs of the advanced western countries are said to consume 100 times as much energy per capita as those in India. Even in Japan which has a low energy consumption per unit of GDP, because the GDP is so high, the energy consumption per capita is 62 times that of Bangladesh. The discharge volume of carbon dioxide, that causes the warming of the globe, is 63 times. However, even if average per capita incomes are low, China and India by virtue of their populations represent vast markets. More than 30% of India's 1 billion people now have access to TV. The emerging urban middle class is variously estimated to be 80 or 200 millions people – who are increasingly becoming motorised.

18.3.9 Conclusion

For much of Asia, the relaity is that many millions more people will have to leave rural homes and agircultural livelihoods, and migrate to the existing cities and new cities. The conditions of many of these cities are already often very poor, with overstretched infrastructure. But Asian cities are energetically fairly efficient. In governing the growth of existing cities and promoting the growth of new ones, planners are presented with new probelsm and new philosophies – how to stimulate and integrate industry, residence and commerce, not how to separate them.

18.4 Present Condition and Problems of the Urbanisation in Japan

Here we look over some lessons on the city planning from the formation of cities in Japan.

18.4.1 Rapid Growth of Japanese Economy and National/Urban Policy

In the atermath of World War II Japan concentrated in the first decade on the reconstruction of industry around Tokyo. From 1960 during the period of rapiid economic growth, Tokyo's centrality was complemented by the development of population and industry in the Pacific Belt Area. This formed the prototype model of the national and urban policy of today's Japan, that is to say the centralisation of administrative functions in Tokyo and the decentralisation policy for the industrial development. This is the policy that divides the Japanese cities into the two types of the "administrative cities", whose principal role is the function as the "administrative center of economy and society", and the "industrial cities", whose principal role is production activities, (see Table 18.2). This is the policy that tries developing the economy efficiently by pursuing the most appropriate urban structure and environment for each type. This is shown in the several comprehensive national development plans after 1962.

The administrative cities are represented by Tokyo, Osaka, Nagoya and another four central cities of four regional blocks, Sapporo, Sendai, Hiroshima and Fukuoka. It has enabled the extension and development of Japanese economy in a short time, through the efficient and centralized administration of the national capital, labor, resource and information. On the other hand, the "industrial cities" of Kitakyushu, Kawasaki, Yokkaichi, Amagasaki and so on produced large amounts of high quality industrial products most efficiently, through the arrangement of the urban structure that aims for the efficient production and trasnportation.

Next, we investigate merits and demerits, achievements and failures of these cities, from the "administrative cities", dominated by Tpkyo, to the "industrial cities", represented by Kitakyushu.

Table 18.2 Administration City and Industrial City

	Administration City	*Industrial City*
Image of City	sky scrapers, flowery, white colour, trip to work with subway, decadent culture,	middle and low buildings, steady, blue colour, trip to work by personal cars, fresh and vivacious culture,
Merits	international city (advanced finance city), specialised in Management, capital (money), information	industrial city of high efficiency specialised in Manufacturing, work (human), resource (goods), technology,
	variety, very attractive, vitality, anonymity, freedom,	warm-hearted human feelings, housing of reasonable price, cars, safety,
Demerits	problems of too concentrated urban spaces (housing, trip to work, disaster prevention) cold-hearted human feelings (lack of community), irresponsibility, hard to live for persons,	problems of pollution simplicity (mono-culture), unrefined, boring, too many old persons, social restrictions (gossipy neighbours),

18.4.2 Administrative City

Tokyo, Osaka, Nagoya and the central cities of the four local blocks are the administrative cities that dominate the network of society in Japan. The principal role of these cities is to collect the information of the "industrial cities" efficiently, and then to respond. In this manner, they have achieved the role of the pulling power to develop the Japanese economic society. In the same time, they have provided good urban services to the citizens and represented the vitality and wealth of Japan. However they still have problems to face.

Quantitatively speaking, the number of houses in Japan had already exceeded the number of families in 1973, although there remains some concern over quality. The most important problem is the price of houses. The selling price in lots of a mid-to-high-rise apartment house is 43,540,000 yen, that is 5.1 times of the annual income. The ratio between income and a flat (67 square metres) in Tokyo is on average 4.8 These are the worst ratios in any advanced country. They are several times worse than prices in New York, London and Paris.

The problem of long commuting times, the result of increasing separation between work and home, is one of the most serious problems of larger cities. Looking at commuters in the three wards of the central Tokyo area who use rail or bus (about 2,040,000 in 1990), the average journey time is 69 minutes. 70% of commuters travel more than 60 minutes each way, and even 20% more than 90 minutes. The congestion rate (number of passengers carried compared with designed capaicty) of railways in the peak hours remains at197% in the main 31 sections (1993). There are some sections where it is more than 250%. Road congestion is also bad. In Tokyo average speeds are 12 kph (8 mph), 20 kph in Osaka and 13 kph in Nagoya.

18.4.3 Dense Urban Area and Disaster Preparedness

The Hanshin earthquake of January 1995, that recorded an intensity of 7 on the Japanese seven-stage seismic scale for the first time, reminded us of the terrors of an earth quake. 6,308 persons died and about 320,000 were made homeless. 210,000 houses were completely destroyed or heavily damaged. Comparing with the damage caused by the North Ridge earth quake of Los Angels in January 1994, the numbers of injured persons were 5 times higher, and the number of persons killed was 110.7 times higher. Many of the deceased persons were killed by falling houses. This was the result of high denisty housing, with many small plots and narrow spaces between, with little open space and few roads, and often also with old wooden structures. Dangerous, densely built-up areas of this type are distributed widely in the larger cities in Japan.

18.4.4 De-Urbanisation of Inner-City

The cores of larger cities in Japan have become hollow just like many Western cities, because the excess centralisation of commercial and business functions, and the high land prices make it hard to live for persons. There is no longer a living central community. In Tokyo, where this phenomenon has progressed furthest, the central residential population density is 35 persons per hectare. The policy of living in city centers (recalling population to city centers) is being implemented today. The low density of population of Tokyo is remarkable compared with New York or Paris, where people still live in many parts of the centres.

18.4.5 Industrial City

The industrial cities of Kitakyushu, Kawasaki, Yokkaichi and Amagasaki are examples of the industrial base of Japan's economic wealth.

These cities have striven for the improvement of land use and transport systems to ensure the efficiency of production and distribution of their industrial goods. They reflect (1) the collective location of factories in the coastal areas that have the merit of marine transport for export and import, (2) the improvement of highways that ensure the convenience of transportation to and from other cities and the inner cities and (3) the development of housing estates to secure the workers. In step with the improvement of production infrastructure, the urban social facilities have also been improved to ensure the quality of urban life

18.4.6 Shadow of Environmental Contamination

The production activity of industrial goods is inevitably accompanied by some degree of terrestrial, atmospheric and water pollution; problems which we have tried to address, as the following example illustrates.

In 1963, five medium cities with a population from 100 thousand to 300 thousand, that were located in the northern part of Kyushu, were incorporated to form Kitakyushu city, the seventh largest city with more than one million population in Japan. The component towns had been in decline after the 1950's because of (1) the shift of energy from coal to petroleum, (2) the shift of from rail to road (3) the post-Fordist shift of industrial structure from emphasizing sheer scale to stressing compactness and flexibility with technological innovation and (4) the shift to the Intense Mass Consumer Society and the change of the values of people. However the consolidation of 1963 improved administration and finance. The new city tried hard to invest in (1) transportation and distribution facilities such as high ways and harbors, (2) industrial sites and industrial treatment works, (3) life infrastructures such as housing, sewage system and parks, (4) cultural and convention facilities of art museums, libraries and international conference halls, as well as (5) the 'soft' institutions such as education and welfare. As a result, Kitakyushu was judged to have the best living environment among Japanese cities with a population of more than one million in 1988 .The city was awarded the "Global 500 Prize" by the UNEP (United Nations Environmental Plan) in 1990 and the United Nations Commendation for Local Governments at the "Global Summit" in 1992. Today, Kitakyushu city is striving for even more urban improvement toward the "International Technology and Information City" that provides advanced urban services, while improving the existing technology and nurturing the information technology. This example shows that where giovernment, citizens, industry and planners cooperate, successful change can be induced.

18.5 Conclusions

Urban planning in the west has developed its own prioities. In Asia there are diverse patterns of urbanisation and diverse lebvels of development. Those countries that are still at low income levels have cities with mixed land-use and low levels of energy use per capita and low service provision costs. Japan, as an example of a high income Asian country, has large cities which have shown some of the costs and diseconomies of scale associated with mega-ctites in the west: but Japan has also instituted a pattern of discrimination between adminstrative and industrial cities, and in the latter at least has made some progress towards healthier and more sustainable living. There may be some lessons to be learnt here which can be applied to the newly industrilsing cities of low income Asia.

The economic development of Asia, especially of East Asia is remarkable, and it accounts for 10% of the world GDP. Despite the current recession, its share of world GNP is set to grow in the next century. Accompanying this economic growth, there has been a rapid increase in urbanisation. In some cases this is creating problems because of the development of primate cities.

To cope with this rapid urbanisation, it may be useful to learn from Japan's experience, and to distinguish between adminstrative and industrial cities, and to plan and encourage enough of the latter that they can grow as alternatives to the single primate city, providing the new urban dwellers with a better environment, in a city can accommodate mixed land-use, short journey times, and happy 'chaos' – all the antithesis of the western model.

Acknowledgement

This study was supported by the Kitakyushu City Academic Research Promotion Fund from 1995 to 1996, and the Monbusho International Scientific Research Program-Joint Research-from 1994 to 1997.

References

Daiwa Research Institute (1997), "*Necessity of New International Airport*", Investment Information of Daiwa 3.

Economic Planning Agency (1996), "Asian Economy 1996".

Economic Planning Agency (1996), "National Life White Paper 1996".

Esho, Hideki (1996), "Future of Sleeping Big Elephant Depends on Closer Connection with Asian Economic Zone", *Nippon Steel Monthly* vol.61, pp. 7 - 8.

Hagishima, Satoshi, et (1996), "*Comparison between Japan and Korea concerning Development of Industrial Provincial City and City Planning*", Proceedings of 4th Asian Urbanisation Conference .

Hokkaido Tohoku Development Finance Corporation (1988), "Formation of Research & Development Base in Hokkaido".

Igusa, Kunio (1994), "Asia and Japan Observed from Data", *Fujitsu Hisho* No.17, pp.34-37

Ministry of Construction (1997), "Re-Improvement of Densley-Built Up Area", *Move of Today* vol.987, pp. 44 - 47.

Muneta, Yoshifumi (1996), "Asian Cities Made of Original Power over Problems" *Nippon Steel Monthly* vol.63 , pp.9-10.

National Land Agency (1996), "Report on National Land 1996".

National Land Agency / Japan Real Estate Estimation Association, (1997) "Second Investigation on Land Prices on World".

Shiraishi, Kaoru (1992), "Traits of Large Scale Urbanisation of East Asia", Research Report on the Urbanisation and the Change of regional society in East Asia 1992 No.1, ICSEAD, pp. 19-20.

Watanabe, Shunichi (1996), "New Paradigm of Asian Cities", *Nippon Steel Monthly* vol.63 , pp.3-4.

19 Aspiration and Reality in Tehran: Land Use Plans and their Implementation

Vincent F. Costello

Editors' note:
Tehran has grown from 1 million in 1950 to 6.5 million in 1995. In order to prevent sprawl and congestion, attempts have beem made since 1965 to plan its expansion. These plans have included a vision to re-orient the city from a north-south axis to an east-west one. The planning has continued through revolutionary shifts in the country's power structure.

19.1 Introduction

The production of city plans may exist at two levels. One level is that of the plans themselves: these may perhaps be the result of detailed analysis of urban problems and contain realistic intentions to solve those problems within the constraints of resources and knowledge, or at another extreme they may be purely statements of political intention, visions of a shining future which are as much propaganda as plan. The second level is that of physical implementation, how the plan is worked through on the ground (Sutcliffe,1981) This chapter looks at both these levels in the past and present land use planning in Tehran. It examines what these plans tell us about Iran and Iranian political processes over the past half century, and what light the current implementation of plans sheds on general processes of urban development and on the particular circumstances which are found in the Islamic Republic. It will concentrate on the Master Plan of 1969 and ask the question as to was the plan, what were its aspirations, and what is the contemporary reality.

The chapter is based on field work and documentary study of Tehran. Past land use planning and demographic change can be studied through maps, census material, photographs, and the aspirations expressed through earlier plans. Contemporary change can be logged by all these means, by field work and increasingly by remote sensing.

19.2 Before Tehran's Master Plan

The old core of Tehran is some twenty kilometres from the foot of the Alborz mountains. The city expanded from its preindustrial core in the second half of the nineteenth century and in the early 1870s an earthen rampart and a fosse were added, consciously modelled on Thiers' Paris fortifications of the early part of the century. They encompassed much open space. In the middle decades of the twentieth century the preindustrial core was largely dismembered by the vigorously secular and nationalistic Pahlavi regime. Analysis of this process can show some parallels with European experience and also with other Iranian cities (Costello,1973). In the 1930s the new Pahlavi regime drove avenues (*khiabans*) through the fabric of the old city. The effect was to impose a plan which was very much in the spirit of Baron Haussmann. The sketch (Bahrambeygui,1977) with the accompanying plan (map 19.1) shows how a grid was imposed on the existing city.

The buildings of the old Qajar palace complex (marked as the *Arg*) were largely destroyed and a number of new ministries were built on the site. The external style of most public buildings was Achaemenian or even Sasanian, as the regime sought to distance itself from the more immediate Islamic past and associate itself with what it saw as pre-Islamic imperial glories.

The grid was eventually extended to most of the city and two boulevards constructed to link the city with the attractive foothills town of Shemiran. The grid has a number of potential disadvantages for modern traffic management but there are reasons for its use in Tehran other than a simple desire for order (Costello, 1993 and de Planhol,1968).

By the early 1940s Tehran's population was growing at more than twice the rate of other cities in Iran to reach 700,000, assuming a position of primacy. After the war it continued to grow as government, commercial, educational and other institutions concentrated there. The First National Census of 1956 gave the total as 1,512,082. Urban growth out from the core was much influenced by environmental contrasts between north, towards the mountains, and south towards the desert. The modern commercial sector and suburbs for the wealthier sectors of society developed mainly northwards, while the poorer suburbs were to the south. There was a free market in land.

There was at first very little land use planning control of the physical expansion of the city which accompanied all this. Before 1951 unused land on the periphery of the city was considered to be owned by whoever developed it. Much was seized by speculators and,as the demand increased, enormous profits were realised. In 1951 such land was placed in the hands of a government controlled bank (Bahrambeygui, 1977). A number of directives were issued by the municipality but there was no land use planning as such. The lower price of land at a distance from the city centre encouraged developments to leap-frog, rather like much of suburban America, leaving unused sections to await development. Low fuel prices made commuting an option for the wealthier classes.

Government intervention in the process was largely confined to pro-

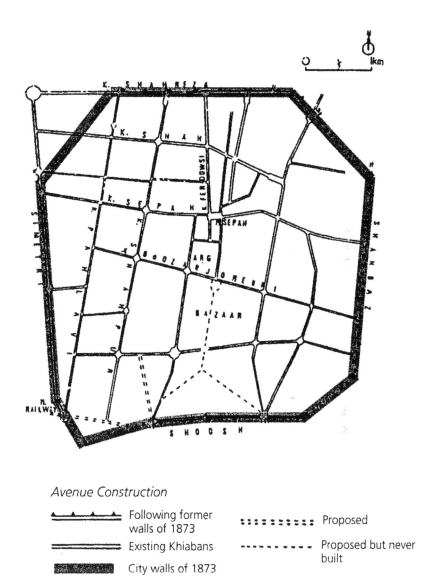

Map 19.1   Sketch-map and plan: Tehran

viding infrastructure facilities - improving the road network and the water supply, with limited participation in housing. The regime was by the 1960s heavily dependant economically on oil revenues and on efforts at import substitution for durable and non-durable goods. Politically, it looked to the United States for support, after the coup engineered by the CIA in 1953. It is hardly surprising then that the Iranian government would turn to a U.S.-based model to try to solve its physical planning problems.

19.3 The Tehran Master Plan

Iranian planning in general was dominated by a series of Five Year Plans beginning in 1949, but these were overwhelmingly sectoral in content and had "insignificant spatial impact" (Amirahmadi,1986,p.524). It was increasingly apparent that some form of comprehensive spatial planning was needed during the 1950s and the 1960s. Eventually a High Council for City Planning was created in 1965 to guide and prepare master plans more all major Iranian cities. However these concentrated on physical planning, particularly the planning of new road networks. There were problems:

the limited number of specialized and experienced engineers in urban development, lack of proper statistics and data concerning towns, the absence of concrete regional development policies at the national level, the failure to determine how to implement comprehensive projects, and the lack of performance guarantees (Plan Organisation, 1968).

A Los Angeles firm, Victor Gruen Associates, in association with Abdul-Aziz Farman-Farmian of Tehran, was commissioned to prepare a master plan for Tehran in 1966. The plan was of the type that was being exported throughout the developing world by American and European planning consultancies. As Lowder puts it:

> These...conditions are reflected in the hallmarks of most urban plans dating from the 1960s: their draughting from the top and from an engineering stance, and their focus on formal land use in isolation from local economic or social circumstances. Not only was there a huge gulf between the `technocrats' who devised them (many were overseas consultants) and the local officials, but rarely was adequate attention paid to the process of implementation—the financial implications, the division of responsibilities and the actual current sequence of events involved in producing the built environment in the specific city..Urban planning was reduced to a technical exercise concerned with assigning land uses in isolation from the economic and social characteristics of the city in question (Lowder, p. 1243)

In Tehran the planning teams carried out comprehensive social, economic and physical studies of the city. As requested, legislative proposals were made as to how the plan should be implemented. A number of implicit and explicit assumptions were made by the master plan. Explicitly, it was assumed that the population of Tehran would rise to twelve or sixteen million by 1991. But it was also assumed that Iran in general and Tehran in particular would by then be through the demographic transition and would begin to show a demographic profile like those of the already industrialised countries. Implicitly it was assumed, naturally, that the sociopolitical framework within which the future development of the city would take place would be substantially the same. With hindsight these assumptions looked reasonable at the time, but the second one was overtaken by events.

The final document ran to 1500 pages in several volumes. A variety of alternatives for the city's future growth were evaluated. The planning literature consulted and quoted was largely, and perhaps surprisingly, British, although works such as Kevin Lynch's *Image of the City* (1960) were referenced. The plan finally proposed a linear extension of the city largely to the west along the line of the Alburz mountains incorporating a number of satellite towns to give focus to the settlement pattern and relieve pressure in the present city. Tehran's development would thus be redirected from its north-south axis to an east-west axis, guided by a new superhighway and subway

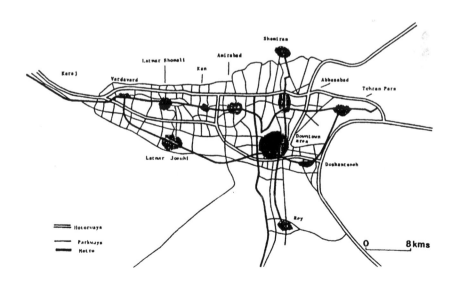

Map 19.2   The Tehran master plan

network (map 19.2).

A survey of major capitals throughout the world looked for examples of what constituted a recognizable and admirable capital image. The plan also made a number a number of recommendations for improving the appearance of landmarks, neighbourhoods and parks in the short term and pointed to opportunities for good design in the long term.

19.4 Implementation of the Plan

The plan was to be implemented both through extensive public investment in the major infrastructure such as the metro and the highways but most development and also through the private sector. This development would be controlled by the establishment of a five year service line, to be moved outward every five years until the outer limit of the city was reached after twenty-five years. An interim development pattern was proposed, with a series of fixed minimum plot sizes, ranging from 120 square metres around the old city of Rey in the south and up to 350 square metres in the more favoured northern suburbs.

Most of Tehran's residential development was to be concentrated on proposed satellite communities to avoid urban sprawl. Each community was to have a mixture of low-, medium, and high density housing, with provision for office employment, retailing, entertainment, and recreation. The three-tier hierarchy of service provision was base at the lower tiers on primary and secondary school catchment areas. A major centre in each community was to be served directly by the superhighways and the metro, when complete.

Standing back from the comprehensive plan now we can see parallels with other master plans of the period. The supergrid bears some comparison with the contemporary plans for Milton Keynes in the United Kingdom, though the scale of the total enterprise may have more in common with the great Paris Regional Plan of 1965. The conscious attempt to create a hierarchy of service provision and neighbourhoods goes back even as far as the first Garden City in the U.K. at Letchworth.

Work was begun on a number of the satellite communities in the early 1970's and from the start the intention was not to integrate different income groups within each community, as was the goal in, say, the post-war New Towns of the U.K., but to segregate them. This was seen at its most extreme in the plan for a wholly new city centre on an almost empty tract of land a few kilometres north of the city centre. This was to be named after the Shah himself, and called Shahestan Pahlavi.

19.5 Shahestan Pahlavi

This was to be a multiuse centre that would bring together government, cultural and commercial functions in a monumental setting. The plan
called for the best elements of formality, order and design which could be
learnt from a number of European capitals and combine them with the best
examples of Iran's own traditional urban culture. It was to be the modern
equivalent of the Persepolis of the ancient Achaemenian kings of Iran or the
Esfahan of the Safavids.

The scale was remarkable. The projected residential population was
fifty thousand, but 200,000 would be employed on the site. The promotion of
this new centre was in direct contradiction to the need perceived by other
planners in Iran to restrict Tehran's growth and try to limit its disproportionate consumption of the nation's resources (Costello,1993). However, as
F.Halliday (1979) put it: "the only kind of planning in Iran is what the Shah
wants."

Already by the mid-1970s there were growing tensions between the
modern and the traditional sectors of the country's economy. Much of the
vastly increased oil revenues which followed the OPEC price rise of 1973
was spent on consumption and speculative housing in Tehran. Iranian planning combined entrepreneurial, corporate, and private planning and industrial and real estate development, with the public entrepreneur acting at the
behest of private interests. The priority afforded to government sponsored
schemes meant that there were significant shortages of materials for private
sector developments. The private sector concentrated on providing middle-
and high-cost dwellings while the state provided for its employees. The military was not constrained by the plan and built apartments blocks outside the
line of the city limits envisaged in the 1969 Master Plan.

The traditional bazaar sector was to some extent marginalised, and there
were plans to develop a vast new commercial centre away from the old core,
and at the same time pull down the old core which was described as requiring
"total renewal" in the 1969 Master Plan. Conservation in the old core was not
seen as feasible, except perhaps for the larger monuments. The long term
intention was to complete the work of destruction begun by Pahlavis in the
1930s. Of course this was standard planning doctrine in many countries at the
time. The U.K. finally stopped wholesale demolition of the central part of its
cities in the 1970s. But in Iran it had also implications for the conflict between the traditional and modern corporate sectors. The states's failure to
address the problem of housing the urban poor was one of the factors that led
to revolution and the downfall of the Shah.

The introduction noted that a plan starts as a mental project, and perhaps may become a physical reality. In this case, the more real it became, the more there was political resistance to the image behind it.

19.6 Planning After the Islamic Revolution

When the revolution came in 1979 the traditional bazaar sector was one of the strongest supporters of the new Islamic regime. Ayatollah Khomeini established his first headquarters in the old bazaar - a clear spatial statement of where the new priorities would be. But clear spatial policies and an awareness of the potential significance of spatial planning in ordering the city did not emerge. Planning remained largely sectoral. Take the case of housing planning: in the years after the revolution attempts were made to end speculation in land through a number of land ownership laws which were based on Islamic principles. The priorities ostensibly at first were towards socially redistributive policies. But in discussions of social justice in the Islamic Republic the concept of property rights was central from the start:

> Repeatedly, it is issues touching on property and the state's power to tamper with private property that have aroused the strongest sentiments among the religious leaders. The controversy also reflects on the inability of Iran's religious and secular leaders to agree on interpretations of Islamic law that can simultaneously satisfy the concern for social justice and the concern for property rights as articulated by different factions and individuals within the leadership. (Bakhash,1989).

In the early years of the Islamic Revolution free housing was promised to the urban poor and access to suitable housing was declared a 'right' of the Iranian people. The expectation of getting a house encouraged further rural to urban migration. Major economic problems, including the consequences of the eight year war with Iraq, meant that these promises could not readily be fulfilled. In recent years the policy has been to rely on the private sector to fulfil these needs. The rapid rise in Iran's population and contraction in both private and public housing investment have, if anything, worsened the housing situation since the revolution. Far from being through the demographic transition, Iran has one of the highest rates of population growth in the world.

19.7 Conclusion

How does Tehran compare now with the original vision of 1969? Firstly, much of the major road infrastructure envisaged by the Master Plan has been put in place. The construction of the highway network was begun before the Revolution continues still. Also the metro has been begun, though there have been acute engineering problems and a shortage of capital which have caused delay after delay.

Next, the reorientation of the city from north/south to east/ west has been followed through as has the construction of many of the satellite new towns. In recent years much more housing has been high-rise and some aspects of the Le Corbusian image have come to pass.

Shahestan Pahlavi was never built, and the site is now partly occupied by a forest park and there are plans for a religious centre with associated retail and recreation functions. The symbols of the Islamic regime have replaced those of the Pahlavis and the major monumental work of recent years has been the construction of the tomb of the Ayatollah Khomeini. The tomb of Reza Shah Pahlavi was destroyed.

The bazaar remains much as it was before the revolution, and attempts by the Tehran municipality to introduce modern services within it have been resisted by bazaar merchants, some of whom may fear the sort of destruction that occurred in the past. However some wholesale functions have been moved to new sites on the edge of the built up area such as the new fruit and vegetable market in southern Tehran.

To summarise:

• At the scale of the city as a whole the there is a good match between 1960s aspiration and 1990s reality.

• At the local scale there is little match. Vast numbers of people do now live in apartment blocks, but they do not pursue the westernised lifestyle envisioned for them.

• The market still is all pervasive, despite attempts to impose an 'Islamic' version of economics.

Unlike the 1960s, there is little information on what the plans for the future are. Those that emerge are mostly propaganda exercises for internal consumption within Iran only. This is planning retreating from being both physical and mental, to mental only. In addition and perhaps because of this, the current regime publishes almost exclusively in Persian. For overseas academics this makes it difficult to find out what are the current aspirations for Tehran.

References

Amirahmadi, H. 1986. Regional Planning in Iran: a survey of problems and policies. *The Journal of Developing Areas* 20, no 4 (July):501:529.

Bahrambeygui, H. 1977. *Tehran: an urban analysis.* Tehran: Sahab Books Institute.

Bakhash, S. 1989. The Politics of Land, Law and Social Justice in Iran. *Middle East Journal* 43, no. 2, Spring.

Costello, V. 1973. Iran: the Urban System and Social Patterns in Cities' (with B.D. Clark), *Trans. Inst. Br. Geogr.*, July.

Costello, V. 1993. Planning Problems and Policies in Tehran, in Amirahmadi,H. and El-Shakhs,S.(eds) *Urban Development in the Muslim World.* New Jersey: Rutgers University.

Halliday, F. 1979. *Iran: Dictatorship and Development.* New York, Penguin.

Lowder, S. 1993. The Limitations of Planned Land Development for Low-income Housing in Third World Cities. *Urban Studies* 30, no. 7, 1241-1255.

Lynch, K. 1960. *Image of the City.* Cambridge: M.I.T. Press.

Plan Organisation. 1968. *Fourth National Development Plan.* Tehran: Plan Organisation.

Planhol, X. de. 1968. The Geography of Settlement, in Fisher, W. B., ed. *The Cambridge History of Iran, Volume I. The Land of Iran.* London: Cambridge University Press.

Sutcliffe, A. 1981. *Towards the Planned City.* London: Blackwell.

20 The Historical Development of Riyadh and its Low-Rise Housing

Ali Bahammam

Editors' note:
Riyadh (see also Chapter 11) has within the recent past invented and experimented with its own urban forms, integrating aspects of an older life style with the materials which new wealth and technology have brought. The adaptation between life-style and buildings works both ways.

20.1 Introduction

The present site of Riyadh dates back to the pre-Islamic town of Hager, then the capital of Al-Yamamah. Riyadh itself was founded on the ruins of several communities about 1740. The city is located far inland in the middle of the region of Najd on the Arabian peninsula, and is situated on a plateau which rises about 600 meters above sea level. The climate of Riyadh is characterized by dry, relatively hot summers and dry, cold winters. Daytime air temperatures may range from 27°C to 49°C in July, but temperatures may drop to as low as 10°C between November and January at night. The average relative humidity is around 40% to 50% between November and February, during periods of colder temperatures and rainfall, and 15% to 16% during the summer from June to August i.e., during hot and dry periods. The annual average rainfall is 59 mm, occurring mostly during December to May, thus leaving June to November completely dry.

This is the geographical setting for what is now the national capital city of Saudi Arabia. This paper examines the historical and contemporary development of Riyadh with an emphasis on the evolution of its housing.

20.2 The Historical Development of Riyadh

The city assumed little prominence until Abd Al-Aziz Al-Saud became its independent governor in 1902 and began his campaign for the consolidation of modern Saudi Arabia. From that point on, Riyadh was the permanent residence of the king, and eventually became the Saudi capital. During the first thirty years of Abd Al-Aziz's reign, the city retained its size inside its fortifications as well as its traditional type of dwellings.Only after the consolidation of the kingdom and the end of the campaign in 1938 did the king himself take the first step towards affecting the city's physical development by deciding to move outside the old city of Riyadh. Two kilometers north of the center of town, he built Al-Murabba— a large complex of palaces and administrative buildings for himself and his entourage. It was built out of dried mud bricks and other local building materials. Each building or dwelling unit was built around one or more courtyards (Fig. 20.1). Thus, the departure of Al-Murabba from the traditional urban pattern lies mainly in the larger size of its components and the scale of the building program rather than in design and layout.

By building Al-Murabba, King Abd Al-Aziz established a precedent for Riyadh. People now felt they could live and build outside the city walls, thereby expanding the size of Riyadh and establishing the direction of its physical growth. The expansion meant that the walls would no longer be a barrier in the way of urban growth and that the preferred direction for development was northward toward Al-Murabba. Although the design concepts and the building processes remained the same as those of the traditional structures inside the old city of Riyadh, a new version of the traditional dwelling

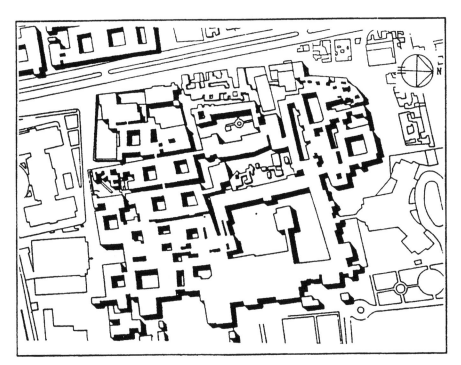

Figure 20.1 The courtyards of old Riyadh

and a new physical pattern evolved as a result of the availability of land and the introduction of the car as a new means of transportation.

When King Saud succeeded his father King Abd Al-Aziz to the throne in 1953, he made two decisions that were to have significant impact on the physical growth of Riyadh. The first was his decision to transfer all government agencies from Makkah to Riyadh and to begin a building program along the road to the [old] airport to house them. The second decision was to expand and rebuild Nasriyah, a country estate 3 km west of town, as his royal residence.

In the early 1940s, Nasriyah was a small estate owned by King Saud, then Crown Prince. However, when the decision was made in 1953 to make Nasriyah the royal residence, the grounds were extended to approximately 250 hectares and plans were drawn up to include modern, luxurious palaces, boulevards, and gardens laid out on a grid pattern (Al-Hathloul, 1981). It was the first to be built with new building materials and different planning and design concepts.

From this point on, a conflict between old and new was consciously felt by the city's residents. In contrast to the traditional pattern, Nasriyah was orthogonally planned. Instead of traditional building materials and technology, cement blocks and reinforced concrete were used. As for its impact on Riyadh, it demonstrated an alternative way of planning, designing, and building techniques. Because it suggested new architectural possibilities, it had a clear effect on Al-Malaz, the first housing project. However, its immediate impact on the city inhabitants was the introduction of new, more durable, building materials as well as new building techniques that accelerated the construction process. Some of the inhabitants used the new materials and techniques instead of the local alternatives to build what came to be known later on as *Al-bait Al-shabai* - the transitional dwelling.

In 1953, when the government decided to move its agencies from Makkah to Riyadh and to build its ministries, the need arose for housing for the transferred government employees. The site of Al-Malaz, 4.5 km northeast of the city center was chosen, and a housing project was initiated by the Ministry of Finance to satisfy this purpose. In 1957, when the move took place, the project was in operation, and some sections of it had been completed.

Al-Malaz project consists of 754 detached, outward-looking dwelling units (villa type) and three apartment buildings. Al-Malaz also includes the required support facilities and services for what came to be known as the "New Riyadh". The physical pattern of the project follows a gridiron plan with a hierarchy of streets, rectangular blocks, and large lots which, in most cases, have a square shape (Fig. 20.2).

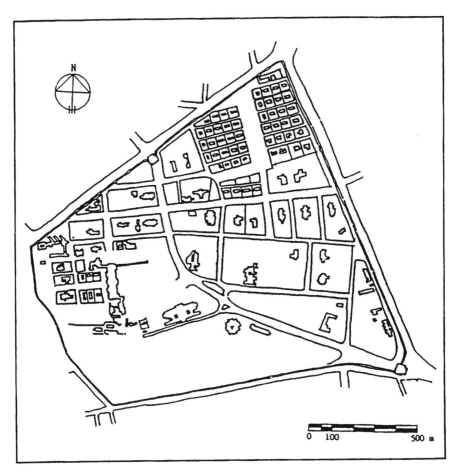

Figure 20.2 Part of the Al-Malaz project

The impact of Al-Malaz on the size of Riyadh can easily be seen. The project covers an area of about 500 hectares, and in fact is a city in itself, as the name "New Riyadh" implies. What was not foreseen at the time of its design, however, was the effect the project would have on the pattern of physical development in Riyadh and in the country as a whole. Al-Malaz introduced new patterns of street layout and new types of dwelling (Fig. 20.3). The gridiron street pattern and villa-type house both became models for the later physical development of every city and town in Saudi Arabia. Riyadh now covers an area of more than 800 square kilometers, with an estimated population of over 3 million people (ADA, 1996). Almost 80% of this expansive area are low-rise, low-density housing follows the grid pattern system and utilizes the villa-type dwelling.

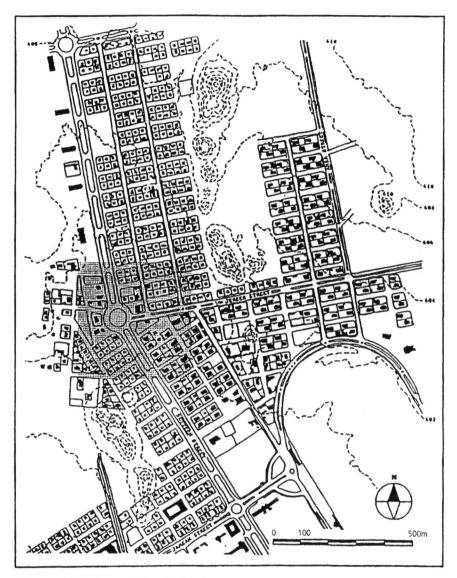

Figure 20.3 Part of the Al-Malaz project

20.3 Low-rise Housing

The evolution of Riyadh low-rise housing includes the traditional dwelling, the transitional dwelling, and the contemporary villa-type dwelling. The traditional dwelling was built from local materials according to the users' own needs and norms in an incremental building process over any conceivable time-scale by the local master builder with the help of the users themselves. The transitional dwelling was built from some new building materials by inexperienced contractors but was no more an incremental process, it was a final product , even if the design concept of the traditional dwelling was still applied. The contemporary villa-type dwelling unit is the product of a new building materials and techniques and a new design concept that satisfies municipal rules and regulations. These three major types of low-rise housing will be presented via a study of their design concept, the organisation of their spaces and elements, the rules and regulations that affect them, and their design and construction processes.

20.4 Traditional Housing

Riyadh's traditional housing can be classified into two types: *The former traditional housing* which was built inside the fortification wall before 1938, and *the later traditional housing* which was built after 1938 outside the city's fortification wall. The main differences between the two types are summarized in the following points:
• Most of the dwellings of the former type have irregular geometric plans. Their shapes were influenced by the organic growth of the neighborhood urban fabric (Fig. 20.4). Conversely, the majority of the dwellings of the latter type had clear rectangular shapes, and were larger.
• The neighborhood layout of the former type, as a part of the organic pattern of the town, was identified by solid masses of connected houses broken up by narrow roads which would branch out irregularly into alleyways and cul-de-sacs that provided access. Furthermore, the urban spaces of the former pattern were organized in a hierarchical sequence from public open spaces to main roads and to the semi-private cul-de-sacs. On the other hand, the introduction of the car as a new means of transportation very clearly affected the neighborhood layout of the later pattern. The streets of the later pattern are wider and straight, featuring long, large blocks of attached houses.
One of the best examples of the later traditional dwelling is King Abd Al-Aziz Palace, which was built as part of the Al-Murabba complex (Fig. 20.5). In general, the dwelling of the latter type has the same design concept and is composed of the same spaces and elements as the former type. Moreover, they were built out of the same local building material and with the same construction technique.

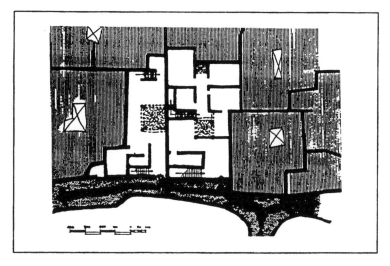

Figure 20.4 Irregular organic growth of traditional housing

Design Concept

Riyadh's traditional dwellings are inward-oriented buildings. This design concept embodies the best response to both the socio-cultural needs of the users and the harsh climatic conditions of the region. The dwelling consists of a lumpy adobe structure built around one or more rectangular courtyards with thick walls, small openings to the outside, and large openings to the inside. Traditional units, when grouped together, share as many as three walls with each other with only narrow streets in between, shading each other throughout the day and thus creating an environmentally consistent solution.

Organisation of the Traditional Dwelling

The internal organisation of the traditional dwelling is the result of an attempt to satisfy the privacy requirements of the society. The dwelling is divided into two main sections. The family living section, designed for use by the family, female guests, and the **maharm** (male relatives, who, according to Islamic teachings, cannot marry females in the house, e.g. their brothers, uncles). Generally, for reasons of privacy, this section is always located away from the entrance.

All rooms within this section of the house are used for a variety of functions, such as sleeping, eating, family socializing, and household work.

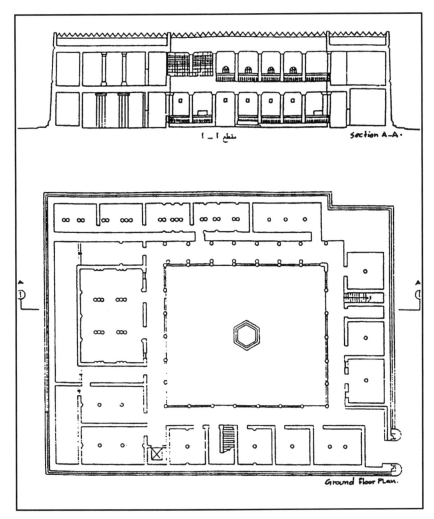

Figure 20.5 King Abd Al-Aziz Palace

The rooms in the traditional dwelling tend to have very little and simple furniture. The entire floor of any room is covered with oriental rugs or mats and, since everyone walks barefoot inside the rooms, the room is really like a large couch and one can sit, lie or walk about as one pleases. At night, a foldable mattress is placed on the floor for sleeping, which in the morning is folded away for daytime use of the room. Another daily function for some of these rooms is that of a place for eating. Meals are served on a circular or rectangular mat, which is set on the floor. Heavy, large furniture, which is difficult to move and which commits a space to a specific purpose does not exist in any room. Therefore, these rooms can meet various needs dictated by the immediate circumstances.

The inner rectangular-shaped courtyard has always been a part of the family section. It is the focal point of the house, usually surrounded by paths for circulation and the rooms of the family section. The inner courtyard performs an important function in the traditional dwelling. It answers the residents' socio-cultural needs and fits the climatic conditions of the region (Talib, 1984). It is an architectural device for maintaining privacy as well as providing an enjoyable extension of the house, especially during the pleasant morning and evening hours.

The second section within the traditional dwelling is designed for receiving male guests. This section is designed in such a way as to allow male guests easy access to the men's reception room without disturbing the privacy of the family section. This section contains the reception room, a separate bathroom, and, in some dwellings, a separate staircase leading to a separate section on the rooftop or upper terrace which is used as a sleeping area for guests during hot summer nights.

The men's reception room is the main element of the guest section. It is usually the largest room in the house. This room is used for many functions, such as a men's sitting and gathering room, as a male-guest bedroom, or as a dining room.[1] The reception room tends to be located adjacent and directly accessible to the entrance lobby. It is usually longitudinal and parallel to the street.

All rooms in the traditional dwelling have full potential for a variety of uses as a result of their simple and multipurpose furniture. All of the furniture that is used in the traditional dwelling can be rolled up and stored away when not in use. The absence of cumbersome furniture lends higher flexibility to the use of living spaces.

Traditional Rules and Regulations

The rules which governed the architecture of the traditional dwellings of Riyadh are the same Islamic rules as were applied in other Muslim towns and settlements. Islamic Teachings from the *Qur'an,* the Holy Book of Muslims, and the *Hadith*, the sayings of the Prophet, are the bases for these rules.

---

1 Most of the large traditional houses have an additional room (adjacent to the reception room) called muqallt. *Although* muqallt *is a term used to designate an eating place, this room also serves other functions e.g., providing additional reception space or a bedroom for guests or visiting relatives, when it is not in use for eating.*

In accordance with the tenets of their Islamic faith, Saudi families traditionally have built their dwellings with considerable care following special designs to respect the rights of the individual neighbors as well as the whole
community.

In general, anyone was free to build as he or she wished as long as one
was not causing any damage or harm to others. However, if constructing a
new building or the expansion of an old one caused any physical damage to
the neighboring buildings, caused intrusions on the neighboring households,
or conflicted with the community's interest, the action is, by Islamic law,
considered a *darar* i.e., a harm or damage to the individual and the community. Islamic law will then insist on the removal of the damage. When the
Prophet says: "*La-darar-wa-la-dirar*"(Ibn Majah, 1953), he is prohibiting
the cause of damage.

These rights were guaranteed in the urban areas by the power of the
gadi (the Muslim Judge) and the *muhtasib* (the official in charge of exercis-
ing *hisbaah*). In its widest sense, the world *hisbaah* means ensuring that the
precepts of the *shariah* (Islamic law), particularly those of a moral and religious nature, are observed within the town and especially in the market as
part of Muslim obligations to society (Lewis et al, 1971)

In Islamic history, there are numerous cases in which judges and jurists
stated that damage had been inflicted upon the community's interest, for example, a new structure blocked or narrowed public roads or infringed upon
individual dwellings, if, for example, a new building was built higher than
another or a new door put in front of or too close to a neighbor's door. In all
those cases, removing the damage involved removing the cause of it (Ali
Bahammam, 1987). However, such violations were rare in the traditional
built environment of Riyadh because the local master builder understood and
honoured all the rules as a part of his Islamic heritage.

Design and Construction Process

It is said that the traditional dwelling was never complete when built.
It grew over a period of time around one or more courtyards. The flexibility
for change and expansion of the original house can be clearly seen in the
evolution of land configuration over a period of time as family size increased
by marriage or by birth.

The building contract was negotiated directly between the user and the
master builder *Al-astaad*. Any building done on the house would not start
until both parties agreed upon the needs, requirements, costs, and general
design schemes. Since the general design scheme was discussed orally, the

only lines drawn would be those of the construction site itself, showing the floor plan of the house as well as the interior organisation of the house, while the rest of the design gradually appeared as the construction progressed (Fadan, 1983). Traditionally, the building process begins with the construction of a boundary wall around a plot of land selected by the family. Well-defined land ownership laws did not exist in the past (Talib, 1984). Religiously speaking, land that is not owned by anyone can be claimed for ownership by building a wall around it (Akbar, 1980), as long as this act did not harm the interest of other members of the community. During the construction process, it was common practice to find the male members of the family and other male relatives helping the master builder and his team in order to speed up the construction operation.

Building Materials and Techniques

Adobe construction is the most salient characteristic of the traditional buildings. It is indigenous to the Riyadh region, where earth itself offers the best, readily available building material. Straw or animal manure are often added to the mixture of water and clay to give the mortar and bricks extra strength, and to make the walls more water resistant (Mousalli, 1997). These sun-dried brick walls have excellent thermal properties, owing to their high heat capacity. As much as 80% of the outside heat is absorbed, and only 20% transmitted (High Commission, 1978).

Despite its scarcity in the region, wood has always been used as roof beams, frames, doors, and windows. Limestone was sometimes used for constructing the building foundations in later buildings, while lime plaster was used to frame doors and windows and to protect the top edges of the roof parapet from rain damage.

The house was covered by a flat roof built out of the following materials: 1) tamarix aphylla (*athel*) or palm tree trunks, 2) palm-branches *jarid*, 3) woven palm-leaf matting, and 4) a layer of mud. The floors of the second story as well as the roof were generally made by laying trunks parallel to each other, spaced about half a meter (1.5 foot) apart on properly raised walls. On top of these trunks and perpendicular to them, the palm branches were placed side by side. A woven palm-leaf matting was usually placed on top of these branches in order to hold the thick layer of mud. However, because of the short length and the low quality of the available wood trunks, the width of the room in most traditional houses rarely exceeded (2.5 meter) eight feet.

Although the use of traditional materials was the optimal solution of challenging the constant harsh summer heat, the adobe structure was vulnerable to compression and weathering, one of the major problems being water leakage during rainy days and nights. Continuous renovation and maintenance was required requirement for the buildings' survival. Finally, the long construction period of the dwelling made the use of the traditional building materials even less advantageous.

All of the above problems and disadvantages were the motives behind the residents' welcoming the use of new, strong, and durable building materials. The new materials helped them speed up the building process and entailed less maintenance effort, and enabled larger rooms to be built. At the time, however, people did not realize the negative consequences of using new materials that were unsuitable for the climatic conditions of the region. Nonetheless, all in all, this was the beginning of a new stage of housing in the city.

20.5 Transitional Housing

The discovery of oil in Saudi Arabia caused a major migration of the population to Riyadh and to other urban areas. The relocation of large numbers of people from rural areas and from neighboring countries to Riyadh created a sudden enormous need for affordable housing in a very short time. In order to alleviate the housing shortage, people started to use the new durable building materials and new and faster techniques, which, by that time had already been introduced in the country, as a way of speeding up the construction process. These new materials included cement blocks, imported wood, and reinforced concrete. The combination of the use of new building materials with the traditional building conception is the most salient trait of the transitional dwelling.

Transitional housing can be found in many of the neighborhoods around the traditional central area of Riyadh. Manfouhah area, which is located south of the city center, is a good example of the transitional stage of housing. It is primarily a residential district with hardly any other use of the land. Its regular street pattern reflects its origin as a land subdivision sponsored by the municipality (Fig. 20.6) (SCET,1982).

Manfouhah, as most of the transitional neighborhoods, has high density housing as a result of small, single family houses occupying most of the lots. All lots have square shapes with an average size of 100 m^2. Most of the dwelling units are one story high, only few of them are two stories high, depending on the location of the neighborhood and the residents' wealth.

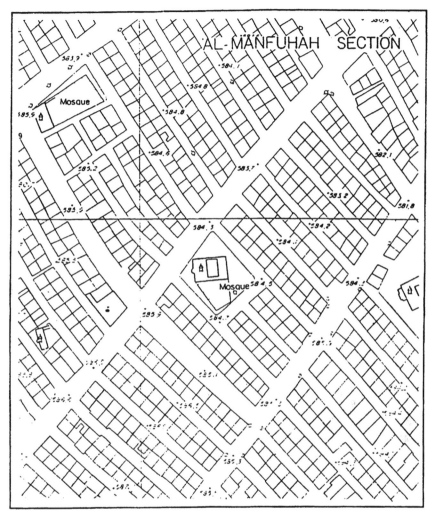

Figure 20.6 Subdivision of land at Al-Manfuhah

Design Concept

The transitional dwelling is still an inward oriented building similar to the traditional dwelling. It is an attached unit that consists of a number of rooms built around one open space. However, in a misapplication of the traditional concept, many of the inner open spaces are no more than a light well which is too small to serve the same functions as in the traditional courtyard. All dwelling units of this type are no more than two stories high. They are grouped together in large blocks sharing as many as three walls with each other, shading each other throughout the hot summer days and preventing the exposure of large parts of the building to solar radiation (Fig. 20.7). This housing pattern helps create a comfortable indoor climate although some of the building materials on the outside walls and the roof of this type of dwelling are unsuitable since they feature bad thermal properties, as in the case of the older cement blocks.

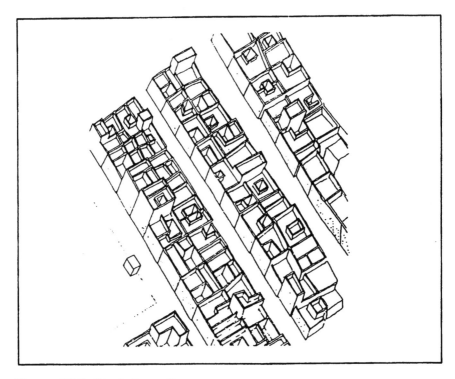

Figure 20.7 Shade from the sun

Organisation of the Transitional Dwelling

The internal organisation of the transitional type of dwelling unit is similar in concept to the traditional dwelling, but many of the elements of the traditional dwelling, such as the brown dates storage room and the animal service yard, cannot be found in this type of dwelling as a result of the smaller size of the lot and the different needs of the users. Except for the men's reception room and a separate bathroom in some of the dwellings, the rest of the house is family domain. All rooms within the family section are furnished with simple and easy-to-move traditional style furniture, similar to the traditional type. These rooms were also used for a variety of functions, thus allowing for high efficiency and flexibility. The men's reception room is the largest room in the house. It has always been situated on the ground floor adjacent to the entrance hall and parallel to the street.

A lack of knowledge of the traditional rules and regulations among some foreign contractors and builders was one of the main reasons for problems with provisions for privacy; one example being the design of the entrance way as a direct access from the street to the center of the family section with no change of direction to block the view from the outside into the house. Large windows in the outside walls and the placement of external doors of the neighboring dwellings in front of each other are other examples of privacy clashes in the transitional type of dwellings.

Rules and Regulations

A number of municipal offices were already established in many of the large cities in Saudi Arabia by the time the transitional dwelling evolved. Therefore, people who wanted to build new dwellings in Riyadh were required to get construction permits from the municipal office. They have had to respect the number of floors permitted by the municipality for their district and to maintain the construction size of the building within the border of the lot. With the exception of these few municipal requirements and regulations, builders and contractors were free to build as they wished without any specific rules. However, the imitation of the traditional design concept implied the use of some of the traditional rules and regulations.

Design and Construction Process

Many of the contractors were foreign immigrants from neighboring countries, who had originally come to Saudi Arabia as low-skilled workers. After a few years of experience, however, they would often start their own business as contractors.[2] Many of them had much less experience with building practice than the master builders. Instead of a long-time vocational

---

2 *Those contractors were able to draw and read free hand sketches and translate them, with the help of their work team, into three-dimensional, physical form.*

commitment, the building profession became an easily-acquired occupation.

Building Materials and Techniques

Sun-dried mud bricks and the mud mortar were abandoned as construction materials. A large variety of local and imported, natural and manufactured, old and new building materials were in use in the transitional type of dwelling. Although it was mainly new materials, such as cement blocks, cement mortar, reinforced concrete, imported wood, and steel doors and windows that were being heavily used, some of the local traditional materials, such as limestone and local woods also still found application. Although the reinforced concrete slab roof could now be used for roofing, in some of the transitional dwellings of the transitional type, local and imported wood tended to be used for roofs in a similar way to traditional building methods.

Two different techniques are used in constructing the dwellings in the transitional stage of housing: 1) the traditional load bearing technique where the loads of the roof and floor of the dwelling are carried to the foundations by the walls, and 2) the new frame construction technique which had come along with the new building materials, and which had brought about a drastic change in the local building industry. Frame construction has become a new phenomenon in the construction of housing in Riyadh.

Finally, although the construction and finished quality of the majority of the transitional dwelling units are not outstanding, the use of the many newly-introduced building materials presented a stage in the housing development which affected the peoples' attitudes in favor of accepting the contemporary type of dwelling from the beginning.

20.6 Contemporary Housing

The contemporary stage of housing represents a completely different dwelling unit and street layout from traditional housing. The differences range from the tiny construction details of the building and the organisation of internal spaces to the external appearance of the single dwelling and the whole arrangement of the neighborhood. It is the adoption of a completely new type of dwelling unit along with a new residential neighborhood street layout, which called for new rules and regulations grounded in foreign concepts.

The detached, outward looking dwelling unit was first introduced in the cities of the Eastern province by ARAMCO (The Arabian-American Oil Company) in the forties (Fig. 20.8) (Fadan, 1983). However, this type of

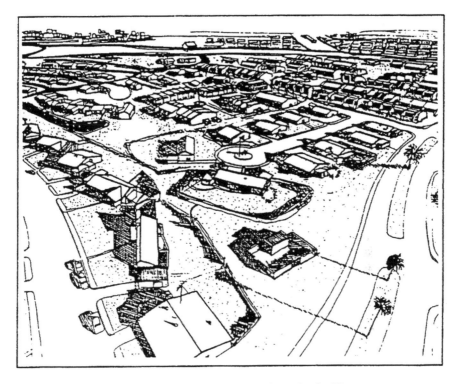

Figure 20.8 The imported concept of the detached villa

dwelling began to have a major impact on the whole country after the conclusion of Al-Malaz housing project in the city of Riyadh, for which 745 villa-
type dwelling units had been built as part of the whole project.

When the government decided to build Al-Malaz housing project, there
were no local architects and the local master builders were ignored because
of their lack of experience in building such large projects in a short time.
Direct involvement of foreign architects in the Al-Malaz project influenced
the design concept of dwelling, cluster, and neighborhood. A new type of
housing was introduced, one more related to Western suburban housing than
to Saudi traditional housing.

Since its introduction, the villa-type house has become the prevalent
type of dwelling. During the last five development plans, thousands of these
models were built in Riyadh and all across the country and thousands more
are scheduled to be built. Two major factors have caused the continuous use
of this model: one, current municipal regulations and two, the strict
conditions of the Real Estate Development Fund (REDF), which have

provided long term, interest free loans to Saudi landowners for the express purpose of private homes construction.[3] The REDF loans have ensured conformity to the municipal building regulations. To provide the loan, the REDF requires two copies of the building permit be submitted together with the application as part of the legal documents.

Design Concept

The contemporary dwelling is a detached, outward-oriented box. It has wide glass window openings in the four outer façades. It also has a maximum wall area exposed to direct solar radiation (Fig. 20.9). This outward design concept is at cross purposes with the hot, arid climate of the region. The private inner courtyard of the traditional and transitional dwellings has disappeared, only to be replaced by open spaces surrounding the building and to be enclosed by an eye level wall. These outdoor spaces fail to answer the

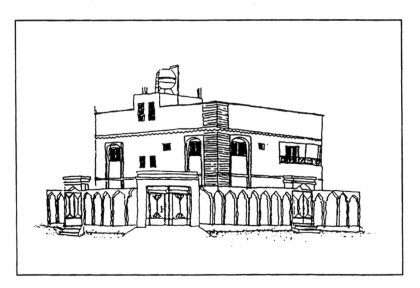

Figure 20.9 A contemporary detached villa

3 *Financial assistance to the private sector was one of the housing policies of the Saudi Government . In 1975/76, at the time the second development plan was beginning to be implemented, the REDF was granting earmarked interest-free loans to private individuals and private organisations for real estate in general loans of SR 300,000 ($ 80,000). The loan must be repaid in 25 annual installments. In addition, there is a reduction of 20% of the principal installment if repaid regularly and on time.3 By the end of the fiscal year 1995/96, REDF had provided loans for the construction of more than 550,000 dwelling units.*

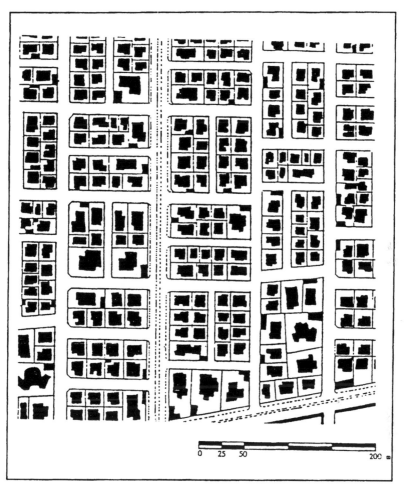

Figure 20.10 Neighbourhood grid-iron

residents' need for privacy and do not take into account the harsh climatic conditions of the region. Finally, almost all contemporary neighbouhoods follow the gridiron plan with a hierarchy of streets, rectangular blocks, and square shaped lots (Fig. 20.10).

Organisation of the Contemporary Dwelling

The rooms of the contemporary dwelling are arranged in one closed block. The early models of the contemporary dwelling was divided only into a living section and a sleeping section. However, most of the later models are two stories high, where the first floor is divided into two domains with two separate entrances; the men's entrance which opens to the male guest domain i.e., the men's reception room, the guest dining room, and the guest separate bathroom on the one hand; and the family entrance which opens to the family living section and the women's reception room on the other hand.

Unlike those of the traditional dwelling, the rooms of the contemporary dwelling are designated for specific domestic uses. Most of the rooms incorporate hard-to-move furniture committed to a special function in each room, which restricts the potential range of uses for the room.

The Contemporary Rules and Regulations

Municipal rules and regulations have played a major role in the rapid spread throughout Riyadh of the villa as the contemporary model of dwelling. After the Al-Malaz project, the setback requirements of the amended municipal zoning regulations were made into a law. One of the reasons why setbacks and special building line requirements have developed are the anticipation of future street-widening and the accommodation of aesthetic interests.

In the city of Riyadh, as in other Saudi cities, the setback and building line requirements cannot be logically justified; the wide streets, the open view, and the green gardens on both sides have proven to be a violation of traditional Saudi privacy. Consequently, home-owners have erected above-eye-level masonry fences (which must not be higher than 3 meters due to municipal regulations) on both sides of the street. In some cases, however, an additional high fence out of steel frame and canvas, plastic, or metal-corrugated sheets has been added onto the masonry fence. The purpose is to safeguard privacy and block the open view into the premises (Fig. 20.11).

By the late 1960s, the municipality of Riyadh, as well as other municipal town planning offices in Saudi Arabia, had enforced several regulations pertaining to building on a plot of land, namely:

- A built-up area generally should not exceed sixty percent of the land area, including attachments;
- Front setback should be equal to one-fifth of the width of the road and should not exceed six meters;
- Side and rear setbacks should not be less than two meters, and extensions should not be permitted within this area.

These regulations and others were issued in the form of a circular by the Deputy Minister of Interior (A-Hathloul, 1981)

Figure 20.11 Metal screen to add privacy to a walled compound

20.7 Riyadh Master Plans

In the late 1960s, the government of Saudi Arabia felt the need to control the growth of urban areas. Riyadh was the fastest growing city in the country and, at the same time, the country's capital. It was therefore the first city to attract the attention of authorities and, in 1968, the task of planning the capital was assigned to Doxiadis Associates of Athens. They were to formulate a master plan and program that would guide the development of the city.

In 1971, the final master plan for Riyadh was submitted; it followed the existing building regulations. Except for buildings already put up, the Doxiadis proposals reaffirmed the existing setback requirements in practice since the late 1960s in all residential areas of the city. This mater plan has set the base for a low-rise low-density housing growth which encourage the horizontal expansion of the city.

By the mid 1970s, Riyadh grew beyond the boundaries laid out in the Doxiadis plan—a result of the economic boom and concomitant pressure for development experienced by Saudi Arabia after 1973. In 1976 the task of revising the Doxiadis master plan and preparing execution and action master plans and development studies for Riyadh was assigned to SCET International/SEDES of Paris.

By this time, it had become clear that some of the proposals and regulations of the Doxiadis plan did not fulfill the socio-cultural needs of Saudi society. SCET proposed revised zoning regulations, one of whose aims was to protect the privacy of individual homes and private grounds (SCET, 1982).[4]

In 1989 the Council of Ministers' issued a resolution which established a metropolitan boundary for Riyadh the outer limits of which are referred to as the Urban Environs. Within the Urban Environs two phase limits— Urban Limits Phase 1 and Urban Limits Phase 2— Were delineated.[5] In 1996 Arriyadh Development Authority (ADA) started the Metropolitan Development Strategy for Arriyadh (MEDSTAR). MEDSTAR is a process-oriented approach to planning and development for the future of the city. It is not a static master planning approach whereby a long-term plan is prepared without consideration for potential future change of the city. MEDSTAR seeks to build consensus for a Planning Vision for Arriyadh, to identify actions to enable planners to monitor and respond to change, and to seek government support in the establishment of policies, structure plans, and plan implementation mechanisms by way of using regulatory an budgeting powers. As an end product, MEDSTAR is expected, when completed, to provide a 50-year Vision for Riyadh, a 25-year Strategic Planning Framework, a 10-year Implementation Program, and a Plan Implementation Strategy (ADA,1996).

However, as of summer 1997, the municipality of Riyadh continues to apply the grid land subdivision and the setback requirements in all low-rise residential areas of the city as formulated in the 1971 Doxiadis master plan.

20.8 Conclusions

Any society can in part be characterised by its predominant cultural and religious norms, by its wealth and the strength of its outside contacts, and by the technologies which history and geography have introduced to it. In this case there has been a clear progression from housing norms which

---

4 *To achieve this objective, the regulations were to bring about two major changes: (1) Side and rear setback requirements in residential areas were abolished; and (2) Owners who wanted setbacks were permitted window openings from the second floor up, however certain conditions and standards had to be followed to protect the privacy of neighbors. The property owner either had to maintain a certain distance between any window and the property line of the neighbors, or the windows had to be designed to prevent direct sight lines into neighboring premises.*

5 *Urban Limits Phase 1 encompasses an area of approximately 632 km2, which are set aside for urban development up to the year 1995. Urban Limits Phase 2 encompasses an area, exclusive of Urban Limits Phase 1, of approximately 1,194 km2, which are set aside for urban development from 1995 up to the year 2005. Lastly, land that falls between the Urban Limits Phase 2 and Urban Environs boundary (approximately 3,120 km2 exclusive of Urban Limits Phases 1 and 2) is set aside for future development after the year 2005. The total land area including the Urban Environs is 4,900 km2 (490,000 hectares).*

fitted both a traditional Muslim society and the fierce climate of central Saudi Arabia. But Saudi Arabia's resources have attracted foreign investment and given it great wealth, while also introducing it to new technologies. These new technologies have not arrived without some of the cultural baggage of their homelands attached – and whether it is at the scale of the house or the scale of the city, they have caused change in traditional design conceptions. The changes are both positive – large rooms and freer spaces – and negative – the loss of privacy, the internal court, and adaptation to the climate.

In the future perhaps instead of seeing a one-way evolution away from traditional styles trhough trnasitional ones, to imported styles, there can be a synthesis of the best of the two, which would create a housing stock once again accommodated to the new society (but old climate) of urban Saudi Arabia.

References

Acton, Maurice (1982) Journal of Royal Institute of British Architects, July.

Akbar, Jamel A.(1980) *Support for the Courtyard Houses, Riyadh, Saudi Arabia.* (Unpublished M. thesis, MIT, School of Architecture and Planing), p. 19.

Al-Hathloul, Saleh A. (1981) *Tradition, Continuity and Change in the physical Environment: the Arab-Muslim City.* (Dissertation, MIT, Department of Architecture,), PP. 159-162, 205.

Ali Bahammam, (1987) *Architectural Patterns of Privacy in Saudi Arabia.* (Unpublished M. thesis, McGill University, School of Architecture), pp. 27-37.

Al-Saati, Abdulaziz.(1987) *Residents Satisfaction in Subsidized Housing: Real Estate Development Fund in Saudi Arabia.* (Dissertation, University of Michigan), p. 101.

Arriyadh Development Authority (ADA) (1996), *Synthesis Report* , MEDSTAR Phase 1. (Arriyadh Development Authority, Riyadh), pp.1-3.

Arriyadh Development Authority (ADA)(1996), *The Investment Climate in the City of Riyadh - 1416H.* (Arriyadh Development Authority, Riyadh), (Arabic), p.7.

Fadan, Yousef M.O. (1983) The_*Development of Contemporary Housing in Saudi Arabia (1950-1983) : A Study of Cross-Cultural Influence under Conditions of Rapid Change.* (Dissertation, MIT, Department of Architecture), pp. 44, 84-86, 101-128.

High Commission for the Development of Riyadh (1978) Design Manual of Riyadh Diplomatic Quarter (Riyadh, Saudi Arabia), p. 20.

Ibn Majah. (1953) *Sunan.* Vol. 2. (Cairo: Dar Ihya Al-Katob-Al-Arabiyah), p. 784.

Lewis, V.L. Menage, C.H. Pellat, and J. Schacht.(1971) *The Encyclopedia of Islam.* Vol. 3. (Leiden: E.J. Brill,), p. 485.

Mousalli-Shaker-Mandily. (1977) *An Introduction to Urban Patterns in Saudi Arabia—The Central Region.* (London: AARP), p. 24.

Real Estate Development Fund. Lending Guide. p.14.

SCET International/SEDES (1982a). Riyadh Action Master Plan. *Technical Reports* No. 9: Planning Regulations, p. 4.

SCET International/SEDES.(1982b) Riyadh Action Master Plan. *Technical Reports* No.11. (Riyadh, Saudi Arabia: Ministry of Municipal and Rural Affairs, Deputy Ministry for Town Planning), pp. 19-20.

Talib, Kaizer (1984) *Shelter in Saudi Arabia* (London: Academy Editions,), P. 45-54.

21 Urbanisation and Housing Problems in the Seoul Metropolitan Region

Seong-Kyu Ha

Editors' Note:
South Korea has gone through three phases of economic development; which have been re-flected in its urban forms. Seoul reveals all of them. In the first phase - before 1960 - an economically underdeveloped base supported a pre-industrial (traditional), partly colonial city form. The second phase - 1960s and 1970s - characterised by rapid economic growth with industrialisation and relying on a labor-intensive base; gave rise to an early industrial city form. The third phase - mature industrialisation in the 1980's and 1990's - made shifts to-wards capital intensive high technology industry; and forms charatceristic of the mature in-dustrial city emerged. The fourth post-industrial city form has yet to emerge and will do so when the Korean economy reaches the post-industrial phase like Japan and western countries. Faced with the rapid trowth of Seoul, the planners thought that the establishment of five new towns around the city would resolve many of its problems – it was part of the Asian uptake of the new town movement that started 100 years before in Europe. Delhi, Singapore, Hong Kong and Bombay are other cities that also planned new towns. Seoul's new towns did relieve some of the housing problems and housed residents in decent housing - mostly apartments - but they primarily remained dormitory towns as they did not generate self-sufficiency in employment.

21.1 Introduction

The economic growth of South Korea (hereafter Korea) has often been called an "economic miracle". For instance, in 1960, per capita GNP in South Korea was US$ 79. This figure increased to US$ 10,079 in 1995. During 1965-1989 its per capita GNP increased 7.0 per cent annually. This annual growth rate was the second highest in the world: the first was 8.5 per cent in Botswana and Singapore shared second place with Korea. Korea is also one of the first of the Asian economies to show signs of recovery from the financial collapse of 1998. In any event, few observers believe the pattern of development of the last few deacdes will be reversed: at worst, growth has been put on hold.

The growth of the Seoul Metropolitan Region(hereafter SMR) coincided with industrialisation of Korea which began in the early 1960s. At the same time, industrialisation stimulated a nationwide migration of people from rural into urban areas. Seoul, in particular, claimed more than the lion's share of the growth in urban population. With the explosive growth Seoul has experienced over the last 30 years, coupled with its near absolute dominance of the nation's economic and political scenes, it is not surprising that Seoul

Metropolitan Region is often referred to as the Seoul Metropolitan Republic.

There are a multitude of problems degrading urban life in the SMR. At the top is the housing problem – the stock of housing being completely inadequate for the size of population. Accordingly government efforts have been concentrated on the expansion of housing production and development of residential land in the SMR.   In the late 1980s, the dispersion of population and economic activities in Seoul has been stimulated by the construction of 5 new towns within the capital region. A prime factor of the new towns decision of the government was to solve housing problems in the SMR. They were intended to cope with the speculative increase of housing prices.

Thus the purpose of this paper is twofold: first, to examine the urbanisation process in the SMR since the 1960s, and second, to explore housing issues and problems in the capital region.

21.2 Urbanisation in the Seoul Metropolitan Region

21.2.1 The Trend of Urbanisation in Korea

Modern capitalist society has been developed on the basis of urbanisation which has proceeded in hand with industrializaltion. Modern Korean urban society has accumulated physical and social wealth, and urban people seem to enjoy their oppulence. We cannot then deny such positive effects of urbanisation. But it also should be admitted that urbanisation process has produced negative effects as well. Most Korean cities face serious socio-spatial problems such as lack of land, housing, education, transportation, health care, and so on.

Korea, as is well known, is small , densely settled country . More than 45 million people live in less than 100,000 km^2 - an average density of 450 persons km^2 , exceeded only by Taiwan , Bangladesh and a few island states. In the thirty years between 1960 and 1990, the percentage urban population increased from 28.3 percent to 79 percent in 1990, and is heading for 86.2 by 2001. The largest increase shown is for the most recent two decades 1960-1980.

21.2.2 Concentrated Growth and Urbanisation in the Seoul Metropolitan Area

The SMR includes the city of Seoul itself, Kyonggi province and Inchon

Table 21.1 Growth of urban population in Korea, 1915-2001

Year	Total Population ('000)	Urban Population (%)	Density (person/km²)
1915	16,278	3.1	-
1920	17,289	3.3	-
1930	20,438	4.5	-
1940	23,547	11.6	-
1950	20,167	18.4	-
1960	24,954	28.3	253.1
1970	31,435	47.1	320.4
1980	38,124	60.0	378.2
1990	43,411	78.9	435.8
2001	47,150	86.2	-

Notes: 1)Until 1940, figure refers to total population of the whole of Korea and thereafter to South Korea only.
2)The urban population in 2001 is estimated in the 3rd Comprehensive National Land Development Plan by the Ministry of Construction in 1992.
Sources: E. Mills and P-N, Song, Urbanisation and Urban Problems, Cambridge, Harvard University Press, 1979, p.9.; EPB, Population and Housing Census Report, various years; Ministry of Home Affairs, Municipal Yearbook of Korea, various years.
The following points are of major importance: first, the population growth rate in urban areas has increased greatly both in absolute and relative terms. Since 1955, the disparity between rural and urban population growth has been increasing.
Secondly, the growth rate of larger cities has been higher relative to smaller ones for most of the period, and so their share of the total population has been increasing.
Thirdly, the 1990 census revealed the rapid growth of urban areas which since the 1960's had received relatively heavy investment for industrial estate development .
Fourthly, the 1990 census revealed the extent of the spill-over growth from large cities to adjacent small cities and counties
Except those surrounding the largest cities, almost all counties showed a population loss in the last census. A significant feature of rural to urban migration in Korea is a high degree of mobility in the young age group(10-29) and in people of high educational attainment.

city.[1] It implies in political and administrative context the whole area covered by metropolitan policies. The Kyonggi region which is located around Seoul has been the object of envy from other regions because it has always enjoyed privileges as the nation's capital region. The total area of SMR is approximately 11.726km² and accounts for about 11.9%, of the whole country. Table 21.2 illustrates the dominance of the SMR in many aspects of national affairs.

---

1 *It was suggested from the early part of the 1980s and acquired an official definition as the Capital region. This area covers the entire Kyonggi Province in addition to the government jurisdictions of Seoul and Inchon at the land of 11.726km² in 1994.*

Table 21.2 Degrees of concentration for Seoul Metropolitan Region

	Nation (A)	SMR (B)	Concentration (%) B/A
Area (km2)4	99,394.00	11,726	11.9
Population (000s)4	45,512	20,445	44.9
GRP (billion Won)1	267,449	126,267	47.2
Total industry2			
No. of establishments	2,118,247	944,812	44.6
No. of employees (000s)	11,356	5772	50.8
No. of manufacturing firms4	91,372	50,810	55.6
Secondary industry2			
No. of establishments	301,143	136,74	45.4
No. of employees (000s)	4294	2270	52.9
Tertiary industry2			
No. of establishments	1,815,984	793,864	43.7
No. of employees (000s)	7036	3493	49.7
No. of universities4	131	56	42.7
No. of hospitals3	27,008	11,293	41.9
Bankings4			
Deposits (billion won)	135,190	88,517	65.5
Loans and discounts			
(billion won)	135,850	82,315	60.6
No. of offices for public services2	12,657	6571	51.9
No. of vehicles5	7,403,347	3,600,798	48.6

Notes: 1) 1993, 2) 1992, 3) 1992 4) 1994, 5) 1991.

Sources: Ministry of Labour, 1993; Economic Planning Board, 1962-1994; City of Seoul Metropolis, 1992-1993; and Korean Statistical Association, 1991; Joochul Kim and Sang-Chuel Choe, Seoul: The Making of a Metropolis, Chichester, Joan Wiley and Sons, 1997.

The metropolitan area has not greatly increased in size, but the population has tripled in the last 30 years, reaching nearly 45% of total population as of 1994(Table 21.2), and it is now one of the world's megacities - 10.79 million in 1994. In the last six years, there has been a decline in the population of central Seoul, but continued growth in surrounding Kyonggi province. From 1970 population growth of Kyonggi province and Inchon city started to outpace that of Seoul. Net migrants into Kyonggi province and

Inchon city from Seoul have continued to be about 100,000 persons per year from 1970 till the present. The highest peak of immigrants into Kyonggi province was recorded in 1986 as 256,297 persons. These phenomena are interpreted as the population suburbanisation in SMR (Y. Kown and J. Lee, 1996).

It is estimated that the population of the SMR will increase to about 31.9 million by 2020 (about 63% of the whole population) (Kong, 1996). It is also estimated that most of the increased population of the SMR will concentrate in Kyonggi province.

Table 21.3 The regional share of population

(Unit: 1000 persons)

	1960	1970	1980	1990	1994
Nation's Total (A)					
SMR (B)	24,989	31,435	37,445	43,441	45,512
Seoul (C)	5,195	8,894	13,302	18,785	20,445
	2,445	5,536	10,613	10,613	10,799
B/A (%)	20.8	28.3	35.7	43.2	44.9
C/A (%)	9.8	17.6	22.3	24.4	23.7

Sources: Economic Planning Board, *Population and housing Census,* various years.

Table 21.4 Annual population growth

	1960-'70	1970-'80	1980-'90	1990-'94
Whole nation	2.58	1.91	1.60	1.19
SMR	7.12	4.96	4.12	2.20
Seoul	12.64	5.11	2.68	0.43

Sources: Economic Planning Board, *Population and Housing Census,* various years.

The share of SMR in the national economy has become ever greater, increasing from 25% in the early 1960s to 47.2% in 1993 – nearly half of the national wealth concentrates on the SMR. The education, culture, and convenient living facilities also concentrate into SMR; 43% of universities, 49% of vehicles, and 42% of hospitals of whole nation are in the SMR.

The industrial structure of SMR depends upon mainly secondary and tertiary sector, while the portion of the primrary industry is relatively small. Among them service industry is the fastest growing sector because of its central role and function such as politics, administration, economy and culture. Manufacturing industry is another growing sector as well. As early as 1980 over 80% of hi-tech industries, notably micro-electronics, resided in the SMR. Because of many opportunities such as good labour supply, well developed market system and good accessibility to communications, factories continue to be built, despite policy to avoid congestion

The government has announced a series of policy measures to control the growth of Seoul and Capital Region, including Population Dispersal Plan(1975), Population Redistribution Plan of Capital Region (1977), and Capital Region Readjustment Plan (1984). In addition, there have been two National Land Development Plans, during the period of 1972 to 1992, also emphasizing the decentralisation of Seoul / Capital Region and regional balance. Furthermore, the government introduced an area-wide restrictive zoning system under the Capital Region Growth Management Planning Law in 1982. A variety of restrictions have been imposed on the expansion and construction of buildings in the Capital region. A typical one is to forbid the construction of manufacturing establishments with more than 16 workers and a floor space of more than 200 square meters.

The decentralisation policy, heavily relying on direct regulatory measures, has proven not very successful during the last three decades (Sung-Woong Hong and Heung-Soo Kim, 1995). Although these measures have contributed to reducing the level of concentration of population and factories in Seoul, Some critics doubt whether or not the effectiveness of the measures was significant (Kwang-Sik Kim, 1995). They argue that the Korean experience shows that direct government interventions for growth control are likely to create negative effects rather than to contribute to the balanced spatial distribution of population.

21.3 Housing Problems and Policies in the Seoul Metropolitan Region

21.3.1 Housing Problems

During the 35 year period from 1960 to 1995 the Korean population increased 2.24 percent annually. But the annual increase of population in the SMR was 8.23 percent. As Table 21.5 shows the growth in households in the SMR in particular has outpaced the growth in dwellings. There is also a higher rate of household increase than of population growth, and in addition there is an upward trend in the annual rate of household increase, despite the downward trend in the annual rate of population growth. Between 1960 and 1995 the annual increase of households in SMR was about 14.75 percent, while housing stock grew from 966,000 to 4,015,667 an annual growth of about 9.0 percent.

Table 21.5 Population, households and dwelling stock, 1960 and 1995

	whole nation			SMR		
	1960	*1995*	*Increase*	*1960*	*1995*	*Increase*
			%			*%*
1.Population (000)	24,989	44,551	2.24	5,194	20,157	8.28
2.Households (000)	4,378	12,961	5.60	940	5,794	14.75
3.Dwelling Stock (000)	3,464	9,579	5.04	966	4,016	9.02
4.Dwelling stock/ household	0.79	0.74		1.03	0.69	
			1960 - 1995			*1960 - 1995*
5, Absolute increase in number of house holds			8,583,138			3,049,452
6, Absolute increase in dwelling stock(' 000)			6,114,712			206,648
7. (6) / (5)			0.71			0.50

Source: EPB, *Korea Statistical Yearbook,* 1962, 1977 and 1983: KNHC, *Collection of Housing Statistics,* 1981; KNHC, *Handbook of Housing,* 1996; National Statistical Office, *Population and Housing Census,* 1997.

As a result, the ratio of dwelling to households in the SMR was 1.3 in 1960 to 0.69 in 1995 (Table21. 5). Korea has thus experienced a rapid increase in house prices. Apartment prices in Seoul were more than doubled between December 1987 and March 1989 (Mai'il Kyungje Sinmun, 1987). The housing price to income ratio (PIR) measured for the city of Seoul was estimated at 9.38 in 1991, which is consistently higher than the world wide average figure of 5. The rent to income ratio(RIR) is also extremely high, being measured at 35%, as compared to the world wide average value of 18%(J H. Kim, 1994; the World Bank, 1992). The prices of the Chonsei had continuously increased even though housing prices slowed down between 1991 and 1995. Chonsei is a rent system in which the tenant pays a lump sum to the landlord and gets the money back when he/she leaves. The earned interest on the Chonsei constitutes the landlord's rental income.

Owner occupancy is the traditional form of housing tenure in Korea. The proportion of owner occupancy in the country as a whole has declined from 79% in 1960 to 53.4% in 1995. This decline was mainly limited to urban areas where it fell from 62% in 1960 to 46.4% in 1995 (Table 21.7). One reason for the drop in owner-occupancy was urbanisation.

The rapid urbanisation in the SMR and the migration form Seoul to Kyonggi province have created a big gap between housing needs and supply. Despite the decentralisation of manufacturing employment to outer ring of SMR, part of the population shift is merely residential, with approximately half of a million suburban residents continuing to commute daily to Seoul.

Table 21.6 Housing tenure type (1995)

(Unit: %)

	Whole nation	*SMR*
Owner-occupancy	53.4	46.0
Chonsei	28.1	36.6
Monthly rent	15.5	15.2
Free rent	3.0	2.2
Total	100.0	100.0

Source: National Statistical Office, *Advance Report of 1995 Population and Housing Census*, 1995(Based on two percent sample tabulation).

Table 21.7 Ownership of dwellings, 1960-1995

(Unit: %)

Area	*1960*	*1970*	*1980*	*1990*	*19951*
Whole country	79.1	69.0	58 6	49.9	53.4
Urban areas	62.0	48.4	42.9	40.5	46.4
(Seoul)	(56.5)	(48.1)	(44.5)	(38.0)	-
Rural areas	86.0	84.3	80.7	77.3	77.3

Notes: 1) Based on two percent sample tabulation, *Advanced Report of 1995 Population and Housing Census*, National Statistical Office, 1996.
Source: EPB, *Population and Housing Census Report*, 1960 – 1990.

Therefore such closer interactions between Seoul and its satellite cities attracts policy attention. The structural pattern of suburbanisation in Korea is different from that of the developed countries in many respects. In the western countries suburbanisation may be characterized as an exodus of the middle and higher income groups to suburban communities, due to their increased accessibility by automobiles on new expressways (John M. Levy, 1990, pp.18 20). The suburbanisation in Korea, however, consists of a relocation of the lower income groups to the outskirts of the city where cheaper housing is available. As a result the living environment within the suburban and satellite cities still remains unsatisfactory. Finally, Korea's basic housing strategy, particularly in the SMR, has essentially been based on the filtering concept: expansion of the supply of housing for moderate and high income households will eventually improve the housing available to lower income households and reduce the rate of increase in the price of housing services. It is, in fact, hard to demonstrate that filtering strategies in Korea encourage distributional equity. Joseph Chung has pointed out that:

Korea's higher income biased housing policy can he summarized as follows. First, for a long time the average size of public sector dwellings which account for more than 40% of total production has been too large for low income people. Second, policies regarding housing finance has not been very useful for lower-income housing either. Third, the housing related taxation system has also penalized low-income housing (Chung, 1995: 322-23).

The government formulated a five year housing supply plan for the purpose of constructing 2 million dwelling units between 1988 and 1992, or 400,000 units per year, so that the housing supply ratio could increase from 69.2% in 1987 to 72.9% in 1992. of total 2 million new dwellings, 900,000 units (45%) will be public sector housing, while the remaining 1,100,000 houses will be private sector housing. Table 21.8 shows that SMR enjoys the highest housing construction of the country, generating more than 47 percent of new housing starts during 1988-1992.

Table 21.8 Housing construction, 1988-1992

	1988	*1980*	*1990*	*1991*	*1992*	*Total*
Whole nation A	316,570	462,159	750,378	613,083	575,492	2,717,682
SMR(B)	151,215	209,288	378,797	274,685	282,983	1,296,968
(Seoul)	54,443	76,273	120,371	103,497	106,441	461,025
(Inchon)	31,473	44,441	62,451	28,227	32,391	198,983
(Kyonggi Province)	65,299	88,574	195,975	142,961	144,151	636,960
B/A(%)	47.8	45.3	50.5	44.8	49.2	47.7

Note: 1) Based on National Housing Starts.
Source: Ministry of Construction and Transportation, 1994.
2) New Towns and Housing Production.

In the late 1980s, the government recognised that a permanent and the most feasible solution would be to expand housing production on a massive scale. The government formulated a five year housing supply plan for the purpose of constructing 2 million dwelling units between 1988 and 1992. In order to expand housing construction in the capital region in particular, the government announced five new town construction plans in 1989. The prime factor behind the new towns was to solve housing problems in the SMR.

Although real household income increased, the increase in housing price was far higher than that. For instance, between 1975 and 1988, the average household' s income increased by 2.9 times and the consumer prices increased by 3.5 times, whereas housing prices increased by 4.7 times (KNHC, Housing Statistical Year book; EPB, Major Statistics of the Korean Economy; S-K Ha, 1991b). Enormous fortunes were made from housing which led to more speculation which in turn accelerated housing price increases. The influx of people to the SMR meant that urban housing become scarce and expensive. The second factor was the lack of developable land in Seoul. during the 1970s and 1980s. Even though the government launched several large scale housing project such as Sangaedong and Mokdong in the form of "new town in-town", there were still housing shortage and chronic house price inflation. This forced the government to move outside of the green belt zone to acquire cheap land for housing.

Third, one of the reasons for the new towns in the SMR was the increased demand for new dwellings and suburban living. Partly due to the rise in income and the deterioration and crowding of the old inner city of Seoul, more and more Seoul residents have been turning their attention to the quality of the environment they live in. For those condemned to the stuffiness of the old inner city, modern accommodation with a well preserved natural environment in suburban areas could be an alternative (Ministry of Construction and Transportation, 1995).

Finally, the government expected that the five new towns would alleviate the congestion in Seoul. Recently, an increase in number of automobiles has made the situation even worse. Land and housing prices in the five new towns would be kept low enough to induce many Seoulites as well as private firms and public organisations to move out of congested areas of Seoul.

Table 21.9 shows some critical features of the five new towns. The five new towns are located within a 25km radius from the city center. Pundang is quite close to the Southern part of capital city, the newly growing urban center(Kangnam) and Ilsan is close to the old central business district of Seoul. Owing to this locational advantage, these two new towns have become middle and upper income residential areas. The other three new towns are also located adjacent to the capital city and the existing satellite cities of Anyang, Gunpo, and Puchen.

The size of the land area of the new towns ranges from 419 ha in Sanbon to 1,894 ha in Pundang. The planned population for each of the smaller new towns is 170,000, whereas the largest one, the Pundang new town is planned to hold 390,000 residents. The development of the new towns will provide about 294,000 housing units and also provide a great chance for improving the housing situation in the SMR.

To achieve the target of mass production of new housing, above all, a vast amount of residential land has been needed. The new towns was a means by which the plan for 2 million new housing units could be implemented. The government has played a decisive role in the supply of residential land, through the public agencies such as the Korea National Land Corporation(KLDC) and Korea National Housing Corporation(KNHC). The public purchase and development(PPD) method was applied to the five new town developments. In a PPD project, the public land development agency purchased land in a project area from landowners at valuation.

Table 21.9 Five new towns development plan

New Town Project	area (ha)	No. of Unit (1,000)	Land Development Project body	Planned Population	Location From Seoul
Pundang	1894	95.7	KLDC	390,000	25 SE
Ilsan	1573	69.0	KLDC	276,000	20 NW
Pyongchon	495	42.5	KLDC	170,000	20 S
Sanbon	419	42.5	KNHC	170,000	25 S
Chundong	543	42.5	KLDC/KNHC/ Local gov't	170,000	20 W
Total	4924	294.0		1,176,000	

Notes: KLDC(Korea Land Development Corporation), KNHC(Korea National Housing Corporation).
Source: Ministry of Construction, 1989.

The most important impact has been the stabilisation of housing price in the SMR, because the housing market experienced a supply of new town housing units in a large scale within a relatively short period of time. According to the Korea Housing Bank (KHB), the rise in the housing prices index has slowed down. Table 21.10 shows that the housing price index of nation as a whole has been declined since 1990. Seoul has experienced a slight decrease housing prices as well. In contrast to the late 1980s, the decline between 1991 to 1995 was largely due to the new town development in the SMR and 2 million housing construction drive.

The housing conditions of new town residents have improved significantly in terms of per capita floor area and housing quality. The Korea Research Institute for Human Settlements survey result shows that the residents increased their individual space in the range of from 30 up to 150 per cent, depending on the tenure of their previous residence(KRIHS, 1993c).

Housing construction in the five new towns generates jobs and income. Its employment impact is significant because the construction industry is basically labour-intensive. In 1990 somewhere near the peak of the housing construction cycle, gross housing investment was 21% of the total fixed capital investment and contributed 8.4% to the nation's GNP. According to a recent study it was found that a 10 % increase in housing investment contributed to 1% increase in GNP, 1.4% increase in money supply, 1.5% increase in employment, and 2% increase in fixed capital formation (Ministry of Construction and Transportation, 1996; and Seong-Whan Suh, 1995)

However, even though the five new town development has achieved a major goal of mass housing production in the SMR, the new town development has been criticized by many housing and urban analysts in Korea.

It has not stopped, indeed it has reinforced, the increasing concentration of population and economic activities into the SMR. The nation's economic activities have been spatially polarized around the capital where tremendous locational advantages exist. Accordingly the SMR has rapidly built up growth as the "circular and cumulative causation" process has continued over a long period of time.

In Korea, it is generally believed that low density is one step towards improving the residential environment. The average gross density of the five new towns is 235 persons per hectare, much higher than that of Seoul, which stands at 181 persons per hectare. Net residential density figures are more dramatic. Net residential density is higher than that of the previous "new town in town and Seoul. The average for the five new towns is 686 persons per hectare, while that of Seoul is 364 persons per hectare(Table 21.11). These are the reflection of government 's rigid land use regulation which has limited the supply of residential land, and in turn enhanced high density development.

Table 21.10 Trend of housing prices (1990-1995)

Year (month)	1990 (Dec)	1991 (May)	1991 (Dec)	1992 (Dec)	1993 (Dec)	1994 (Dec)	1995 (Dec)
Nation as a whole	100.0	105.2	99.5	94.5	91.8	91.7	91.5
Seoul	100.0	106.1	97.8	92.5	89.6	90.0	89.5
(Apt.)	100.0	109.5	95.5	91.3	88.8	89.9	89.9

Source: The Korea Housing Bank, *Housing Finance Bimonthly,* 1990-1995; Sang-Ho Nam, "The Socio-economic Impacts of the Five New Towns in the SMR", *Kyonggi for the 21st Century,* No 18. 1996, p57.

Table 21.11 Density for population and housing

Specification	Five New Towns In SMR	Seoul	Mogdong(New Town in-Town)
Gross Population Density (persons/ha)	235	181	264
New Res. Pop. Density (persons/ha)	686	364	495
Floor Area Ratio For Res. Area (%)	184	-	122

Source: Ministry of Construction and Transportation, 1995; Seoul Municipal Government, 1995.

Although the five new towns belong administratively to Kyonggi province, about 66 percent of the residents commute to Seoul for jobs and other purpose. It seems to be that the five new towns are functionally dependent on and just fill the role of Seoul's suburban dormitories. Unfortunately the five new towns have not been designed as complete living and working entities.

The decision for the five new town development was made in a matter of a few weeks. There were no sufficient socio-economic survey and studies for the new town development in the SMR before inception. One of main purposes of new town development is to avoid the further congestion of the capital city and to provide opportunities for life close both to nature and to urban activities. In order to achieve these goals, the new towns should demonstrate innovative techniques in planning and design. However, there have been very little changes in house types and building designs. About 90% of the housing in the five new towns is apartments, lacking diversity and variety. It seems to be that new towns are a copy of large cities. Price control is responsible for uniformity in housing development, since there is no incentive for the housebuilders to improve the quality of design for the apartments. On the other hand housebuilders have sought to profit under the constraints of price control by building at low quality. They have tended to use cheap building materials and a cheap unskilled labor force (Il Seong Yoon, 1991; Chosun Ilbo, Feb. 14, 1996). Accordingly many have occurred (Korea Center for Housing and Environment Research, 1995).

Since 1982 the sale prices of new apartment housing have been set by the government and housebuilders have had to sell under the ceiling price. In this system, a flexible ceiling price system is based on construction cost determined by the material cost and labour. In 1983 a bond bid system was introduced. The bond bid system is applied to the limited number of private housing units whose controlled price is less than 70 percent of the market price. To buy in this case, people must purchase national housing bonds, type

2, which have a 20 year repayment period with 2 percent annual interest, in addition to paying the price of the housing in advance the main aim of the bond bid system is to channel a part of the 'premium' into the National Housing Fund (KRIHS, 1987, p. 33). The aim of introducing these price controls was as follows: first, to give new house buyers the chance to buy cheap apartments, and second, to stabilize the price of existing housing, which could be achieved by price control on new apartments.

21.3.2 Social Goals of New Towns

New Towns appear to be a means of equalizing opportunities in the realm of housing and employment, for bringing together population groups of different income backgrounds. This would encourage social interrelationships and integration. The emphasis on socially balanced communities is one of the most important justifications for the five new towns in the SMR. New towns are suitable for achieving social goals since they should provide a variety of housing types and prices that cater for to a wide range of income groups and life styles.

As shown in Table 21.12, the five new towns are generally filled by relatively higher socio- economic status groups than other cities. The average income of household is accordingly higher in the five new towns than in other Korean cities, thus being 1,628,000 Won for the former and 1,356,000 Won for the latter in 1992. In terms of education attainment in the new towns, it is noticeable that in the five new towns more than half of the household heads had a college education.

2) In January 1988 the ceiling price was set at two levels: 1,268,000 Won per pyong for apartments whose size is less than 25.7 pyong(85 square meters), and 1,340,000 million Won per pyong more than 25.7 pyong apartments (middle and large apartments). 1 pyong is equivalent to $3.3m^2$.

3) Won is the Korean currency unit. In June, 1997, US $ I was equivalent to 890 won and £ 1 was equivalent to 1451 won.

Filtering Strategies

In Korea many planners and government officials tend to believe that trickle-down is advantageous to the poor. It means that the total housing supply in the SMR increases and vacated middle income accommodation becomes available for the low income group, thus, easing housing pressure. However, the KHB loans are earmarked for those who contribute to the funds including the subscribers to the state housing pre-emption subscription deposits. The subscribers are the only ones eligible to purchase state developed new housing. It means that only those who have the ability to save can obtain loans. This policy prevents the low income group from obtaining loans and maintains inequalities. Second, the administration controlled dwelling price is another aspect of Korea's higher income biased housing policy. In the Capital Region where the housing shortage is severe, newly built dwellings are sold at the administrative controlled price which is sometimes only 70% of the market price. To be eligible purchasers have to have made a given number of deposits at KHB's housing saving deposits. Moreover, they have to have been tenants or non-owner occupiers for a stipulated period of time. In spite of administration controlled prices which are much lower than the market price, the dwellings are still too expensive for the lower income group. Seen in this light, the policy is actually a subsidy to the wealthier groups.

Table 21.12 Socio-economic characteristics of New Towns

	Pundang	Ilsan	Pyongchon	Sanbon	Chungdon	Five New Town as a Whole
Age of Household Head	43.0	43.8	42.2	40.7	41.4	425.0
Household Size	3.9	3.8	3.8	3.9	4.1	3.9
Monthly Income (10,000 won)	194.0	147.7	146.6	143.9	149.4	162.8
Education Attainment 1) (%)	60.6	47.2	46.7	53.4	44.0	51.8

Notes: 1) the household heads with a college education.
Source: KRIHS, "A Study on Housing Conditions and Household Mobility in the SMR New Towns", 1993.

21.4 Conclusions

The country has yet to solve its housing problems. The most serious one is the shortage of housing stock in major metropolitan areas, particularly in the SMR, where the increase in housing stock fell short of the household increase due to continuing in migration and the emergence of the small and single economic unit of the nuclear family. In response to such problems, the government has launched 2 million housing units construction plan of 1988-

1992. The key strategies were development of new towns in the metropolitan region, supply of a large amount of residential land, and expanding of housing credit. Mass production of housing and the five new town development in the SMR have contributed to the alleviation of housing shortages, particularly for the middle income households. It seems to be that the five new towns are functionally dependent on and just fill the role of Seoul's suburban residential area for the middle and higher income group.

The SMR remains the biggest concentration of population and economic activities. A stark contrast has appeared in Korea, between backwardness and stagnation of the remoter small and medium cities and the dynamic growth of the agglomerations metropolitan areas and some industrial cities. These trends are producing a new geography of Korea - a geography of stagnation and growth, of 'have not' regions and 'have' regions.

References

Bourns, L.S. *The Geography of Housing*, 1981, London: Edward Arnold.

Brandt, V.S.R. 1982. "Up Ward Bound: A Look at Korea's Migrant Squatters", Korea Culture Service, *Korea Culture*, vol., No 4.

Bratt Rachel G. Chester Hartman and Ann Meyerson(eds.), 1986. *Critical Perspectives on Housing*, Philadelphia: Temple University Press.

Chosun Ilbo(daily newspaper), February 14.

Chung, Joseph. H. 1995. "Economic Development and Housing", Gun Young Lee and Hyun Sik Kim(eds), *Cities and Nation*, Seoul, Nanam.

Economic Planning Board(EPB), *Population and Housing Census*, 1960-1990.

Economic Planning Board(EPB).1990. *Major Statistics of the Korean Economy*.

Ha Seong-Kyu(ed.). 1987. *Housing Policy and Practice in Asia*, London: Croom Helm.

Ha, Seong-Kyu. 1989a. *"Spontaneous Settlements and Urban Redevelopment in Seoul"*, paper presented at the Conference on [Dynamic Transformation: Lessons in Planning and Development from Korea and Other developing Countries], University of Maryland at College Park.

Ha Seong-Kyu. 1991b. *The Study of Housing Policy*, Seoul: Pakyoungsa.

Ha Seong Kyu. 1992c. *"Housing Needs and Co-operative solutions in Korea"*, *Housing~Studies Review*, Vol.1, No.2.

Ha Seong-Kyu, 1993d, The Improvement of living Conditions for the Development of Small and Medium Cities, *Journal of The Korean Regional Development Association*, vol.5 No.1.

Ha Seong Kyu. 1994e. "Low-income Housing Policies in the Republic of Korea", *Cities*(International Journal of Urban Policy and Planning), Vol. 11, No.2.

Ha, Seong-Kyu, 1995f. "Policy Alternatives for the Urban development Programme", *Community Development Review*, Vol. 22, No. 1.

Ha Seong-Kyu. 1995. "Housing Crisis and Perspectives of Housing Policy in Korea", *Housing Studies Review*, Vol.3, No.2.

Hankook Ilbo, April 25, 1994.

Hankook Chuteak Shinmun(weekly newspaper). 1994. *A Report on Resident.s' Response on Apartments in the Five New Town*. Seoul.

Hong, Sung-Wong and Kim, Heung-Soo. 1995. ''(Global City in a Nation of Growth: A case of Seoul'', in G. Y. Lee and H.S. Kim(ed), *Cities and Nation*, Seoul: Nanam.

Hur, J-W. 1990. "The Policy Orientation for Housing Price Stabilisation". *Monthly Economic Review*, May.

Kim. J-H. 1990. "*Korea Housing Policies: Review and Future Directions*", paper presented at the international conference on Korean housing polices, November. Seoul.

Kim, J-H. 1994. "*Changing Perspectives for the Korean Housing Industry*", paper presented at the international conference on Urban and Regional Development Strategies in an Era of Global Competition, Seoul: KRIHS.

Kim, Kwang-Sik. 1995. ''Growth Management Measures and Industrial Location Patterns in the Capital Region", in G. Y. Lee and II. S. Kim(ed). *Cities and Nation*, Seoul: Nanam.

Kim, Manjae. 1992.*The State, Housing Producers, and Housing Consumers in Tokyo and Seoul*, Ph.D thesis, Brown University.

Korea National Housing Corporation(KNHC), *Housing Statistical Yearbook*, 1989.

Korea Research Institute for Human Settlements(KRIHS). 1981a. *Studies in Housing Speculation*, Seoul.

Korea Research Institute for Human Settlements. 1987b. *A Study on the System of Multi-Family Apartment Sales. with Particular Focus on the Bond Bidding Practice*, Seoul.

Korea Research Institute for Human Settlements. 1993c. *A Study on Housing Conditions and Household Mobility in the SMR New Towns*, Seoul.

Korea Research Institute for Human Settlements. 1994d. *A Study on the Provision of Elderly Home*, Seoul.

Korea Research Institute for Human Settlements. 198~3. *A Case Study on Seoul Housing Market - with Particular Focus on Filtering* Seoul.

Korea Research Institute for Human Settlements. 1989. *A Study on Policies for the Urban Poor*, Seoul.

Korea Housing Bank(KHB). 19~37. *Housing Finance Bimonthly Review*, Seoul.

Korea Center- for City and Environment Research. 1995. *Korea's Welfare as Viewed by the Urban Poor and a Search for Grassroots-Government Collaborations*, Seoul.

Kong. S. K. "*Population of the Capital Area: Trends and Policies*", Paper presented at Seminar of the Kyonggi 21C Committee, 1996, April.

Kwon, Yongwoo and Lee, Jawon, "Residential Mobility in Seoul Metropolitan Region in Korea", paper presented at the 20th International Geographical Congress Held in the Hague, Netherlands on August, 1996.

Lee, Gun Young and Kim, Hyun Sik. 1995. *Cities and Nation*, Seoul: Nanam.

Lee, H. and Kim, Y-T. 1994. "Price Control Policy of Newly Construction Apartment: The Case of Korea", paper presented at the 14th EAROPH World Planning Congress, Beijing.

Levy, John. M. 1990. *Contemporary Urban Planning*, New York: Prentice Hall.

Lewis, O. 1966. "Culture of *Poverty", Scientific American. Vol. 21, No.4, pp. 19-25*.

Lewis, O. 1967. La Vida: *A Puerto Rican Family in the Culture of Poverty San Juan and New York*, Panther.

Lim, D.H. 1992. "A Scheme For the Deregulation of Lotting out Price", *Housing Studies Review*, Vol.l, No.2.

Mai'il Kyungje Shinmun (Daily Newspaper), Dec. 12, 1987.

Ministry of Construction and Transportation. 1995. *National Report for Habitat II*, Seoul.

Ministry of Home Affairs, *Municipal Yearbook of Korea*, 1972, 1977, and 1981.

McAustan P 198. *Urban Lad and Shelter for the Poor*, International Institute for Environment and Development. London.

Murphy Denis. 1990. *A Decent Place to Live*, Bangkok: ACHR.

Nam, Sang-Ho. 1996. "The Socio-economic Impacts of the Five New Towns in the SMR. *Kyonggi for the 21st Century*. Vol.8. Suwon.

Stafford, D.C. 1988. The Economics of Housing Policy, London: Croom Helm.

IJNCHS, 1996. *An Urbanizing World. Global Report on Human Settlements,* Oxford: Oxford University Press.

World Bank. 1992. *The Housing Indicators Program. Extensive Survey/Preliminary Results,* Washington D.C.: the World Bank.

Yoon, IL.Seong. 1994. Housing in a Newly Industrialized Economy, Aldershot: Avebury.

22 Urban Landscapes of Japan

Cotton Mather and **P.P. Karan**

22.1 Introduction

Japan today is one of the most highly urbanized nations in the world. Four-fifths of its population now reside in urban areas and the landscapes of the cities, towns, and villages are a vivid portrayal of both growth and modernity.

In the Edo Period (1603-1867), Japan's population was almost stable at 30 million. The country then was largely dependent upon agriculture, and many scholars believed that this was the pinnacle population under this type of economy. But Japan's current population represents a quadrupling of that of the Edo Period, and a transformation of a rural to an urban society. Japan now is one of the ten most populous nations in the world. Amazingly, this strikingly large population is on an area of just 1/26 the size of the United States. But the Japanese problem of living space is far more constraining than these figures imply.

Contrasts in scale on the Japanese landscape are the rule. The areas of flat land are very small and are separated by large areas of rough topography. Mountains dominate the terrain of all the main islands. Indeed, Japan is mostly mountainous, mostly forested, and mostly sparsely inhabited. This is a stunning aspect of Japan, so populous and so urban a nation of such small area yet with much of the land essentially devoid of population. Only 14 percent of the land is arable. Moreover, the arable land is patchy in its distribution, and is poorly accessible. In yesteryear, Japan was rural in nature and dependent upon agriculture. The cities, towns, and villages grew upon this base, a base of relatively small parcels of flat land - flat land that is mostly the deltas of streams. These deltas are at the seaward end of stream courses, so most of Japan's population has been coastal oriented.

The challenge of modern and urban Japan is basically to organize society compactly, three-dimensionally, efficiently. and interconnectedly. This Japan has done. Japan now has the highest average life expectancy in the world, the world's largest source of investment capital, and the highest per capita income in Asia. Furthermore, Japan is now the second largest economic power on the planet. One out of five Japanese individuals now resides in the eight largest cities Tokyo, Osaka, Yokohama, Nagoya, Kyoto. Kobe, Fukuoka. and Sapporo. In contrast to the rural reality of the Edo Period, the present reality is one of most Japanese living in jammed urban environments where private space per person is drastically constricted.

The apex of this urbanisation is *Tokaidopolis*, the spectacular metropolitan sprawl that stretches along southern Honshu from Kobe and Osaka through Kyoto and Nagoya to Yokohama and Tokyo. This amazing agglomeration, 300 miles long, is an urban system of gigantic dimensions and intricacy. Tokaidopolis, Japan's megalopolis, embraces most of the nation's residences, all of the central government, the headquarters of most of the large industrial and financial institutions, most of the prestigious universities, and all of the great organisations of publication and communication. This concentration of human beings and their institutions has developed into a world-renowned complex famous for its unparalleled productivity, but one with stressful social and environmental aspects. Mainly they pertain to limitations of space for the individual and the family. So Japan's population is crowded upon the nation's patchy distribution of flattish land, which in aggregate is only one-eighth of the entire country - an area smaller than Costa Rica!

Thousands of individual scenes in urban Japan meet the eye of the observer, but there are recurrent characteristics that represent the distinctive cultural impress of the Japanese people upon their land. The Japanese landscape is a vivid portrayal of Japanese ideas and their value system of spatial organisation. An identification of the urban landscape lends an understanding to the cultural refinements that have evolved on such a physical base.

22.2 Primary Characteristics of the Japanese Urban Landscape

The primary characteristics of Japanese urban landscapes are:
1 . Paucity of Idle Land
2. Interdigitation
3. Compactness
4. Meticulous Organisation
5. Immaculateness

These features of the urban landscape of Japan are related to the severely limited land and the endeavours to organize and maximize the use of that land.

Paucity of Idle Land

The largest of Japan's lowlands is the Kanto (or Tokyo) Plain which is only about 5,000 square miles in area - about the size of Connecticut. Yet upon this single small lowland is a population equal to the entire population of Canada! Small wonder that land in Japan is used intensively, is sometimes sold by the square meter, and that idle land in urban areas is virtually non-existent

Land is precious throughout Japan. more so in the cities than in the towns, more so in the towns than in the country, and more so in old Japan than in Hokkaido. But nowhere is land more precious than in the large metropolitan centers. Not only are these centers without idle land, they are used so intensively that they are three dimensional.

Interdigitation

The Japanese response to severe areal constriction in urban areas has been to leave no land vacant even temporarily, and to have no exclusion of any major type of land use. This evolved as the population increased and with the cultural progression of experience with this extraordinary shortfall of space.

Two significant aspects that evolved on the urban landscape should be borne in mind. One is that it is interdigitated, and the other is that the interdigitation pattern is finely textured. That is, the land is subdivided into relatively small parcels. What does this mini-parcelisation denote?

Japan is mainly a nation of mini-sized units. Most of the retailing, for example, is done in small shops. The nation does have huge apartment buildings, but most urban dwellers live in small houses or upstairs over shops or other commercial establishments (Fig 22.1). Great corporations exist such as Sony and Nissan, but over two-thirds of all Japanese industrial workers are in companies with fewer than 300 employees. Over half of Japan's factories have less than ten employees. And Japan, on a per capita basis, has almost twice as many wholesalers and retailers as does the United States. The urban landscape mirrors this preponderance of mini-sized units, or fine texturisation. And each type of land use is interdigitated with another type (Fig. 22.2). Also, there is almost no unused land awaiting speculative development. Japan does have land speculation, but since the land is so extraordinarily expensive it is already in some type of interim land use. Empty lots in America's urban areas may exist for speculative reasons, but in Japan such lots will be used *now* even though they are subject to higher economic use tomorrow.

Compactness

Compactness is a fundamental characteristic of Japanese culture. Indi-

Figure 22.1 Living above the shops

Figure 22.2 Interdigitation of fields, homes and factories

viduals without historical perspective may assume that this is simply a response of so many people on so little space. Let us bear in mind, however, that the Japanese population has quadrupled in the last 125 years. And in 1870, Japan was not straining for space. Yet at that time they did venerate their long established miniature verse form known as haiku, they had their scaled-down garden, they had the bonsai, their box within a box, their folding fan, their anesama doll, their tatami mat which could be folded, and their tiny tea house. So the Japanese predilection for compactness is not simply a modern accommodation to so many people on so little space. And compactness is an intriguing aspect of the national mindset on the modern urban landscape. While Hokkaido has more space per person than Honshu, Shikoku and Kyushu, throughout the nation the Japanese have contracted their space in much more refined and confined terms than Americans would have done in similar circumstances.

The pressures of the Japanese people on place are most accute in the huge metropolitan agglomerations such as in Nagoya, Osaka, and Tokyo. Those three urban entities alone embrace 43 percent of the national population. Such an astonishing concentration of human beings on such limited space has been possible only with an extraordinary sense of compactness. Huge apartment buildings stack living unit upon living unit, and each of these units have internal space refinements that overwhelm the American mind. Americans also have large apartment complexes, but the individual units in the United States are much larger, and they lack the subtle internal arrangement of confined space that typify the average Japanese apartment.

And where do the urban residents of Japan shop? Huge shopping complexes have been constructed, but much of the retailing occurs in the immediate neighbourhood where each shop operator is a specialty retailer. These shops are small, varied, and numerous. Moreover, they are nearby. Sidewalks are absent and the narrow street serves as a passageway for cyclists and pedestrians. And for them, no parking space is needed. Thus more compactness! Their home refrigerator is necessarily small, so their contact with the shopkeeper and his family is frequent. The relationship is both commercial and social. Both they and the members of the shopkeeper family sleep in the same neighbourhood and are members of the same community.

Many Japanese do not own an automobile. Those who do may park it in a multi-level, steelgirded structure with an elevator that minimizes the utilisation of space. School grounds are confined. Residences are on minimal lots. Everywhere in urban Japan there is exemplification of the premium of space. This compactness of the urban landscape is statistically manifested by much higher population densities than found in the American urban scene.

Meticulous Organisation

The urban landscapes are meticulously organized in terms of both time and area. This perhaps has been an outgrowth of a rural heritage where most Japanese farmers owned tiny plots of land. So land was precious even in the era of rural dominance. At that time *multiple cropping* - the following of one crop after another in rapid succession so that two or more crops could be grown each year on the same piece of land - was practiced. And, long ago, the farmer practiced *interculture* - the growing of two or more crops on the same land at the same time. Thus fast-maturing crops such as vegetables could be interplanted among slower-growing tree crops, another type of meticulous organisation of time and space as well as an illustration of spatial interdigitation upon the landscape. These concepts from the rural era have in effect been transmitted into a meticulous organisation of the modern urban scene. This has evolved into an interculture or interdigitation of land use for the function of the urban neighborhood.

Americans, in contrast, ever generous with space, have focussed on the simplicity of broad urban areas, each relegated to a single type of land use, thus the zonation of space into industrial, wholesale, retail, or residential areas. The American urban landscape focusses on functional economic zones rather than on the social concept of the neighbourhood.

This meticulous organisational aspect of the Japanese as it pertains to the urban neighbourhood is linked with their predilection for reduction. Thus the small Japanese rock garden, the pioneer development of the small automobile, the small transistor, and the small computer.

Immaculateness

No modern industrial nation seems more immaculate than Japan. The clean waterfronts, the neat machine yards, the litter-free factory grounds, the spic-and-span public areas, the swept streets. and the debris-free homes both back and front set Japan far apart from the trash-laden lands and those with gaudy billboard-lined thoroughfares and with ugly back alleys and sprawling zones of urban shacks. Americans especially may well maintain that immaculateness is a fundamental characteristic of the Japanese landscape.

Yet there is a curse, a dark and somber side, one common to all industrial nations - severe pollution. Kogai, or pollution, is menacing in absolutely every Japanese city. A yellow pall overhangs all Japanese urban centers. And the streams, lakes, bays, and the surrounding areas are laden with poisonous liquids and solids. Japan is an advanced industrial nation, and one with an advanced stage of toxicity.

22.3 Secondary Characteristics of the Japanese Urban Landscape

Secondary aspects of the urban landscape of Japan are:
1. Gardens with Sculptured Plants
2. Lack of Lawns
3. Profusion of Aerial Utility Lines
4. Walled Areas with Gates
5. Waning of Traditional Architecture

Gardens with Sculptured Plants

The public garden of Japan is internationally renowned, but it is the garden of the home that constitutes an omnipresent landscape element of Japanese culture.

The public garden is the model. It is the idolized representation of the aesthetic element from which the economic and spatial compromises must be made for the home. The public garden is a subjugation of nature in which the scale and form of spacious panoramas are miniaturized. Nature in the public garden is in a controlled setting with paths that proffer ever new and glorious perspectives. At its best, the public garden soothes the soul, invites inspiration, and leads to meditation. It is indeed an exultation of nature and the human spirit.

The home garden, however, is typically in a very confined space, separated from the public thoroughfare by a masonry enclosure (Fig. 22.3). It is arranged with meticulous attention to perspectives gained through the fenestration of the home. This represents an integration visually of the interior and the exterior, so often lacking in the Occidental world. And in Japan, the lawn is absent as it denotes a bland undeveloped element of space. Space is ever precious in Japan; it can be reduced, enriched, and controlled by the sculpturing of plants. Gardens of the home in Japan are a careful endeavour to enhance the aesthetics of the home environment, and they are very private.

The Japanese garden emphasizes nature controlled by the human hand, whereas the English garden is one of studied naturalness, and the French garden represents a rational order with a geometric aspect that is imposed on disorderly nature.

Figure 22.3 The domestic garden

Lack of Lawns

A striking aspect of the Japanese urban landscape is the absence of lawns. Even in Hokkaido, where space is least crucial, lawns are lacking. The lawn is not a feature of the private residence, of the great public garden, nor is it even a component of the Japanese cemetery (Fig. 22.4). Every parcel of flattish land in the nation is prized; it is not to be squandered on a monotonous swatch of sward. In the Japanese garden, the ultimate objective is to produce a contrived. constricted, controlled and inspired interplay of rock, water, plants and light for every minute parcel of place. In Japan, that is truly the finest manifestation of "landscape architecture." And it is verily an absolute rejection of filling in empty areas with splotches of grass!

This lack of lawns in Japan is to Americans an especially intriguing landscape characteristic. Upon reflection, Americans may wonder why they merely fill in their open space with just grass. It is certainly noteworthy that for most of the regions in America imported types of grass are planted, the lawns occupy more land than any single crop, their 26 million acres of turf grass is an area larger than the state of Indiana, that most of the water in their cities is used for watering lawns, that urban householders use far more chemi-

Figure 22.4 A local cemetery

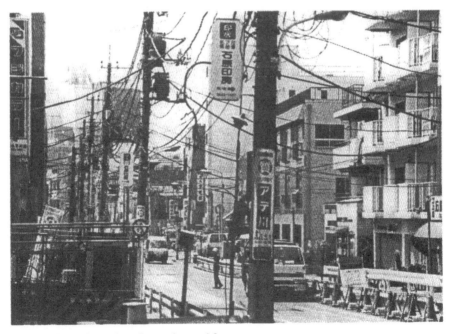

Figure 22.5 Overhead service cables

cals than do the farmers, that most of the pesticides used on the lawns have been untested for long-term effects on humans, that they expend most of their lawn-time just cutting the growth that they stimulated with fertilizers, and that the common lawn scene in the country is working the lawn, not enjoying it! Seemingly stuck in America's cultural craw are the 1872 words of Senator John James Ingalls that "grass is the foregiveness of nature - her constant benediction."

Profusion of Aerial Utility Lines

Among the most conspicuous elements on the urban landscape of Japan are the utility lines. They are strikingly noticeable for four main reasons: there are so many, they are above ground, they spread in every conceivable direction, and they are at many elevational levels. The Japanese are fascinated by technology and employ power and communication lines in myriad ways; hence their extreme dependence upon these wired connections.

But why are the Japanese utility lines so profuse, so complex in pattern, at so many elevational levels, and why are they solely above ground?

The Japanese are a frugal people, they waste no space, they are slow to discard the yet useful, they are quick to add and adapt, and lastly they have an "inside" not a flamboyant "outside" perspective.

Japan's utility line system evolved with one more wire or one more cable for every new line needed. Old lines were retained, new lines were added; they were not combined, nor were they buried (Fig. 22.5). If they were serving different elevational needs, so they were placed, and they ran in new directions wherever they were to be used. The appearance of this awesome maze bypasses the Japanese perspective. Their system was not based on appearance or on one grand organisation. Rather, it just grew.

The "outside" perspective ranks high in the Occidental world. Their cities favour broad boulevards, grand traffic circles, and heroic monuments and statues. So in the Occidental realm are such examples of grand exterior display as the Champs Elysees of Paris, celebrated for its impressive breadth, its tree-lined beauty, and the fountain display at its center. Vienna has its imposing Ringstrasse, a magnificent 150-foot-wide boulevard planted with four rows of trees and lined with splendid edifices and enormous monuments. Rome has its St. Peter's Church with its elliptical piazza bounded by quadruple colonnades and the monumental avenue leading to the piazza. In Buenos Aires there is the Avenida de Mayo and the Avenida 9 de Julio, the latter reputed to be the world's broadest boulevard. Even Washington, D.C. has its magnificent mall flanked by great avenues and dominated by the capitol, the

Washington Monument, and the Lincoln Memorial.

Tokyo and the other Japanese metropolitan centers have no counterpart. The Imperial Palace has an "inside" not a flamboyant "outside" perspective. Indeed, the actual palace is walled and surrounded by moats. So the streets of Japan have one main characteristic, be they the ordinary thoroughfare in a residential-commercial sector or in a major downtown area: they are functional. They are thoroughfares, along and across which are utility lines, pedestrian and vehicular traffic. Whereas the Occidental streets have an "outside" or display aspect. That is of major significance to their cultural psyche.

Walled Areas with Gates

Residential life in Japan in the 20th century underwent enormous changes, and during this period the idealized was far from the realized.

In the early part of the century, most of the population was rural although the settlement form was in a strassendorf or street village pattern where neighbourly social relations were strong. The houses were usually flimsy, one-storied structures with only three or four rooms, with no basement, no attic, and no continuous foundation. The building was framed upon wooden poles and roofed with thatch or tile. Unlike the Chinese preference for clay, brick or stone, the Japanese were partial to wood. They appreciated the patina of weathered wooden exteriors and the mellow tones of hand-rubbed wood on the interior. It was also in plentiful and local supply.

The homes were much adapted to a subtropical clime with sliding panels on the south side that opened to the southern breezes of summer and to the sun in winter. Sliding panels with translucent panes facilitated the multi-functional use of rooms and the passage of light. The broad roof overhang beyond the walls added shelter to the open rooms and related to the changing angle of the sun from winter to summer. A fire pit was used for cooking. Room heating was inadequate, so during cold periods the occupants hovered near the fire. This was the realized.

The wealthy homes were in bold contrast. Their houses had roofs of tile, metal or composite; they were larger, often two-storied, and were set back from the street behind a walled area with a locked gate for privacy. Behind the wall was a garden with weathered rocks, sculptured plants, and a pond arranged meticulously to make it an integral component of the home. These relatively spacious residences, with their walled-in gardens, represented the substance and style of the idealized. They still do.

Most of Japan's population resides now in a new reality. Life is largely in crowded urban areas with drastically confined private space per individual.

Figure 22.6 Local urban road

The culmination of this is in Tokaidopolis. But this world-renowned, massive urban complex, so famous for its productivity, has produced extremely stressful social and environmental consequences. They pertain mainly to the long daily commutate via public transport from residence to work and to the limitations of space for the individual and the family. So the residential unit is compact, children go to parks or commercial playgrounds, automobiles are a luxury, and much of daily life transpires in public and semi-public places. Lacking a guest room in the house, most friends are entertained not in the home but at restaurants and coffee shops. The idealized home for the majority is just a dream, but one that embodies both the space and style of an elegant and re-fined tradition.

Waning of Traditional Architecture

Linking the urban centers , and a vital part of the landscape within the urban areas , are the road and railway systems. The road system is low in quality and quantity compared to those in other major industrial nations (Fig. 22.6). Japan does have a superhighway system, but it has limited outreach. And the rights-of-way are mostly shoulderless and strikingly narrow. About 84 percent of Japanese roads are classified as "other roads" which are suitable only for slow-moving traffic; a third of them are unpaved. Japan has far fewer cars per 1,000 population than any other leading industrialized nation.

Japan, however, has one of the world's most efficient and intricately organized railway systems. Significant are the Shinkansen, known as "bullet trains," in addition to the Limited Express Trains, and the local trains. The nation's railways carry over 22 billion passengers annually: most of these are daily cornmuters. The high price of space does not accommodate huge movement by automobile. But the railway is a vital landscape element in both rural and urban Japan.

The nation has precious jewels on the urban landscape from the past such as the renowned ones in Nara and Kyoto (22.7), but most of the constructs on the present landscape are similar to those found throughout the modern industrial world. Traditional architecture has been waning, and the

Figure 22.7 A small Shinto shrine in Kyoto

loss of old displaced structures has been hastened by both natural and human disasters. In the past half-century alone, typhoons, tsunamis, earthquakes and fires have exacted a dreadful and recurrent toll on the nation of Japan. Indeed, few old societies have so few structures from bygone eras. This has been accentuated not only from disasters, but also by Japan's striking pace of economic development. The capital city of Tokyo, for example, rightfully places high regard and pride on its Imperial Palace, but few primary cities in the world are so overwhelmed by modern edifices and with so few symbols of the past. It is conspicuous even in the small outlying urban nodes of Japan that most of the buildings are modern in age, form and function.

22.4 Conclusion

It may thus be possible to typify Japan's new urban landscapes as both a general representation of all our possible futures, and also as the outcome of its own particularistic cultural history. Whichever, it repays detailed study of its social methods of the organisation of urban space.

References

Arnold, Edwin. 1892. *Japonica*. New York: Charles Scribner's Sons.

Association of Japanese Geographers. 1980. *Geography of Japan*. Tokyo: Teikoku-Shoin.

Christopher, Robert C. 1983. *The Japanese Mind*. New York: Fawcett Columbine, Ballantine Books.

Collcut, Martin, Marius Jansen, and Isso Kumakura. 1988. *Cultural Atlas of Japan*. Oxford: Phaidon.

Henderson, Harold G. 1958. An Introduction to Haiku: *An Anthology of Poems and Poets fiom Baho to Shiki*. Garden City, N.Y.: Doubleday.

Itoh, Teiji. 1984. *The Gardens of Japan*. Tokyo: Kodansha International.

Keene, Donald. 1971. Landscapes and Portraits. *Associations of Japanese Culture*. Tokyo and Palo Alto, California: Kodansha International. Reprinted *as Appreciation of Japanese Culture*. Tokyo and New York: Kodansha International, 1981.

Kornhauser, David. 1982. *Japan:Geographical Background to Urban-Industrial Development*. New York: John Wiley.

Lee, O-Young. 1982. *The Compact Culture. The Japanese Tradition of "Smaller is Better"*. Tokyo: Kodansha International. Translated by Robert N. Huey.

Minear, Richard H. 1994. *Through Japanese Eyes*. New York: Apex Press.

Noh, Toshio and John C. Kimura. 1989. eds. *Japan. A Regional Geography of an Island Nation*. Tokyo: Teikoku-Shoin.

Reischauer, Edwin O. 1988. *The Japanese Today. Continuity and Change*. Cambridge: Harvard University Press.

Statistical Handbook of Japan. Tokyo: Statistics Bureau, Management and Coordination Agency.

Trewartha, Glenn T. 1965. *Japan: A Geography*. Madison: University of Wisconsin Press.

PART V
TRANSPORT AND INFRASTRUCTURE

23 Accelerating Integration between Hong Kong and Southern China

David K.Y. Chu

23.1 Introduction

To discuss the accelerating integration between Hong Kong (the Special Administrative Region(SAR) of the People's Republic of China) and Southern China, it is important first to review relevant global trends in international transport and telecommunications interactions between Pacific Asia's emerging world cites. Rimmer (1996) pointed out that the flows among the world cities are bound to be multi-layered comprising of low-speed transport, high-speed transport and telecommunications. Secondly, he spells out the importance of international container movements among the Asian world cities operating through hub/feeder structures as the key support to physical distribution within Pacific Asia. "Spurred by the globalisation of manufacturing, an annual growth rate of 10% was experienced."(p.68) with the result that the busiest container ports in the world are mostly Asian container ports, with only Rotterdam as a notable exception. However, he also warned, "Given the huge potential for trade between Pacific Asia's emerging world cities, shipping lines have flooded routes with surplus capacity , driving rates below the costs of providing the service"(p. 68) ; "Shipping lines' strategies coupled with economic growth have sparked marked port development in and around world cities" and " finally there is a glut of expansion projects at lesser ports... which have been designed to attract cargo previously handled by the ports of neighbouring world cities."(p.71). He projects the significance of this for future developments in the Asia Pacific transport scene:-

> Looking ahead, all forecasts show that growth rates for transport and communications flows in Pacific Asia will be above world averages. The logical extension of this pattern of development is to recognize that linkages and interaction between Pacific Asia's world cities will create development corridors...Inevitably, new transport and communications superhubs will have to seek locations outside these corridors (p.95).

The last sentence is of particular interest to me but it conforms to the theoretical stand that "goods transportations (are) increasingly decentralized

419

because of the decreasing costs of transport, the introduction of robots, and computerised production control" (p. 49).

Indeed there is now a general consensus that the future pattern of economic activities in the information era could be more dispersed, more flexible, but closely networked. Future superhubs would tend to choose less congested locations with more room for future expansion. The existing development corridors are just too congested for the development of superhubs. Economic restructuring is the key mechanism in driving the world to this new "order".

23.2 Economic Restructuring

The emergence of the East Asian container ports is a response to the cargo traffic generated by the export-oriented industrialisation of the Asian NIEs which successfully integrate their economies within a globalised production system, depdendent the new international division of labour (NIDL). Under this, the "sunsetting" industries of the metropolitan countries are relocating themselves in selected developing countries, whilst the metropolitan countries become the major service providers and locations of R & D activities (Frobel, Heinrichs and Okreye, 1980).

In pre-war years, the mainstay of the Hong Kong economy was entrepot trade. That ceased in 1953 when the United Nations put an embargo on trade with the Peoples' Republic of China. From very poor beginnings, Hong Kong commenced its industrialisation which peaked in the 1980s when about 850,000 people in Hong Kong were employed by manufacturing. This success recruited Hong Kong as one of the four Asian NIEs (also known as the four little tigers). However, now less than 400000 people are employed in manufacturing, and the mainstay of the economy today is services which employ over a million employees. The rapid growth of the tertiary sector (including retail, wholesale , finance and transport sectors) is illustrative of the transformation of the Hong Kong economy.

The process of economic restructuring in Hong Kong has been driven by two main forces. The first is its failure to upgrade its industries. The second is open policy launched by the Peoples' Republic of China since 1978. The manufacturing industries owned and run by the Hong Kong entrepreneurs did not close down but simply relocated northward. In most of the cases, they even expanded by hiring more Chinese workers, whilst the retained Hong Kong employees became the supervisors or technicians of the relocated factories. The cross-border division of function is equally important. On the Hong Kong side are the office headquarters and the show room.

Some retain the quality control function, whilst others keep highly skilled functions like product development. The other end (or ends) concentrate on production according to the specifications set in Hong Kong. The finished products will normally be shipped back to Hong Kong before they are transshipped or re-exported to the Euro-American destinations. All this set the stage of the functional integration of Hong Kong and southern China even before the conclusion of the Sino-British Joint Declaration in 1984, which eventually led to the political accommodation of the two territorial entities in 1997.

23.3 The Physical Bottlenecks

Spatially, the outcome of the economic restructuring at Hong Kong has extended Hong Kong beyond the existing border into the entire southern China with the Pearl River Delta as the immediate hinterland. For those familiar with the traditional port literature, the hinterland of a port can be classified into immediate hinterlands, intermediate hinterlands, distant hinterlands and beyond. This seems to apply to Hong Kong. The only difference is that a border equivalent to an international one exists between the city and the immediate hinterland, which doubles the efforts on customs inspection between the port and its hinterlands in general. Even after the reversion, the border has been maintained and Hong Kong remains a separate customs entity from PRC and border controls on the movement of people will be equally stringent in order to stop an illegal influx of migrants. In a certain sense this is undesirable because it increases the costs of transportation between the port and its hinterland. In theory this is not insurmountable, provided there is political will and determination plus the necessary technical support to keep the costs and inspection to the minimum while keeping smuggling and illegal migration under control.

In reality this is not the case. The crossing of the borders between Hong Kong and Mainland has become the nightmare of all the container lorry drivers. Passenger transports, though better regulated, also suffer from congestion and queues – e.g. at the railway station of LoWu. Disagreements are aired on the meeting points of the Northwest Rail corridor and the Shenzhen rail system. A year after reversion, a solution has still not been worked out. Other supplementary modes of transport like the ferries and coastal shipping are basically very underdeveloped. Bridges or tunnels are proposed but few of these have reached the stage of serious discussion with sincerity. The current situation is definitely undesirable. How has it come about?

23.4 Planning Philosophies of the British HK Government (1840-1997)

From the perspective of the colonial government of Hong Kong, the New Territories were very different from the ceded Kowloon Peninsula and the Hong Kong Island. The New Territories were leased areas for 99 years from 1898 whilst the latter would be British forever unless the Treaty of Nanking was replaced by other diplomatic agreement. The planning philosophy derived from this was to crowd all the infrastructure and investment in the ceded area -" the metropolitan area" so called - and the New Territories were left alone as far as possible, providing farm land for vegetables, and sources of fresh water. The colonial policy was to "respect" the local customs so that the villagers and their land of the New Territories were not used for urban development except that deemed absolutely necessary when no other alternative was available. The fact that most new towns in Hong Kong are built on reclaimed land can be traced back to this philosphical underpinning. The Boundary Street that separated the ceded area and the leased area thus had great implications for the siting of infrastructural projects at least up to late 1960s, and it is believed that its shadow remained even after the signing of the Sino-British Joint Declaration. According to personal information, mainland China was not considered as a factor in the Town Planning Department until the 1980s, and the previous Territorial Development Strategy only considered China as a "passive" factor (Zhu , 1997). It is fair to say that, for the government planners under the British administration, Hong Kong was a 'borrowed' place with a limited time horizon. Therefore town planners took twenty years to mean the 'long term'. Hong Kong would not be a partner to its mainland neighbouring counties and cities. This is the root of the underdevelopment of the infrastructure and institutional arrangements between Hong Kong and mainland China. Technical difficulties are only secondary to the colonial planning philosphy and not the primary cause of the problem. The tradition and the inherent planning philosophy and practices are not going to change overnight after a smooth transition with minimal changes in the bureaucracy.

23.5 Conflicting Interests in the Local and Regional Setting

It will never be easy to ensure that different localities share similar development strategies because of competing interests and conflicts. This is even so in the case of one country and one system. The inherited "colonial system" of Hong Kong and the socialist system of Mainland China are different both in kind and in degree. The consequent developmental strategies in transportation infrastructure are therefore miles apart. This is not just figurative but real. The meeting points of the rail system between Shenzhen and Hong Kong, the landing site of the Zhuhai-Hong Kong bridge, the number and location of airports and the southern China aviation monitoring system are all cases in point. At a regional level, the difference is equally distinctive. Hong Kong regards herself as the dominant node in the regional transport network whilst other neighbouring cities, except Singapore, are its small brothers. Most Hong Kong citizens and officials share the view that to maintain the prosperity of Hong Kong it is necessary to keep Hong Kong's dominance even at the expense of others — to underdevelop the neighbours by competition, for example — is perfectly acceptable. Others take the opposite view — to build up their own cities to compete with Hong Kong. What is at stake is the cost-effectiveness of all these projects, particularly when all these are put together. This is exactly the case if the planned capacities of both Lantau and Yantian ports (in Hong Kong) are added together with the existing facilities.

Coordination at higher levels is sometimes the only way out of situation of this kind. Cooperation between localities with conflicting interests is always paid lip-service, which is mostly insincere. Politicking and lobbying with the highest level at the power hierarchy is inevitable. This is the newest rule of the game, and Hong Kong seems quite unprepared for it, at least for the time being, as if it is still stuck with its pre-1997 mentality.

With the SAR government takes up its office, one would expect many changes. One can predict that Hong Kong under the one country - two systems formula will have to take care of the national interests as well as Hong Kong's interests alone. Better coordination in transport, hardware and software fits into the category of regional and national interests. Accelerated integration in planning and construction of regional rail and road networks and bridges, cross border links would seem logical and inevitable. With the influence and tradition of Hong Kong's practices, these projects would be open to the world for tendering and participation, promising good business for the construction and engineering companies especially those with state of the art technologies.

23.6 Conclusion: the Timing and Phasing of Integration

For most internal and domestic affairs, Hong Kong's autonomy is satisfactory. Transportation, especially with external linkages, is bound to be bilateral and easily linked up with Defense and Foreign Affairs which are beyond the jurisdiction of the Hong Kong government. But there are many grey areas too. The understanding of the wishes and desires of the whole country before setting priorities in these areas is essential. To say the least, unilateral decisions by Hong Kong in these areas should not jeopardise the national view or else the decision will back fire against Hong Kong's own interests.

In another paper (1994), the author argues that it is essential to maintain the market in transport services as competitive, open and fair to all players. The author still maintains this stance and would like to see service providers from some Chinese cities, including seaports operators and shipping lines, sharing Hong Kong's trade, expertise and services. Equally it is important to ask them to open the Chinese market to the Hong Kong players and their related companies. In the short run, possibilities of win-win situations which one can perceive include:

1. more landward connections between Shenzhen and Hong Kong — add 2 to 3 more crossings and computerise the entire system, if possible, and share data base on both sides;

2. customs inspection on one side only (or joint inspection) on entry, not on exit transit containers should be inspected at destination not on route;

3. separate the freight rail link and passenger rail link— redesign the proposed northwest rail corridor. A freight rail line for Hong Kong is not as urgent as suburban train to link up Tuen Mun, Tin Siu Wai and Kam Tin. Passenger rail can have a much steeper gradient so that there is wider choice in terms of routing therefore it is less expensive;

4. support the Shenzhen ports, including Yantian and Shekou ports in diverting some port traffic from the extremely congested Hong Kong terminals, for this Hong Kong should upgrade the Taipo to Shataukok Road to highway standard and a bridge or tunnel to connect Shekou with Yuen Long plus a new highway.

In the longer run, Hong Kong ought to study and re-consider its stance on the Zhuhai bridge carefully. It will help Hong Kong to open up the west wing of the Pearl River Delta and western Guangdong. The bridge is originally Zhuhai's initative but it is difficult to understand Hong Kong's reluctance to support such a beneficial venture. Many of these may not be entirely government ventures. They could be backed by governments but built by publicly listed companies. Besides, the MTR and KCR should be privatised and make themselves listed public companies so they can participate in financially sound projects in cities other than Hong Kong. Public buses and coach companies could do the same too. One would like to see more transport companies in Hong Kong go international, whilst transborder operation, be it jointly with Chinese or with other interests, or solely owned by Hong Kong entrepreneurs, is a good starting point.

Furthermore, integration is a concept that sometimes sounds worrisome to some Hong Kong people. The sensitiveness of the issue calls for slow but steady progress to alleviate unnecessary fear and resistance. In front of Hong Kong , there are a couple of time horizons— 2047, for example is an important one that nobody could ignore. From now on 2008, 2018, 2048 are 10, 20 and 50 years that we should look into and beyond. What kind of scenarios for Hong Kong and southern China that Hong Kong people and the mainlanders would like to see and accomplish by these dates?

References

Chu, D.K.Y. (1994) " Challenges to the Port of Hong Kong before and after 1997" *Chinese Environment and Development* Vol. 5, No. 3 , pp.5-23.

Frobel, F, Heinrichs, J and Kreye, O (1980) *The New International Division of Labour,* Cambridge University Press.

Planning Department, Hong Kong Government (1996) *Consolidated Technical Report on the Territorial Development Strategy Review '96,* Hong Kong Government Printer.

Rimmer, P. (1996) "International transport and communications interactions between Pacific Asia's emerging world cities" Ch.3 in Lo, Fu-chen and Yeung Yue-man (eds) *Emerging World Cities in Pacific Asia,* United Nations Press, pp. 48-97.

Zhu, Jian-ru and Wang, J (1997) "World City Spatiality and Transport Infrastructural Construction in Hong Kong" *Academic Sinica, Vol. 52 Supplement,* pp. 62-70. (in Chinese)

24 Challenges to Sustainable Transport in China's Cities

Andrew H. Spencer

Editors' Note:
Bicycles are good for sustainable urban development because unlike automobiles and buses they emit no atmospheric pollutants and they are quiet. In China the bicycle is the main means of urban transportation. For most people, the journey to work is conducted by bicycle. Nonetheless, as Chinese become more affluent, they are likely to demand the use of automobiles, with all the threats of pollution and congestion which this entails. It is still not certain how much of this can be prevented by the socialist government.

24.1 Introduction

The bicycle is to China what the car is to America. For many, that is part of the attraction. Compared with other Asian countries, Chinese cities appear to offer (almost literally) a breath of fresh air:

Tianjin, which relies on non-motorised vehicles for four out of ten person-trips, instead has high mobility, few traffic congestion problems, very low traffic accident rates, very low public and personal cash expenditures with only modest time expenditures for transport. Thornhill (1991), quoted by Replogle (1992, p.20).

Replogle himself (1992, p.xiii) credits China with having developed 'the most resource efficient urban mobility systems in the world'.

All the same, motor traffic and congestion are growing. China stands not so much at a crossroads as at a T-junction: having approached by way of the bicycle, the diverging arms beckon to mass motorisation on one hand and mass transit on the other. How far down any of the roads must, or can, China travel? This chapter attempts to highlight some of the challenges.

24.2 Trends in Bicycle and Private Vehicle Ownership

Table 24.1 shows that in 1988 China had, nationally, 272 bicycles per 1000 population. While below the level for the industrialised countries shown here, it is well above those for other developing or even newly industrialising countries. Conversely, China lags behind other Asian countries in ownership of motor vehicles. Although pre-war photographs show that bicycles were fairly common in cities like Nanjing, most people will associate them with the post-Liberation order when policies of social equity and a drive to conserve oil supplies encouraged its wider adoption An expansion of cycle production during the 1950s, the provision of travel allowances for workers in state enterprises even if they cycled (Wang Z.H., 1989) and the creation of cycle lanes on many urban roads (Replogle, 1992, p.42) all encouraged its use.

Table 24.1 Bicycle and motor vehicle ownership in selected countries

Country	Year	Bicycles/000 pop.	Motors/000 pop.
China	1988	271.7	1.1
India	1985	58.8	2.0
Indonesia	1985	100.0	9.0
South Korea	1982	153.8	14.0
Thailand	1982	51.0	8.2
Malaysia	1982	178.6	64.3
Japan	1988	491.8	251.6
Netherlands	1985	785.7	350.0
USA	1988	420.4	567.3

Note: The 'Motors' figures for Indonesia and South Korea are for 1981.
Source: computed from Replogle (1992, Table 1.1); Sinha, Varma nd Faiz (1990).

The economic reforms have been an additional spur to ownership, in both urban and rural areas. Between 1980 and 1990 the number of bicycles in Guangzhou rose by 157 per cent; as a result three-quarters of the city's population now have a bicycle. Similar trends can be seen in the municipalities of Beijing, Shanghai and Tianjin. As the 'municipalities' contain extensive rural areas, figures for the urbanised area of Tianjin are included in Table 24.3 for comparison

The changing demand for bicycles has been documented by Zhang (1992) who suggests three main determinants: employment, income levels and price. Employment status has changed rapidly, particularly in rural areas where the setting up of new township industries has increased the need to commute. Incomes have also risen rapidly (Table 24.2) and bicycles are readily affordable. Prices were for a long time around 160 yuan (currently US$19), and even before the urban reforms began a worker could save this sum over a year or two. Although average prices had climbed to 213 yuan (currently US$26) by 1987 the percentage increase was well below the rise in incomes over the same period.

Table 24.2 Per capita income, and bicycles per 100 households, 1978-1990

Year	Urban areas		Rural areas	
	Income (yuan)	*Ownership*	*Income (yuan)*	*Ownership*
1978	316.0	102.3	133.6	30.7
1990	1387.3	188.6	629.8	118.3
Percentage change, 1978-1990	339.0	84.4	371.4	285.3

Source: from Zhang (1992, Table 2).

Table 24.3 Trends in bicycle ownership per thousand population, major cities

	Beijing M	*Shanghai M*	*Tianjin M*	*Tianjin U*	*Guangzhou*
1978	321.5	141.3	169.1	-	145.8
1980	-	152.1	-	-	346.5
1982	408.7	189.1	415.8	635.4	-
1987	-	-	529.4	809.8	-
1988	-	451.6	-	872.0	-
1990	837.0	-	-	-	741.4
1993	-	501.2	-	-	-

Note: M – municipality, U – urban area (smaller than a municipality). The Guangzhou "1978' figure is for 1977. Figures may be overestimated owing to the persistene of abondoned bicycles in the licensing records.

Source: computed for Cai (1994, p.19); Replogle (1992, Table 1.1); Ren and koike (1993, Table 2); Thomas et al. (1992 Table 2 and p.21); Yang (1985, Table 1).

But at the same time as universal adult ownership of bicycles is approaching reality, a new yet familiar trend is becoming apparent. Table 24.4 makes it clear that in Guangzhou motorcycle and car ownership, though still in their infancy, are rising fast. In the first half of 1996 production of bicycles fell by 24 per cent nationally compared with the previous year; much of this fall was attributed to a growing popularity of light motorcycles (Anon, 1996a). A survey of cyclists in central Shanghai has suggested that 64 per cent of them would like to own a motorcycle despite the much higher cost (Yan and Zheng, 1994). Significantly, economic liberalisation in Vietnam has also led to a sharp rise in motorcycle use and a corresponding fall in cycling (Hai, 1994). The drive for personal mechanised transport, whatever its underlying cause, appears to all intents and purposes to be universal and China's cities are unlikely to be any exception.

Table 24.4 Bicycles and motor vehicles per 1000 population in Guangzhou, 1980-1990

Year	Bicycles	Motors	Cars	Motor cycles
1980	346.5	7.0	2.0	1.2
1990	741.4	23.7	7.5	43.8
Percentage change, 1980-90	114.0	238.6	275.0	3550.0

Note: "Motors" includes all forms of motor vehicles except motorcycles; "Cars" means cars, taxis, vans and minibuses.

Sources: Computed from Thomas et al. (1992, Table 2 and p.21).

For the moment, institutional factors are holding these trends in check. Most city governments are restricting the issue of motorcycle licences because of concerns about fuel consumption and safety. Guangzhou allows 6000 new licences per year (Thomas et al., 1992) while Shanghai's annual quota is less than 1000 (Xu and Yu, 1996). Cars, light vans and minibuses are usually owned by work units and access to them is a matter of privilege. A 1990 survey in Beijing suggested that only a quarter of motor vehicle movements involved cars and that 80 per cent of these originated from work units, mostly for business trips (MVA Consultancy, 1993, Table 3.4). The typical car or van driver travels to work by cycle or bus and only then takes out the vehicle, incidentally spreading and flattening the morning traffic peak.

Pressure for change is coming, however. The China Automobile Industry Corporation expects demand for motor vehicles to rise by 8-9 per cent annually between 1996 and 2010, by which time China could he one of the world's three biggest markets. It is the prospect of (government-sanctioned) mass private motoring which makes the need for appropriate urban transport policies all the more pressing.

24.3 Bicycle Use

Table 24.5 gives an indication of bicycle use in urban areas. In the cities depicted, bicycles account for up to 75 per cent of trips (the higher figures for Zibo and Ningbo are for work trips only). It is interesting to compare these figures with Indonesia's 'bicycle city', Yogyakarta, where 23.5 per cent of trips in 1976 were by bicycle and 15.5 per cent by motorcycle (Kartodirdjo, 1981, p.lll). Only in Dalian and Chongqing (a very hilly city) is public transport more important than cycling.

Smaller cities might he expected to be more conducive to cycling than

Table 24.5 Modal splits in selected cities in China (row percentages)

City	Year	Walk	Bicycle	Public	Other
Over 2 million population...					
Shanghai M.	1986	41.00	31.00	24.00	4.00
Beijing U.	1992	14.20	47.40	32.70	5.70
Tianjin M.	1990	10.58	74.63	8.32	6.47
Guangzhou	1992	30.60	33.80	21.80	13.80
Shenyang	1985	29.03	58.65	10.07	2.25
Chogqing		69.20	0.60	26.10	4.10
Median of nine cities		*37.40*	*35.20*	*19.20*	*4.10*
1 – 2 million population...					
Dalian	1990	36.20	17.90	36.40	9.50
Shijiazhuang	1980	33.35	58.65	5.00	3.00
Changsha	1983	39.21	31.39	25.19	4.21
Changchun	1989	37.73	39.63	18.85	3.74
Median of ten cities		*35.90*	*45.50*	*14.70*	*3.45*
Under 1 million population...					
Yantai		30.0	59.7	4.5	5.80
Guiyang	1987	69.74	12.96	11.57	5.73
Zibo	1989	19.03	75.98	2.87	2.12
Ningbo	1990	11.50	82.30	5.80	0.30
Median of ten cities		*38.20*	*53.25*	*4.35*	*2.40*
For comparison: under 1 million population...					
Yogyakarta	1976	45.0	23.54	13.86	17.60

Note: M – municipality, U – urban area (smaller than a municipality).
"Other" includes motor vehicles (in Yogyakarta, mostly motorcycles).
Figures for Zibo and Ningbo are for work trips only.
Note that the percentages in a row do not add up to 100.

Source: Chongqing, Yantai and medians – computed from Mei et al. (1994, Table 1); Guangzhou – from Zhou and Frame (1994); Shenyang and Changchun – from Yang, Paaswell and Rouphail (1990, Table 2); Ningbo – from Jamiesona nd Naylor (1992, Table 1); Beijing – computer from MVA Consultancy (1993, Tables 3.1 and 3.2); Shaghai – from Cai (1994, p.18); Yogyakarta – from Kartodirdjo (1981, p.111); Zibo – Mao (1974); other cities – from Liu et al. (1993, Table 2) with amendments based on other authors.

larger ones but although the median figures bear this out the relationship is certainly not a fixed one - as can be seen hy comparing Shenyang, Dalian and Guiyang. Just as significant as size is urban form; cities dependent on cycling tend to he compact, polycentric and with mixed land use (Replogle, 1992, p.42). Zibo fits into this category; it is a loose agglomeration of five towns spread up to 20 km away from their geographical centre (Mao, 1994). Tianjin is of particular interest. On the one hand it is compact enough for most homes to be within 5.4 km of workplaces: a comfortable 25 minutes' cycling time. On the other hand, central area redevelopment and the building of new housing further out have made walking to work less attractive. Buses are both infrequent and overcrowded. As a result, Tianjin has come to be regarded as the most cycle-dependent large city in China (Ren and Koike, 1993).

The bicycle is first and foremost a means of travelling to work. 62 per cent of work trips by Beijing residents are by bicycle, whereas only 39-45 per cent of other home based trips are (MVA Consultancy, 1993, Table 3.1). In a smaller city like Changchun the proportions are 51 per cent for work and 8-28 per cent for other purposes (Yang et al., 1990, Table 4). In both cities walking tends to dominate for non-work trips, either because they are shorter or because the main income earners appropriate the cycles for commuting and the rest of the household have to manage without. Despite what this may imply, women use bicycles almost as much as men do. Women form a substantial proportion of China's workforce and China seems to be largely free of prejudices against women cycling. Nor is there any major social or class stigma; although the highest income groups tend not to cycle, a survey in the fairly small city of Baoding showed that whereas 14.7 per cent of respondents regarded cycling as a low status mode, 6.7 per cent saw it as high status (Tanaboriboon and Ying, 1994; Kubota and Kidokoro, 1994).

As with ownership, bicycle use is rising. Not surprisingly, Tables 24.6 and 24.7, for Shanghai and Guangzhou, also show a rise in the use of personal motorised modes (concealed in 'other' in 24.6). The trend in public transport use is less clear, though Shanghai seems to be exhibiting a steady decline. Interestingly, figures for Beijing (MVA Consultancy, 1993, Tables 3.1,2) suggest that bicycle use by 'floating' populations is far lower than among permanent residents and that their use of public transport and taxis is correspondingly higher.

Bicycles represent an admirable solution to most people's travel needs, particularly in smaller cities (Shimazaki and Yang, 1992). They can easily negotiate the often extensive narrow lanes or *hutong* which serve the older residential areas. Even in larger cities, where public transport is supposedly more competitive, congestion has often vitiated this. A survey by Beijing Public Transport Corporation in 1986 found that door to door journey speeds by bicycle were almost double those by bus (16.3 as against 8.7 km/hr). Lack of investment and the poor traffic conditions have meant that many bus fleets are unable to keep pace with the growing demand for travel, leading to long boarding times, overcrowding and the familiar vicious circle of declining patronage and worsening service (Jones, 1991).

24.4 Trends in Travel Demand

As economic transformation proceeds, so the demand for urban transport is evolving. The sheer physical growth of cities is steadily increasing not only the number of people travelling hut also the length of their journeys. These impacts, daunting enough in themselves, are being accentuated by at least two further developments: changes in urban land use patterns, and changes in personal lifestyles. The first of these is worth considering more closely.

The traditional Chinese city was compact and displayed a mixed pattern of land uses. Following Liberation the transfer of business occupations to public ownership commonly meant their removal from shopfronts lining the streets to office or factory complexes. Much of the responsibility for providing housing during this period was borne by work units. It was always intended that residential areas should provide a range of communal facilities such as schools, health centres, markets and shops (Wang D.H., 1992; Chiu, 1994). Although some workers might have to commute the idea, however hard to realise in practice, was that residential areas should be more or less self contained and the consequence was that the demand for travel outside the work unit was effectively contained (Bjorklund, 1986).

Table 24.6   Trends in vehicular modal split, Shanghai Urban Area, 1981-1991 (row percentages)

Year	Bicycle	Bus	Other
1981	30.5	67.7	1.8
1986	40.3	58.2	1.5
1991	43.9	53.8	2.3

Source: Cai (1994, Table 2).

Table 24.7   Trends in modal split, Guangzhou, 1984-1989 (row percentages)

Year	Walk	Bicycle	Bus	Car/taxi	M.Cycle
1984	40.9	30.6	21.2	6.9	0.4
1989	36.5	30.6	21.0	10.6	1.3
1992	30.6	33.8	21.8	7.5	6.3

Note: Based on origin-destination surveys in 1984 and 1992 and estimates for 1989 based on vehicle registration data. Trips by ferry boat are excluded.
Source: computed from Thomas et al. (1992, Table 1); Zhou and Frame (1994).

Since the reforms three factors in particular have been at work. One has been the wider range of opportunities for businesses and commerce. A second has been the proliferation, since 1979, of organisations willing and able to invest in housing and other forms of urban developments. A third, more long term in its impacts, is the fostering of an urban land market. In the past, urban land was allocated to work units in perpetuity, at no charge. Since 1990 most land use rights have been made transferable for payment and over the next 20-30 years it is envisaged that all eligible urban land will have been transferred to "paid land use" (Zhou et al, 1993). The result is likely to he a gradual sorting of land uses, with inner-city manufacturing premises cashing in on their assets and moving to peripheral zones where more modern factories can be built, their place being taken by the conventional gamut of high rent-paying Central Business District activities.

Table 24.8   Projected population and employment change in Beijing, 1992-2010 (percentages)

	Population	*Employment*
Four inner districts	-20.3	+0.4
Four other districts	+34.5	+46.3
Whole urban area (eight districts)	+14.9	+26.7
(including floating population)	+24.2	

Source: computed from MVA Consultancy (1993, Tables 2.3, 2.4, 2.6).

In Beijing, for instance, it is proposed to develop peripheral residential areas, to remove manufacturing industries to the suburbs, to redevelop several areas in the central city for commercial, shopping and entertainment uses, and to create separate business nodes both to the east and around the new railway station in the west (MVA Consultancy, 1993, para. 2.4.4). Table 24.8 summarises the projected changes in the distribution of population and jobs up to 2010 and underlines how the city is increasingly mirroring world-wide trends. In the much smaller city of Ningbo (population 430,000) a similar process can he observed:

> ...single-function zones are apparently being adopted...Ningbo plans to redevelop existing central commercial space (after this function's relocation to less-travelled streets) into a financial and administrative centre along the lines of 'downtowns' in Western cities. (Jamieson and Naylor 1992, p.35).

Urban planning, never a powerful force in modern China, appears unlikely to play much part in resisting or shaping these trends (Ng and Wu

1995).

Some estimates of how these changes may affect travel patterns have been offered by the Beijing Transport Planning Study (MVA Consultancy, 1993). Daily trip making is forecast to rise from 1.19 trips per capita in 1992 to 1.68 in 2010 (Table 24.9). Most of this increase will consist of non-work trips fuelled by rising incomes and growing opportunities for private consumption whether of housing, consumer goods or leisure and cultural activities. It must he stressed that the study was a modelling exercise, not a detailed behavioural analysis, but the projected trip rates parallel past trends in Hong Kong. Interestingly, Beijing's forecast trip rate for 2010 will still be below the current rate in Hong Kong, implying that there is still enormous potential for further growth. Although the report gives no estimates for overall trip lengths, the average length of public transport trips is forecast to grow from 10.6 km in 1992 to 11.5 km in 2010 (ibid., Table 4.8) .

Table 24.9   Growth in per capita trip rates, Beijing and Hong Kong (excluding walking)

Year	Beijing	Hong Kong
1973	-	1.27
1981	-	1.58
1992	1.19	1.85
2000	1.34	-
2010	1.68	-

Note: Hong Kong figures are from survey data; Beijing figures are estimates and projections based on surveys in 1990 and 1992.
Source: from MVA Consultancy (1993, Table 4.1).

Table 24.10 Daily vehicular trips in Beijing by mode, 1992 and 2010

	1992		2010	
	Trips (thousands)	%	*Trips (thousands)*	%
Bicycle	3921	47.7	5331	40.9
Public Transport	3334	40.6	5210	40.0
Other motor vehicles	965	11.7	2488	19.1
All modes	8220	100.0	13029	100.0

Source: computed from MCA Consultancy (1993, Table 4.2).

Table 24.10 shows the projected change in modal split over the same period. The study only modelled the effect on modal split of changes in vehicle ownership (whether bicycles or motorised); it took no account of whether public transport improvements such as new subway lines, might themselves he able to attract passengers away from private modes. Although this imparts a slightly conservative bias it is significant that the proportion of trips by public transport is expected to change very little. Motor vehicle trips, on the other hand, are forecast to gain ground at the expense of the bicycle (even though in absolute terms the number of bicycle trips will increase by over a third).

The implications are clear. Even in Beijing, a city with many wide streets considerable lengths of road will be carrying volumes above their capacities in peak periods by the year 2000, while the average speeds of motor vehicles will fall from 30.5 to an estimated 22.5 km/hr between 1992 and 2010. Already long tailbacks are occurring at many junctions. Some of the most severe impacts will be on local access and distributor roads, particularly in the central area (ibid., paras. 4.3.7-10). Hamer (1993) asserts that in most cities the land area currently occupied by roads will need to be doubled. In the foreseeable future this will be almost impossible to achieve, leading to the prospect of development running ahead of transport infrastructure and overloading it, increasing the difficulties of travel. Some of these difficulties, and some associated environmental problems, are described in the next section.

24.5 Bicycles and Urban Transport Problems

China's urban transport problems follow a litany which will be familiar to all researchers of developing cities.

24.5.1 Congestion

Congestion is all pervasive. Buses in central Shanghai average 10 km/hr compared with 25 km/hr in the suburbs (Armstrong-Wright, 1993, Box 3.1). On one of Beijing's widest streets, Chang'an Avenue, the average speed of motor vehicles has fallen from 35km/hr in 1959 to 25 km/hr in 1980 (Yang, 1985). It is easy to blame much of this on the bicycles: after all, they occupy a great deal of roadspace. A World Bank study of nonmotorised transport in ten Asian cities estimated that whereas buses could carry 2700 passengers per hour per metre lane width in mixed traffic, bicycles could carry only 1330 (Kuranami et al., 1994) - though this figure is rather on the low side. At junctions bicycles can cause obstructions either as they start away from the green light in massed formation or as they make left turns across the lines of traffic, commonly waiting in the middle of the junction for a clear path (Wang and Wei, 1993). On the sections between junctions Zhou and Akatsuka (1994) suggested in a regression analysis that a volume of 1750 bicycles per hour per metre lane width, or 6650 in a lane 3.8 metres wide, would reduce motor vehicle speeds by about seven km/hr.

But arguably this is a very modest impact for a flow of over a hundred bicycles per minute! Research by Guo (1996) suggests that cycles similarly cause very little obstruction or delay to buses, even though buses constantly have to 'break into' the bicycle stream at stops. He found that it was motor vehicles which caused the greatest delays by obstructing not only the buses themselves but also the bicycles as they attempted to make way for a stopping or departing bus. Extrapolating to the future, it is not the bicycle which appears as the great consumer of roadspace but its potential successor - the private car.

24.5.2 Accidents

Again, these are a serious problem in China. In 1985 there were 48 deaths per 10000 motor vehicles; a rate similar to India's, almost double those of Indonesia or Thailand and about 19 times those of the UK or the USA (Navin et al., 1994; Spencer, 1989). It is quite easy to show how cyclists form a large proportion of the casualties. Wang Z.H. (1989) argues that 60 per cent of traffic accidents in urban areas involve bicycles, while Xu and Li (1994) point out that Shanghai sees 300 cyclist fatalities a year - half of the city's total road deaths. Yet neither of these studies takes into account the degree of exposure to accident. If there is a large number of cyclists, it is only to he expected that a large proportion of accidents will involve them. Figures for causes of accidents - assuming that these have been correctly attributed - present a fairer picture. Table 24.11 shows how in Beijing and Changchun motor vehicles, despite their smaller numbers, take the largest share of responsibility. Nor do bicycles generally do much damage. Although in Xi'an they were responsible for just over a quarter of all accidents, they caused no more than 16 per cent of fatalities. If the cyclists were to change to other modes accidents, and certainly injuries, could by no means he guaranteed to fall.

24.5.3 Environmental Challenges

Officially, China has fairly strict standards of ambient air quality. Its

Third Class hourly standards, applicable in urban areas, are 20 mg/m³ for CO and 0.3 mg/m³ for NOx. These compare quite well with the World Health Organisation's hourly recommendations for Europe of 30 mg/m³ and 0.4 mg/m³ respectively (Chan et al., 1994; Whitelegg, 1993, p.40). What matters, naturally, is what occurs in practice.

Table 24.11 Percentages of accidents attributed to bicycles and motor vehicles

City	Beijing	Changchun	Xi'an	Xi'an	Xi'an
Date	1986-90	1987-8	1980-1	1981	1981
Type	Injury accidents	All accidents	All accidents	Injuries	Fatalities
Attributed to:					
Bicycles	36.0	14.5	26-29	33.7	16.3
Motor vehicles	48.8	70.0	-	-	-

Source: computed from Navin et al. (1994, pp. 6-7).

Table 24.12 Energy and oil consumption accounted for by the transport sector, selected regions, 1985

	Per capita energy use by transport (gigajoules)		Percentage used by transport	
	Road	*All modes*	*Energy*	*Oil*
China	0.2	1.2	7	13
Taiwan	10.1	11.5	-	-
India	0.8	1.3	-	-
South & Southeast Asia	-	-	24	35
Brazil	12.0	13.3	-	-
Latin America	-	-	33	Over 40
Kenya	1.8	2.7	-	-
Africa	-	-	40	Over 40
USA	66.7	80.8	-	62

Note: Leeming (1993, p.25) suggests that transport accounts for a little under 20 per cent of oil consumption in China (around 30 per cent if agricultural tractors are included).
Source: from Replogle (1992, pp.21-22 and Table 2.3).

In 1990 NOx levels in a sample of northern cities were 47 mg/m³, and 38 mg/m³ in southern cities - well within the Chinese standards. Roughly two-thirds of these emissions could be attributed to domestic and industrial coal burning - sources which will presumably diminish over time (Edmonds, 1994; see also Smil, 1993). More significant is the situation in one of China's most modernised cities, Guangzhou. Here, the growing numbers of motorised vehicles have made these the predominant source, accounting for 67 per cent of NOx and 87 per cent of CO produced in the city. A sample of NOx emissions taken at various locations and various times of day, suggested that the Chinese standards were exceeded in 57 per cent of cases and that there might well be a health risk, particularly to cyclists themselves (Chan et al., 1994), though standards for CO levels were only rarely exceeded.

A related environmental challenge is energy consumption. China's oil reserves are far from unlimited; production in 1989 was not much higher than the UK's and under-investment has meant little provision for exploring new fields (Leeming, 1993, pp. 25-7). Consequently it is reassuring to see that China's transport sector accounts for a remarkably low consumption of energy compared with other parts of the world (Table 24.12).

The question is how long this reassurance can last. It is a feature of developing countries, particularly the more dynamic ones, that income demand elasticities for motor fuel are well above those in developed countries, and price demand elasticities generally lower (Birol and Guerer, 1993). Rising incomes in China will be more than reflected in rising demand for both vehicles and their fuel. The Automobile Industry Policy, mentioned earlier, is calling for the annual production of motor vehicles to rise by over 300 per cent between 1994 and 2010, and that of private cars ('sedans') by no less than 1500 per cent (Huang, 1995). A rise in fuel prices from their currently low level will do very little to offset this. It takes only a small exercise of the imagination to picture the pollution and energy implications; Taiwan's figure in Table 24.12 is perhaps indicative. One needs hardly add that very little pollution or fuel consumption can be blamed on bicycles - and most of what does occur will he during their manufacture, not when they are on the streets.

24.6 Development of Urban Transport Policies

It remains to consider how China's urban policymakers are responding to these challenges. Until the late 1970s traffic was light and urban transport was not seen as a productive activity; hence transport planning was given a low priority. It was only with the reforms and the consequent growth of congestion that research into urban transport received significant official support.

Perhaps inevitably, Chinese transport planners have drawn upon methods developed in the cities of Western Europe and North America. In these countries planning for bicycles was barely taken seriously; the emphasis was on the problems posed by the private car and on means for attracting its users onto public transport. This had two implications. First, China imported techniques and software which were based on Western experience: algorithms for modelling modal splits and the effect of congestion on speeds had been designed around cars and were quite inappropriate for handling bicycles. Second, and in spite of this, the bicycle found itself treated as cars might he treated. Thus a joint research agenda, sponsored by the Ministry of Construction and supported by other government bodies, referred to

"The development prospect of bicycles *and other private transportation means* in our cities and the relative policies." (Xu, 1992, p.36, emphasis added).

While a 1986 symposium on traffic policy concluded that:

> We should...give priority to the development of public transport. Its development should be ensured in policy-making, urban planning construction and financing. Otherwise, private transportation means will continue to grow. (ibid., p.39).

More explicit still were the proceedings of a conference on the role of bicycles, held in Beijing in 1983. These recommended restrictions on private motorised vehicles and 'controls' on non-motorised vehicles in order to achieve a modal split based on city size. Bicycles would only be the major mode in cities of under 100 000 population; in larger cities they would progressively yield their modal share to public transport, which would play the dominant role in cities with over half a million inhabitants (Jamieson and Naylor, 1992).

It would be wrong to infer that Chinese transport planners automatically see the bicycle as a problem. Nonetheless, most appear to be healthily sceptical of its virtues. The general opinion was apparent at a 1994 conference on non-motorised transport in Beijing when some Westerners, enthusing about the city's apparently 'green' transport system and urging its preservation, provoked a response from Chinese delegates who argued that retaining the status quo was simply not feasible. The reasoning behind this view should by now be fairly clear. If today's cyclist is potentially tomorrow's motorist, it follows that it is essential to pre-empt the trend by fostering a public transport culture.

This does not mean that urban roads are not being built - far from it. Undoubtedly there are some who hold that these are the 'infrastructure' needed to create a 'modern' image and attract foreign investment. There have been suspicions that Beijing is following this line and it may he no coincidence that a series of postage stamps for 1995 featured motorway interchanges in that city. Its extensive almost monumental layout makes it something of an exception among Chinese cities and it has been possible to build three orbital motorways and to widen several radial roads without wholesale destruction of buildings. Shanghai is also with World Bank support building a series of bi-level motorways.

At the same time many cities are planning to expand public transport. The Ministry of Construction has called for the number of buses per 10 000 inhabitants to rise by a quarter in cities of over two million people and to be more than doubled in smaller cities. Beijing Shanghai and Guangzhou all have new metro lines under construction and there have been proposals for Chongqing, Shenyang, Nanjing, Tianjin and Qingdao (Anon 1993). Inevitably attracting the needed investment funds has been difficult. In mid-1996 it was announced that the last three of these schemes had been put on hold; Guangzhou has gone ahead relying on a mixture of bond issues and sales of development rights close to stations (Anon 1995 1996b). Even Beijing's main bus operator has found it necessary to enter into a joint venture with the Austrian manufacturer Steyr in order to modernise and expand its fleet.

Regarding bicycles themselves Wang Z.H. (1989) identifies two bodies of opinion. One group taking a fairly critical line would attempt to divert as many cyclists as possible onto public transport in order to release precious road space. Those taking such a view would seek to limit their production introduce licensing with the proceeds earmarked to public transport improvements and regulate bicycle parking more strictly. Others feel that the investment needed to soak up the required number of cyclists would be prohibitively expensive and that its ability to make public transport attractive enough is questionable in any case. Western transport planners will not need to be reminded that many hugely expensive bus and rail improvements have drawn only a disappointingly small proportion of riders away from private transport.

In fact there is a considerable overlap between the two views. Both acknowledge that much could be achieved by segregating traffic types and reducing conflicting movements at junctions. The World Bank study referred to earlier has claimed that providing segregated lanes for both buses and bicycles could raise passenger capacities by 93 per cent and 35 per cent respectively compared with letting the vehicles fend for themselves in mixed traffic (Kuranami et al. 1994). Shanghai has begun to implement a plan for a complex classification of streets open to different traffic types and with a number of them restricted to buses only or bicycles only. Even its elevated motorways will have physically segregated cycle lanes on their lower levels. The aim is to preserve a role for the bicycle, particularly for short journeys of up to 5 km (Cai 1994). A combination of priority measures and a generally conservative policy towards providing road space for cars may yet be the most feasible way of achieving a 'sustainable' urban transport system.

24.7 Conclusion: Pressures for Change

China's urban transport dilemmas may be distinctive in their details but in their underlying nature they are familiar enough. National ambitions for a car industry must be reconciled with accommodating the vehicles which it produces (Spencer and Madhavan, 1989). Its people's aspirations to a 'developed' lifestyle must be reconciled with that lifestyle's costs in terms of paying for the roads and enduring the ensuing congestion, noise, severance and pollution. At the same time please for respecting the ozone layer, coming as they do from countries whose own environmental records are far from impeccable, appear rather hollow. Vice Premier Zou Jiahua stated in 1994 that China could not put environmental protection ahead of economic development, and that developed countries would need to back up their structures with aid (Anon, 1994). (General Motors responded by donating an electric car to China for trials)!

Impressive though mass cycling may seem to the advocate of sustainability, one cannot expect the nouveau-riche Chinese urbanite to live in an ecological museum. All the same, the opportunity for preventive action is there. China's cities do not yet exhibit the motorised mayhem found in Bangkok. Press reports in mid-1996 suggest that some Chinese policymakers are beginning to acknowledge the need to make economic development subject to what is environmentally feasible (Anon, 1996c). Beijing itself is planning to allow cars to travel in the city only on alternate days, according to their number plates (Anon, 1996d). There is still time to take a hard look at urban land use policies. The next few years could be critical for the emergence of a sustainable urban transport policy for China.

References

Anon, 'China will make great efforts for development of public transportation', *China City Planning Review*, 9/1 (1993) 54-5.

Anon, 'GM gives electric car', *Far Eastern Economic Review*, 157/29 (1994) 75.

Anon, 'Subway fare', *Far Eastern Economic Review*, 158/34 (1995) 59.

Anon, 'Sales of bicycles start to yield way to motorcycles in what has been dubbed the 'bicycle kingdom'', *Far Eastern Economic Review*, 159/33 (1996a) 25.

Anon, 'Shanghai metro deal', *Far Eastern Econmic Review*, 159/33 (1996h) 71.

Anon, 'State councillor puts environmental protection above development', *Far Eastern Economic Review*, 159/34 (1996c) 23.

Anon, 'Global transport news: road: China', *Global Transport*, 6 (1996d) 15.

Armstrong-Wright, *Public Transport in Third World cities*. (London: HMSO, 1993).

Birol F. and Guerer N., 'Modelling the transport sector fuel demand for developing countries', *Energy Policy*, 21 (1993) 1163-72.

Bjorklund E.M.. 'The danwei: socio-spatial characteristics of work units in China's urban society', *Economic Geography*, 62 (1986) 19-29.

Cai J.X., 'Bicycle transport in Shanghai: current status and future prospects', *The Wheel Extended*, 90 (1994) 18-24.

Chan L.Y., Hung W.T. and Qin Y., 'Vehicular emission exposure of bicycle commuters in the urban area of Guangzhou, South China (PRC)', *Environment International*, 20 (1994) 1 69-77.

Chiu R.L.H., 'Housing', in Yeung Y.M. and Chu D.K.Y. (eds) *Guangdong: survey of a province undergoing rapid change*. (Hong Kong: Chinese University Press, 1994) pp. 277-300.

Edmonds R.L., 'China's environment: problems and prospects', in Dwyer D. (ed.) *Chin:. the next decades*. (Harlow: Longman, 1994) pp. 156-85.

Guo J., 'A study of traffic behviour in bus stop areas with mixed traffic'. Unpublished PhD thesis, Transport Studies Group, University of Westminster (1996).

Hai L.D., 'Bicycle use in the extraordinary background of urban transportation in Hanoi and new policies for it', in Ren F.T. and Liu X.M. (eds) *Proceedings of the International Symposium on Non-motorised Transportation*, Beijing Polytechnic University (1994) pp 340-51.

Hamer A., 'China urban land management: options for an emerging market economy', The World Bank, Washington DC. Executive summary reprinted in *China City Planning Review*, 9/1 (1993) 17-26.

Huang W., 'Auto industry faces challenges and opportunities', *Beijing Review*, 38/45 (1995) 15-18.

Jamieson W. and Naylor B., 'Planning for low-cost travel modes in Ningbo, China', *Transportation Research Record*, 1372 (1992) 31-9.

Jones, T.S.M., 'Urban public transport in Beijing (PRC) present and future'. Paper presented at the London-Beijing Symposium on Transport and Tourism, Middlesex Polytechnic (1991).

Kartodirdjo, S., *The pedicab in Yogyakarta. a study of low cost transportation and poverty problems*. (Yogyakarta: Gadjah Mada University Press, 1981).

Kubota H. and Kidokoro T., 'Analysis of bicycle-dependent transport systems in China: case study in a medium-sized city', *Transportation Research Record*, 1441 (1994) 1 1-15.

Kuranami C., B.P. Winston and P.A. Guitink, 'Nonmotorized vehicles in Asian cities: issues and policies', *Transportation Research Record*, 1441 (1994) 61-70.

Leeming F., *The changing geography of China*. (Oxford: Blackwell, 1993).

Liu X.M., Shen D.L,. and Ren F.T., 'Overview of bicycle transportation in China', *Transportation Research Record*, 1396 (1993) 1-4.

Mao H.Z., 'Investigation and planning of bicycle transportation in Zibo City', in Ren F.T. and Liu X.M. (eds) *Proceedings of the International Symposium on Non-motorised Transportation*, Beijing Polytechnic University (1994) pp. 96-101 (in Chinese).

Mei B., Wang X. and Xu J.Q., 'Study on the development trends of urban bicycle traffic', in Ren F.T. and Liu X.M. (eds) *Proceedings of the International Symposium on Non-motorised Transportation*, Beijing Polytechnic University (1994) pp. 52-7 (in Chinese).

MVA Consultancy, *Beijing Transport Planning Study: draft final report*. For Beijing Academy for City Planning and Design and Great Britain Overseas Development Administration (1993).

Navin F., Bergan A., Qi J.S. and Li J., 'Road safety in China', *Transportation Research Record*, 1441 (1994) 3-10.

Ng M.N. and Wu F.L,., 'A critique of the 1989 City Planning Act of the People's Republic of China', *Third World Planning Review*, 17 (1995) 279-293.

Ren N. and Koike H., 'Bicycle: a vital transportation means in Tianjin, China', *Transportation Research Record*, 1396 (1993) 5-10.

Replogle M., *Non-motorised vehicles in Asian cities*, Technical Paper 162. (Washington DC: World Bank, 1992).

Shimazaki T. and Yang D.Y., 'Bicycle use in urban areas in China', *Transportation Research Record*, 1372 (1992) 26-9.

Sinha K.C., Varma A. and Faiz A., 'Environmental issues in developing countries', Seminar N, PTRC Summer Annual Meeting, University of Sussex (1990) 37-46.

V. Smil, *China's environmental crisis*. (Armonk, NY: M.E. Sharpe, 1993).

Spencer A.H., 'Urban transport', in T.R. Leinhach and Chia L.S. (eds) *Southeast Asian transport: issues in development* (Singapore: Oxford University Press, 1989) pp. 190-231 .

Spencer A.H. and Madhavan S., The car in Southeast Asia', *Transportation Research A*, 23A (1989) 425-37.

Tanaboriboon Y. and Ying G., 'Characteristics of bicycle users in Shanghai, China', *Transportation Research Record*, 1396 (1993) 22-9.

Thomas C., Ferguson E., Feng D. and DePriest J., 'Policy implications of increasing motorisation in developing countries: Guangzhou, People's Republic of China', *Transportation Research Record*. 1372 (1992) 18-25.

Thornhill W., 'Non-motorised transport in China'. Paper presented at the Transportation Research Board Annual Meeting, Washington DC (1991).

Wang D.H., 'A review on the development of planning and design of China's urban residential areas', *China City Planning Review*, 8/1 (1992) 46-55.

Wang Z.H., 'Bicycles in large cities in China', *Transport Reviews*, 9 (1989) 171-82.

Wang J. and Wei H., "traffic segregation on spatial and temporal basis: the experience of bicycle traffic operations in China', *Transportation Research Record*, 1396 (1993) 1 1 -7.

Whitelegg J., *Transport for a sustainable future: the case for Europe*. (London: Belhaven, 1993).

Xu P. and Li B., 'The study of strategy for bicycle safety in Shanghai', in Ren F.T. and Liu X.M. (eds.) *Proceedings of the International Symposium on Non-motorised Transportation*, Beijing Polytechnic University (1994) pp. 378-87 (in Chinese).

Xu K.W. and Yu T.W., 'Macro-management of Shanghai public transport', *Public Transport International*, 1996/2 (1996) 28-32.

Xu X.C., 'Ten years of urban traffic planning development in China', *China City Planning Review*, 812 (1992) 32-41.

Yan K.F. and Zheng J.L., 'Study of bicycle parking in central business district in Shanghai', *Transportation Research Record* 1441, (1994) 27-35.

Yang J.M., 'Bicycle traffic in China', *Transportation Quarterly*, 39 (1985) 93-107.

Yang Z.S., Paaswell R.E. and Rouphail N.M., 'Growth of urban transportation in the People's Republic of China', *Proceedings of CODATU 5*, Sao Paulo (1990) pp. 91 9-30.

Zhang X.H., 'Enterprise response to market reforms: the case of the Chinese bicycle industry', *Australian Journal of Chinese Affairs*, 28 (1992) 111-39.

Zhou G.Z. and Study Group, 'Policy framework for the development of real estate and the real estate industry in China', *China City Planning Review*, 9/1 (1993) 2-16.

Zhou H.L. and Frame G., 'Study on Guangzhou's bicycle policy', in Ren F.T. and Liu X.M. (eds) *Proceedings of the International Symposium on Non-motorised Transportation*, Beijing Polytechnic University (1994) pp. 90-5 (in Chinese).

Zhou Y.Q. and Akatsuka Y., 'A study on characteristics of bicycle traffic under mixed traffic on a road segment in Beijing, China', in Ren F.T. and Liu X.M. (eds) *Proceedings of the International Symposium on Non-motorised Transportation*, Beijing Polytechnic University (1994) pp. 195-200.

25 Water and the Urban Future of the Ganga Plains

Graham P. Chapman

25.1 Introduction

The Ganga (Ganges) river basin , together with the lower Brahmaputra in Assam and Bengal, constitutes one of the most remarkable natural resource units on earth. This one river system is home to 10 % of the human race. This 10 % includes half of the poorest people on earth in Uttar Pradesh, Bihar and West and East (Bangladesh) Bengal. Their living standards are an affront to the normal expectations of human dignity. Physically this is a river basin of extremes too – part of the world's greatest river plains (linking to the Indus plains in Pakistan), with the deepest riverine deposits (well over a mile deep), one of the world's great aquifer regions for groundwater, adjacent to the world's greatest mountains, and rained on by its heaviest monsoon. Indeed it is the extent of the ecosystem and the energy and material fluxes through it which have enabled it to support so many people for so many – not centuries – but millennia. In all that time, the civilisations the valley has spawned have all been based on agriculture, and within the last centuries, they have been based on better and more extensive use of irrigation, harnessing the perennial waters from the snow-melt rivers that flow from the Himalayas.

Some of the towns and cities of the plains were also founded thousands of years ago. Others are more recent, including the 10 million city of Calcutta. Some are the products of both ancient foundations and new impetus – like Delhi which is one of the fastest growing (and most polluted) of the earth's mega-cities, and flood-prone Dhaka in Bangladesh, which may even be the fastest growing outside of China. But at the moment the valley is only 20% urban – this is one of the least urbanised even if most densely settled major regions of the earth. If it is to develop and to industrialise to give its people a better living standard – indeed as it is doing and as it will – the cities will grow. These cities will demand more water, more food, more power. Just the thought of this ought to make planners start calculating the problems the politicians will face, and the possible solutions they might suggest. I make some rough calculations here to illustrate the nature of the problems.

25.2 The Future Urban Population of the Ganges Plains

For statistical purposes the central Ganga plains have been defined as comprising the state of Haryana, the two Union Territories of Chandigarh and Delhi, the plains districts of Uttar Pradesh and the state of Bihar. In broad terms this is an area 1400 km long (NW to SE) by 300 km across; the total area of the territories just named is 514,000 km^2, slightly less than the area of France. In 1991 the total population of this area was 246m people, or about 5% of the human race, and nearly 5 times the population of France. The growth rate of the population from 1981-1991 was 2.3% annually, meaning that the first constraint on the development process will be the basic necessities of this ever burgeoning total.

The assumptions are that population growth rates will have declined to 1.5% annually by 2031, so the starting rates which for each part of the basin vary around 2.3%, are reduced slightly in each year. The reason why the rate will not have reduced more by 2031 is that the current Indian population pyramid is heavily weighted towards the young, people who have not yet reached their reproductive years. The assumption about urbanisation is that there are two components to growth - one the internal natural increase, the other migration. The natural increase of both urban and rural areas is projected using the same growth rate as for the state, but then the urban growth is augmented by a migration component subtracted from the rural area, which is calculated as a specific multiple of the natural urban increase. The specific multiple changes according to a curve related to levels of urbanisation. At 10% urban the migration component is 2 times the natural increase. At 30% the migration component is equal to natural increase (which is what happened in broad terms in the 10 years 1981-1991 in India); at 90% migration it is only 1% of natural increase.

The results, wholly credible, suggest that in 35 years time the population of the middle Ganga plains will be about 600 m, or 100 m more than for the whole of the Ganga-Brahmaputra in 1991. The population will also be about 40% urban rather than 20% urban as now. The implication is that rural populations will increase by about 80% - nearly doubling the pressure on land. The surest way to increase agricultural production and levels of employment - livelihood creation - has been shown above all to be through increasing irrigation, both from gravitational canals and by powered tube-wells, and then secondarily by other means such as the use of fertilisers and new seeds.

The urban population of the basin will increase from the 1991 figure of 94m to 485m, of whom 140 m will be in the one state of Uttar Pradesh. In passing, we may note that Delhi's size is predicted to be 25m by 2031. These

Table 25.1 Current and projected total population and level of urbanisation for the Ganges and lower Brahmaputra basins

	Total Population 1990	% Urban Population 1990	Annual % Population growth 1981-91	% Population below Poverty Line (India only)	Total Population 2010	% Urban 2010	Total Population 2031	% Urban 2031
Ganges hills								
Up Hills*	5900.9	21.6	2.0	-	8763.7	31.5	13015.4	41.0
Himachai Pradesh	5111.1	8.7	1.8	19.1	7277.8	16.2	10363.0	24.9
Nepal	19379.0	8.0	2.6	-	32084.2	18.8	53119.3	31.3
	30391.0	**10.8**	**2.3**	**-**	**48125.7**	**20.7**	**76497.7**	**32.1**
Ganges Plains								
Delhi	9370.4	89.9	4.2	15.8	15155.5	91.2	24512.2	92.3
Chandigarh	640.7	89.7	3.6	-	1039.4	91.0	1686.1	92.2
Bihar	86338.8	13.2	2.1	47.9	131197.7	23.1	199363.9	33.6
Haryana	16317.7	24.8	2.4	17.2	25907.9	36.3	41134.3	46.8
UP Plains	133130.0	19.8	2.3	42.9	208967.1	31.1	32 004.2	41.9
	245797.6	**20.7**	**2.3**	**-**	**382267.6**	**31.2**	**594700.7**	**41.7**
Brahmaputra hills								
Arunachal Pradesh	858.4	12.2	3.1	-	1572.0	26.6	2878.8	41.3
Manipur	1826.7	27.7	2.5	12.3	3001.3	29.8	4931.2	50.5
Meghalaya	1760.6	18.7	2.8	37.3	3039.2	32.3	5246.4	45.0
Mizoram	686.2	46.2	3.4	-	1314.5	58.5	2518.0	67.9
Nagaland	1215.6	17.3	4.6	-	2949.3	38.9	7155.4	56.9
Sikkim	405.5	9.1	2.5	-	662.5	20.0	1082.3	32.2
Bhutan	1476.0	4.0	2.3	-	2325.7	11.6	3664.7	22.2
	8229.0	**19.0**	**3.0**	**-**	**14864.4**	**33.0**	**27476.7**	**47.3**
Brahmaputra plains								
Assam	**22294.6**	**11.1**	**2.1**	**38.0**	**33939.4**	**20.7**	**51666.5**	**31.2**
Bengal Delta								
Tripura	2744.8	15.3	3.0	47.4	4958.6	29.7	8958.0	43.8
West Bengal	67982.7	27.4	2.2	48.1	105094.1	38.1	162464.5	47.8
Bangladesh	107993.0	15.7	1.9	-	155549.6	24.6	224048.7	33.8
	178720.5	**20.1**	**2.0**	**-**	**265602.3**	**30.0**	**395471.2**	**39.5**
South Asian Ganges- Brahmaputra Basin	**485432.7**	**19.4**	**-**	**36.9****	**744799.4**	**29.7**	**1145812.8**	**39.9**

Sources: Census of India, 1991, Muthiah (1987), and Encyclopaedia Britannica (1992).
* Districts of UP included in Hills: Uttar Khashi; Dehradun; Almora; Chamoil; Garhwal; Naintal; Tehri Garhwal; Pithoragarh.
**All India average poverty

urban populations will make demands on water for direct use, and indirectly through the demands for more power, much of it from hydro-electricity. But they will be in competition with the rural population, increasing from 391 m to 688 m.

In sum, we have to consider what are the likely patterns of direct water demand by the following users:- agriculture, urban non-industrial, urban industrial, and waste/sanitation disposal.

25.3 Urban Water Demand and Urban Sanitation

The water supply situation at the level of the individual household or enterprise in both urban and rural areas is not well known. The Tata Energy Research Institute (Reidhead, Gupta and Joshi, 1996) records an increase in clean water availability from 69% to 86% of the urban population between 1985 and 1993 – but it does not say how many litres per capita are available – the figures relate to the presence of supply points. Many supply points are not known to the authorities –tubewells in particular may remain unreported. Where water tables are high enough, hand-pumps can also be found in urban areas - but within urban areas there is a very high chance that the near-surface water that such pumps can lift will be contaminated (Reidhead, Gupta and Joshi, 1996). It is clear that many households and enterprises are not connected directly to the supply, and there are private water traders meeting some of the excess demand.

TERI's report on the State of India's Environment (Reidhead, Gupta and Joshi, 1996) includes a map (25.1) of the river basins of India showing the status of water stress and water scarcity. The map is based on rainfall, assumed recovery rates, population figures, and assumed coefficients relating to demand. Given water recycling and better water recovery, the coefficients could change – thereby changing the pessimistic conclusions. However, as they stand at the moment, the figures suggest a precarious situation: a little of India faces some water problems, most of it is already in water stress (including the Ganga basin), some of it faces water scarcity, and parts of the southeast in Andhra Pradesh and Tamil Nadu and parts of Gujarat arealready facing what is known as 'absolute scarcity'.

The National Institute of Urban Affairs (NIUA, 1994) provides some figures and maps for four cities - Delhi, in the Ganga basin, and three cities outside - Mumbai, Vadodara and Ahmedabad. In the latter, per capita daily availability peaked at 209 litres in 1971 and by 1991 had declined to 141 litres. Bombay is forecast to have a major deficit by 2011. The National Commission on Urbanisation (GOI, 1988) has provided a sketchy over-view of some of the issues and requirements. It notes that currently per capita water supply in large cities may be over 100 litres per day - but that the distribution is highly unequal, with a majority of households not directly supplied. The connection for sewage is similarly unequal: only one third of households in urban areas nationally are connected to a closed water sewer.

The same source quotes some figures which are slightly confusing - it is difficult to work out what has been stated as target figures by the Planning Commission (note: not the National Commission on Urbanisation) and what are current achievements. It appears that, to provide for a complete coverage of the urban population by the year 2001 with 150 litres per day, the State would have to make an investment of Rs 15000 crores (£3 bn or $4.5 bn)

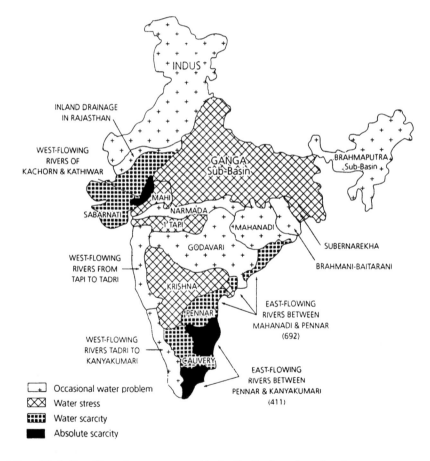

Map 25.1 Predicted water scarcity in India by river basin
Source: Tata Energy Research Institute. New Delhi.

annually (from 1988 to 2001). Noting that this is an impossible financial target, the National Commission suggests a more modest approach - with lower targets to aim for, and a more multi-faceted distribution system. The norms are shown in Table 25.2 - much more modest than the Planning Commission's target of 323 lpcd for Delhi (225 lpcd for domestic purposes, the remainder for industrial and civic purposes.). Even to achieve this and equitable distribution, simple means of distribution - such as neighbourhood holding tanks with tapped outlets, and stand pipes, are postulated - plus private provision from ground water, although clearly with the proviso that either private or public regulators have the ability and mandate to monitor quality.

Table 25.2 Suggested urban water requirements

Purpose	Absolute Minimum litres per capita per day	Desirable lpcd
Cooking and drinking	10	15
Bathing, flushing etc	30	40
Washing utensils and clothes	30	35
Total	70	95

Source: GOI (1988, p 294).

In the past the public water utilities have been owned and directed by urban local government. However, over the last twenty years, the status and power of local urban government, never strong, has been further undermined (see Kumar, Chapter 26), and many of the public water utilities are now run by state boards. Potentially this has the advantage of a wider view being taking of the relation between supply and demand, but it also of course makes the utilities less responsive to specific local problems.

It is also possible to approach these problems from a different statistical base. Statistical evidence from a variety of sources all suggest that about 30-40% of the urban population live in slums - whose inhabitants do not have access to flush sewage systems. Some of these may have pit latrine systems, and some have night-soil collection systems. It would appear that half the remaining population are not properly connected either, although more likely to have some partially adequate alternative disposal system.

The low levels of connection have two 'beneficial' effects: one is that current water demand levels are lower than they would otherwise be, and the second is that much of the potential sewage that could be discharged into the rivers is not. These benefits are of course bought at the cost of hazardous health conditions for the urban population.

The National Commission observes that 80% of water supplied to urban areas passes through the urban area ending as sullage, sewage, or effluent of some kind. Virtually all is untreated, some is discharged simply to sink into the ground, a lot into the drainage system. By the year 2001 Delhi alone will discharge about 4100 million litres a day, against its supply of 5100 million per day The Commission is at pains to point out that proper collection and treatment will not only provide a better habitat, but crucially, also will recycle the water. Indeed it suggests that recycled water costs significantly less than developing new sources, at increasing depth or distance.

25.4 Future Urban Water Demand and Supply

I am assuming that at most the current actual average water supply in urban areas for all inhabitants is at most 50 lpcd. Other estimates (but older) put it considerably lower. Using the projections for urban population to the year 2031 given in Table 25.1, if development were to mean water consumption for all inhabitants at 200 lpcd, then with the urban population quintupling the total urban water demand would rise 20 times in 35 years. Industry that accompanies industrialisation will also in all likelihood increase its demand at a faster rate.

Urban demand is by its nature fairly constant throughout the year. The question is then whether the sources are equally able to face this anticipated high demand throughout the year. The historically used, cheapest and preferred source for most major cities has been surface flow, abstracted from major rivers. We will look at this source first.

Surface flow is so seasonal that it can barely be considered adequate to meet existing levels of provision, let alone demand. Some idea of the seasonality can be judged by the discharge figures for the Ganges at Varanasi - a flow of 13,454 cumecs in the summer, and only 285 cumecs in the winter, when the river has the appearance of a sequence of languid pools, or by the figures of river flow by Messerli and Hofer (1995). Another way of looking at seasonality is to say that for most of this area the four monsoon months of June to September account for 85% of total rainfall. Supposing surface flow is split the same way, then the Yamuna at Delhi has a dry season flow of 8.5 m cumecs per day. This flow (probably an exaggeration of the lowest flows) would only just meet the water demand of 25 million people using 200 litres per day (5 m cumecs). Although clearly a substantial part of the consumption would be discharged, in effect it would be discharged with at least some level of contamination back into a nearly dry channel. (The abstraction rate of water from the Yamuna at Delhi has more than once reached the point where water is being drawn back upstream from what should be downstream sections - into which untreated effluent is poured. In other words, abstraction from upstream of the city is drawing sullage from what should be downstream locations.) Such a scenario is neither impossible, nor on present trends necessarily unlikely, although at present urban areas are minority users compared with agriculture. If enhanced dry-season surface flow is seen as a solution, the finger will point towards building more big dams in the Himalayas, and in the south-bank Deccan hills.

The second source for urban areas is groundwater. The sediments of the Ganges basin form a giant sponge, usually with a water table near the surface, and also many other aquifers in different deposits at increasing depths. The near-surface water levels do vary throughout the year, but not by us much as the seasonality of rainfall would lead one to suppose. Often the variation is measured only in a few metres. It would seem that there are adequate resources in most of the Ganges plains to meet the levels of demand envisaged from this source, but on three conditions. The first is that investment is provided for an adequate number of bore-holes, the second is that there are adequate energy supplies for the extraction required, and the third is that the water is not contaminated, or only to the extent that modest treatment will render it potable . The finger again points towards the provision of more power supplies for the plains - and again towards dams in the hills as the best source of such power.

Logic says that any approach to the problems of water supply will involve storage in the hills and control of the river regimes. From the urban point of view the maximum power demands occur in the hot season (March-May) when river flows are least, so surface flow can be augmented and power provided for groundwater extraction at the same time. On the other hand these patterns of water release do not match the demands of agriculture: the hot season is the low season for agriculture. Irrigation demand in agriculture usually peaks in the winter season, December to February. This logic suggests that increasingly agriculture may lose out to the urban-industrial needs. But it will do so from what is currently a privileged position.

25.5 Competition from Irrigation and Agriculture

In India somewhere near 90% of total human water demand is for agriculture. This is near the average for Asia - but very substantially different from Europe (see Table 25.3), where industrial and urban development is greater (and of course the climate and terrain very different.) Yet it would appear prima facie that economics do not justify such levels of use by agriculture. To quote Fauceys (1992):

> In a year, 1000 m3 of water may be used either to provide water for 80 people, or to grow food for between 1.6 and 3 people. This imbalance suggests...domestic water should win every time. It is far more economic to move food from a rain-rich area to a dry area than it is to move water.

Although the disparity of the two figures suggests that overwhelmingly the equation will work out in favour of urban water use, the equation is, of course, never this simple. The level of agricultural water demand depends on a wide range of variables, most of which are liable to substantial change. At first sight there would appear to be a simple kind of climatic determinism at work in India - with the wetter eastern parts of this plains region a traditional rice-growing area, and the drier western parts traditionally more wheat orientated. But crop breeding during the Green Revolution has produced new breeds of rice which perform better in irrigated conditions in drier sunnier areas than in the moister and cloudier conditions of the east. Given also that there is a substantial demand for rice and a price margin in its favour, there has been a very substantial increase of rice growing in the western areas of the plains, in some cases rice becoming even more important than wheat. The difference in water demand between rice and wheat is substantial: therefore for this reason amongst others there has been a substantial increase in agricultural water demand in the western parts of the plains.

Table 25.3 Water demand in two global regions

	Annual withdrawal m^3 p.c	Withdrawal as % available	% domestic	% industry	% agriculture
Asia	526	15	6	8	86
Europe	726	15	3	54	33

Source: Chisholm in ODA (1992).

Variation in the preferred crop and the associated implicit water demands is therefore one reason why it is difficult simply to speak of areas as "rain-rich" or "rain-poor" – and another reason to show that the coefficients the TERI map (25.1) is based upon can vary substantially. Another is the pattern of seasonality of rainfall. The maximum rainfall in the western wheat areas occurs during the Kharif (monsoon) season, when it is too hot and humid for wheat to grow; and the monsoon is too unreliable this far west for rain-fed rice to be a reliable staple crop. Thus the traditional farming pattern has been to grow wheat in the cooler Rabi (winter) months, using retained soil moisture from the monsoon, and irrigation from open wells where possible. The first of the major nineteenth century canal irrigation schemes, the Upper Ganga Canal, started in the 1850's, used the low winter flow of the Ganges to provide protection and security for this winter wheat crop. The arable acreage could be expanded for a while by taking new land into cultivation, but ultimately further increases could only be accommodated by increasing the cropping index (gross cultivated area/net cultivated area: indicating the extent to which a piece of land is cropped more than once a year). This of course means growing more in the second season. After Independence in 1947 the Upper Ganga Canal was modified to provide for irrigation in the monsoon period and the growing of crops such as rice and sugar cane. There are technical, seasonal, and geomorphological reasons why water could not be provided in the wet season before extra investment was made – in outline the principal difference is that in the low season the river water carried little silt, but in the high season silt flows are very high and can choke a canal system fast. 'Bed-load' extractors had to be designed and installed in the head reaches of the canal.

In sum, the relationship between rainfall, agriculture, and water demand is too complex for agricultural regions to be characterised in a static manner as rain-rich or rain-poor.

25.6 Summary and Conclusions

In the upper Ganga catchment (and also in places in South India) water is already the subject of fierce inter-state rivalries, riots and police firings, particularly in the hot season. Disputes between Haryana, Delhi and Uttar Pradesh are a predictable annual feature. But this is hardly even the beginning of trouble, if urban populations increase by five times and water demand by twenty times in the next three and a half decades. The demand for water for the urban areas may well out-price agricultural water – yet to feed the population agriculture is also set to increase its irrigation demands on current trends .

Resolution in the future is not impossible. The Himalayas are sometimes described as the world's biggest water towers, and there are many possibilities for developing new dam sites (but see the words of caution in Thompson 1995). There is great potential for the extension of extraction of groundwater – but this may need power from the hills. Both urban and rural areas can substantially improve water efficiency - in the urban areas by proper treatment and recycling, in rural areas by stemming the massive waste that occurs in large-scale surface systems, and adhering to different cropping regimes (easier said than enforced.)

The problems of the future need not be impossible to solve – but if solutions are to be successfully implemented, then it is imperative that the sheer scale of the problems is realised *now*, and the first strategic steps for implementing solutions are also taken now.

References

Chapman, G.P. and Thompson, M. (Eds) (1995) *Water and the Quest for Sustainable Development in the Ganges Valley*, Cassell, London

Chapman, G.P. (1983) "Underperformance in Indian Irrigation Systems: the Problems of Diagnosis and Prescription" *Geoforum* Vol 14 no 3 pp267-275

Fauceys, R. (1992) "Domestic Water Use: Engineering, Effectiveness and Sustainability" in ODA(1992).

GOI (1988) "Water and Sanitation " Chapter 14 pp 293-301 in *Government of India: Report of the National Commission on Urbanization*, Vol II, New Delhi

National Institute of Urban Affairs (NIUA) (1994) *Urban Environmental Maps* New Delhi.

ODA(1992) *Proceedings of the Conference on Priorities for Water Resources Allocation and Management* Natural resource Engineering Advisors' Conference, Southampton July 1992, Oveseas Development Administration, London.

Rasheed, S. (1995) "Nepal's Water Resources: The Potential for Exploitation in the Upper Ganges Catchment" Chapter 5 in Chapman, G.P. and Thompson, M. (Eds)

Reidhead, P.W. , Gupta, S. and Joshi, D. (1996) *State of India's Environment (a quantitative analysis)* Tata Energy Research Institute, New Delhi

Seckler, D. (1981) *The New Era of Irrigation Management in India* Ford Foundation, New Delhi, Mimeo

Thompson, M. (1995) "Disputed Facts: A Countervailing View from the Hills" Chapter 6 in Chapman, G.P. and Thompson, M. (Eds).

26 Organisations and Approaches for the Development and Provision of Infrastructure in the National Capital Territory of Delhi

Ashok Kumar

Editors' Note:
The National Capital Territory of Delhi is a bit like the District of Columbia in the USA – it belongs to the federation, and not to any state. But the NCT has both rural and urban components, and the urban component – though of one contiguous conurbation - is divided btewen different municipal corporations. Delhi, as the capital of India and one of its megacities, combines both administrative and industrial activities. The several organizations that manage Delhi's administrative affairs have over-lapping jurisdictions. Lack of coordination between organizations, bureaucratic bankruptcy, lack of policy direction and breakdown of communications affect the smooth organization of the administration. Such problems are common in other cities as well – London has been run by a multitude of local councils since Mrs Thatcher abolished the Greater London Council (though that situation is again beginning to change.) Delhi has opted for convergence and networking approaches for resolving its administrative problems. Whatever may be the approach adopted, it is not easy to have smooth coordination when so many organisations work in the same space.

26.1 Introduction

Infrastructure is the very glue that holds modern urban society together (Graham and Marvin, 1995 : 169; Prasad, 1996; HUDCO, 1996 : 9 - 10). The provision of infrastructure makes economic sense because then activity can flourish in the cities, and it makes environmental sense because it greatly contributes to a better physical environment. The health of urbans citizens is also contingent on the provision of adequate levels of infrastructure (Cotton and Franceys, 1994; Cotton and Franceys, 1991, Dimitriou, 1991). The lack of adequate water supply and sanitation leads to a high incidence of water borne and diarrhoeal diseases in poor urban areas. In a study of the slums of Delhi, an Infant Mortality Rate (IMR) of 112 per 1,000 live births was reported which is scaringly high compared tothe Indian urban average of 66 in 1984 and the Delhi average of 39.69 in the year 1985 (Muddassir and Risbud, 1993 : 40 - 41; World Bank, 1994). In other words a lack of urban basic services makes cities uneconomical and unhealthy (Seabrook, 1996; Stephens,

1996). By contrast, the provision of basic services can "reduce the costs of being poor" (Seabrook, 1996 : 7).

Significantly, the World Bank also perceives that infrastructure development and economic productivity reinforce each other. The Bank's policy of integrated infrastructure development aims at low cost infrastructure investment with a view to using the low income communities for "the city's economic production processes through the making of focused marginal cost investment" (Dimitriou, 1991 : 198). Therefore, it could be argued that "utility infrastructures are important factors in the economic, social and spatial development of cities, regions and space economies" (Graham and Marvin, 1995 : 170).

Significant new infrastructure is needed to sustain the more than 12 million people who are expected to live in the National Capital Territory of Delhi at the turn of the century. A fraction of that population, i.e. just 1.7 million people, lived in Delhi in 1951. During the last three decades, the urban area of Delhi has increased by more than two times. Although the twin cities of (old) Delhi and New Delhi constitute the centre of the connurbation, New Delhi has a separate sub-jurisdiction in the New Delhi Municipal Committee. Efforts to bring about organisational changes have resulted in the establishment of new organisations such as the development authorities, but not the abolition of the old. Overlapping functions and jurisdictions are the inevitable outcome.

In this chapter, which is divided into three sections, unless it is qualified, the word Delhi will mean the whole of the National Capital Territory, including New Delhi. In the first part an analysis of the governance of Delhi has been provided. The second section of the chapter then contains a discussion on the approaches to the provision of infrastructure with a clear emphasis on the National Capital Territory of Delhi. The third draws general conclusions and makes several recommendations.

26.2 Governance of Delhi after Independence

Delhi has been governed by elected governments in sporadic short spells since 1947 (see Table 26.1).

The local government of New Delhi has also remained nominated and unelected, and therefore largely unrepresentative. For example, the Presidential Ordinance promulgated on 25 May 1994 provides that three MLAs, 5 Union Government officers, 2 nominated members such as lawyers etc. and the MPs of New Delhi will oversee the working of the renamed New Delhi Municipal Council. But the Chairperson of the Council will be still an IAS

Officer appointed by central government. Similarly the Municipal Corporation of Delhi has been primarily run by unelected bureaucrats. Infrastructure such as roads, drains etc. in "development areas" is provided by the Delhi Development Authority which operates under the direct control of central government. Major urban development projects have been undertaken by a central agency called the Delhi Development Authority.

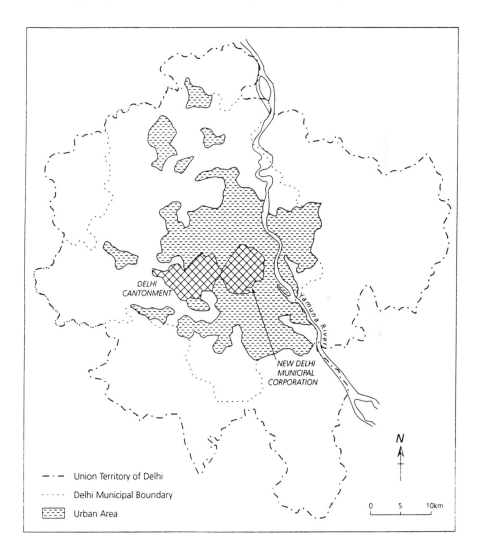

Map 26.1   Delhi: National Capital Territory and urban area

The National Capital Territory of Delhi is much less than a full state of the Indian union. It has a special status under the Government of the National Capital Territory of Delhi Act 1992 (Goyal, 1993). Since 1992 a seventy member Legislative Assembly and seven member Council of Ministers has been created. The Legislative Council is led by the Chief Minister. However, the Legislative Council has no powers to legislate on various important issues including public order, police, land, and matters on the State List particularly Entry 1, 2 and 18 (also refer to 26.1). The authority of the state of Delhi is further curtailed by minimizing its financial powers. Whereas other states have their public accounts in the Reserve Bank of India, the Delhi Government does not have any such account; nor can it, like other states, borrow from the market.

Table 26.1 Governance of Delhi between 1947 and 1997

1947 - 1952	Government of India
1952 -1956	Self Governing Part C State with the Legislative Assembly of 48 Members
1956	Delhi Legislative Assembly abolished
1956 -66	Directly Administered Union Territory and the Birth of the Delhi Administration
1957	Delhi Municipal Corporation created
1966	Metropolitan Council set up under the Delhi Administration Act
1967	First elections held to the Metropolitan Council of 56 elected and 5 Nominated Members
1980	Metropolitan Council dissolved by the Indira Gandhi Government
1983	Metropolitan Council revived
1990	National Front Government dissolved Metropolitan Council
1992 onwards	Elected Government of the National Territory of Delhi with 70 Members' Legislative Assembly headed by the Chief Minister

Source : Compiled from Kumar (1996a : 24) and Goyal (1993 : 16-39).

Major arguments for not according the full status of a state to Delhi have been repeatedly outlined. Firstly, it is contended that since New Delhi hosts important national and international offices, its governance cannot be left to the administrative control of a state government. This underscores the lack of faith in the administrative capacity of local governments. Secondly, it is argued that since 80 per cent of the property in the National Capital Terri-

Table 26.2 Major Organisations providing various services in the NCT of Delhi, 1996

Name of the Service	Organization providing the Service	Administrative Control
Urban Planning and Development	Delhi Development Authority	Ministry of Urban Affairs and Employment
Water and Sewerage	Delhi Water Supply and Sewage Disposal Undertaking	Ministry of Water Resources and the MCD
Electricity	Delhi Vidyut Board*	Delhi Government
Transport	Delhi Transport Corporation	Delhi Government
Drainage	Irrigation Department of Delhi Administration	Delhi Government
Fire	Delhi Fire Service	Delhi Government
Telephone	Mahanagar Telephone Nagar Limited	Ministry of Communications
Milk	Delhi Milk Scheme	Ministry of Agriculture
Law and Order	Delhi Police	Home Ministry
Slum and Squatter Improvement	Slums Department	Delhi Government
Enforcemnet of Planning Controls	Special Task Force	Home Ministry
Urban Form and Aesthetics	Delhi Urban Arts Commission	Ministry of Urban Affairs and Employment
Jhuggi Jhompri Rehabilitation	Delhi Slum Improvement Board	Delhi Administration
Finance for Industry and the Business	Delhi Finance and Development Corporation	Delhi Government

Source : Compiled by the author from various sources (1997).
Note : * DVB was known as the Delhi Electric Supply Undertaking before early 1997.
Slum Improvement Board and Trans - Yamuna Development Board could also be taken into account. Both these boards are also headed by Bhartiya Janta Party MLAs and MPs. However, these are not statutory bodies.

tory of Delhi is owned by central government, it is only pertinent that central government or its own civil service agents look after and develop these lands. Lastly, it is argued that it would be appropriate if central government is involved in the administration of Delhi in order to provide efficient services which will bolster the country's image.

26.3 Who Governs Delhi? - a Problem-Centred View

26.3.1 Maze of Organisations

More than 10 million people now reside in Delhi. The governance of these people within the National Capital Territory is undertaken by more than 13 large and about 200 medium and small organisations. All these organisations look after the total area of 1,483 sq. km. out of which 685.34 sq.km is rural and the remaining 797.66 sq.km is classified as urban area. Delhi has 209 villages, 29 census towns, 3 municipal bodies and two tehsils. Therefore, the governance of Delhi has become almost impossible with the present administrative structures. In addition to the organisations shown in Table 26.2, numerous technical colleges and hospitals are run by the Delhi Government. Each of these organisations is as big as a medium size (in terms of number of employees and the extent of operations) private company.

26.3.2 Centralisation and Decentralisation Versus Fragmentation

The existence of a large number of organisations in the National Capital Territory of Delhi in itself is not a bad situation of governance. However, uncoordinated decentralisation also leads to fragmentation. For example, large scale privatisation of public transport in the late 1980s unleashed the phenomenon of decentralisation and fragmentation. Overnight individuals became owners of buses. No coordination and enforcement mechanisms were put in place to regulate private operators. The result was catastrophic. Thousands of people died on the Delhi roads in the eight years as a result of the chaos created by the private bus operators (Kumar, 1996b) and their use of unqualified drivers. The Delhi Government had to act. In early 1997 the government declared that individual private operators would cease to operate by the end of 1997 and their buses would be run by the DTC on mileage basis.

26.3.3 Fragmentation: More Choices or Simple Chaos

Arguing in favour of centralized provision by a single metropolitan authority, Ellen Brennan concluded that "A metropolitan development authority may be an effective mechanism for managing mega-cities provided that its role is clearly defined. The limited territorial control of existing municipalities and the extensive jurisdiction fragmentation usually mean that metropolitan problems can not be handled at the sub-metropolitan level"

(Brennan, 1994 : 251).

26.3.4 Overlapping Jurisdictions

Take a look at the mess created by the different administrative boundaries in Delhi. According to new administrative set up (see below) Delhi has 27 revenue and police sub-divisions. Delhi is also divided into 12 election zones but only 8 town planning zones. To further add to the confusion, one hundred and thirty four unique election wards have been created. Moreover, there are unique boundaries of the slum and squatter areas. This situation is extremely confusing and results in chronic delays as far as delivery of services is concerned.

26.3.5 New Administration

But now Chief Minister Sahib Singh Verma seems to have established an effective new administrative set up which he claims will be transparent and impartial (also see Fig. 26.1). The Government of the National Capital Territory of Delhi now has nine Deputy Commissioners and 27 Sub Divisional Magistrates with a view to bringing the administration at the doorsteps of the people. The nine districts are West, South West, North East, New Delhi, Central, North, East, North East and South. The offices of the Deputy Commissioners and Sub-Divisional Magistrates have started functioning since 1 January 1997. These offices will work as one window offices for all government departments at the district and sub-division level.

Administrative districts have been created to cover the entire geographi-

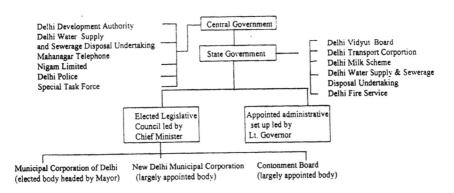

Figure 26.1 The structure of government for Delhi

Table 26.3 District administration of the NCT Delhi, 1997

West District and its Sub-Divisions	West Police District and its Police Stations
West District and its Sub-Divisions • Punjabi Bagh • Rajouri Garden • Patel Nagar	*West Police District and its Police Stations* • Punjabi Bagh, Paschim Vihar, Naggloi • Hari Nagar, Kirti Nagar, Rajouri Garden • Patel Nagar, Moti Nagar, Anand Parvat, Tilak Nagar, Janakpuri, Vikaspuri
South West District and its Sub-Divisions • Vasant Vihar • Nazafgarh	*South West Police District and its Police Stations* • Vasant Vihar, Sarojini Nagar, R.K. Puram • Nazafgarh, Jagarpur Dabri, Delhi Cantonment, Vasant Kunj, Narayana, Mayapuri, Inderpuri
North East District and its Sub-Division • Seelampur/Seemapuri • Shahdara	*North East Police District and its Police Stations* • Seelampur, Gokulpuri, Bhajanpura, Seemapuri, Nand Nagri • Shahdara, Welcome, Mansarovar Park
New Delhi and its Sub-Divisions • Connaught Place/ Parliament Street • Chankya Puri	*New Delhi Police District and its Police Stations* • Connaught Place, Tilak Marg, Parliament Street, Mandir Marg • Chankya Puri, Tuglak Road
Central District and its Sub-Divisions • Darya Ganj • Pahar Ganj	*Central Police District and its Police Stations* • Darya Ganj, Chandni Mahal, Jama Masjid, Kamla Market, Hauz Quazi, I.P. Estate • Pahar Ganj, D.B. Gupta Road, Nabi Karim, Karol Bagh, Prasad Nagar, Rajinder Nagar
North District and its Sub-Divisions • Sadar Bazar • Civil Lines	*North Police District and its Police Stations* • Sadar Bazar, Kashmere Gate, Bara Hindu Rao • Civil Lines, Timarpur, Roop Nagar, Maurice Nagar, Subzi Mandi, Pratap Nagar, Sarai Rohilla • Kotwali, Lahori Gate, Chandni Chowk
East District and its Sub-Divisions • Vivek Vihar • Preet Vihar • Gandhi Nagar	*East Police District and its Police Stations* • Vivek Vihar, Anand Vihar • Preet Vihar, Shakarpur, Kalyanpuri, Trilokpuri • Gandhi Nagar, Krishna Nagar, Geeta Colony
South District and its Sub-Divisions • Kalkaji • Defence Colony • Hauz Khas	*South Police District and its Police Stations* • Kalkaji, Badarpur, Okhla • Defence Colony, Lodhi Colony, kotla Mubarakpur, Lajpat Nagar, • Srinivaspuri, Hazrat Nazamuddin • Hauz Khas, Malviya Nagar, Mehrauli, Greater Kailash, Chittaranjan • Park
North East District and its Sub-Divisions • Saraswati Vihar • Narela • Model Town	*North West Police District and its Police Stations* • Rohini, Jahangirpuri, Shalimar Bagh, Ashok Vihar, Saraswati Vihar, • Keshav Puram, Sultanpuri, Kanjhawala, Mangolpuri • Samaypur Badli, Narela, Alipur • Model Town, Mukherjee Nagar, Adarsh Nagar

Source: Government of the National Capital Territory of Delhi (1997 : 8).
Note: *In all there are 27 revenue and police sub-divisions. Delhi Cantonment, Karol Bagh, Seemapuri and Parliament Street are listed separately.

cal area of the National Capital Territory of Delhi. A Deputy Commissioner or Development Commissioner will head a district which will have more than one sub-division magistrate. Administrative districts will correspond to the police districts. In all Delhi will have nine administrative districts and police districts, and 23 sub divisions (see Table 26.3).

The Development Commissioner is a civil servant who will be in charge of new administration in a district. Each sub-division will have more than one police station. But Delhi Administration's 11 directorates and 24 departments will continue to exist. But nobody quite knows what to do with them. The problem of duplicity and coordination that they create remains unresolved.

The functions of the Deputy Commissioners and Sub-Divisional Magistrates include the day to day administration, implementation of government programmes and schemes, and redressal of the public grievances. They have the following functions:

Magisterial Functions (parts of the criminal code, inquests, human rights abuses etc); *Revenue* Functions (.g. Urban Land Ceiling Act, Panchayat Act, Land Acquisition Act and Delhi Land Reforms Act); the Issuance of various *Certificates* (Scheduled caste, backward caste, birth, death, marriage); *Registration* Function (stamp duty); *Relief* Work (riot victims, relief to victims of natural calamities such as floods, fires, earthquakes and victims of terrorist activities, and payment of solatium to the victims of hit and run accident); *Coordination* and *Supervision* of other Government Departments (cooperatives, employment and labour, excise, environment and forests, food and civil supplies, land and building, medical and health, and transport etc. They will further have a broad supervisory role in respect of the other departments in their jurisdiction such as Delhi Vidyut Board, Municipal Corporation of Delhi, Education, Public Works Department, Social Welfare, Rural Development, Medical and Health, Irrigation and Floods, Fire Services, Industries, Food and Supplies etc. The Sub-Divisional Magistrates will carry out the raids to detect theft of electricity); *Miscellaneous* Functions (Supervision of Rajya Sabha Board, statutory functions as District Election Officer, Electoral Registration Officer, Returning Officer and Assistant Returning Officer for Parliamentary and Assembly Constituencies are also decentralised to district level.)

26.4 Impediments to Effective Governance of Delhi

26.4.1 Lack of Coordination between Organisations

The existence of a large number of small organisations providing the same or different services is not a problem by itself. Committees and commissions can be formed to resolve conflicts and deliberate on issues of mutual interest (but most committees that do exist are advisory in nature.) For example, development plans for rural areas are prepared by the state government whereas the Delhi Development Authority prepares the Master Plan for Delhi including rural areas. As of today, no coordination systems are developed to understand clearly the implications of the two different planning documents.

26.4.2 Permanence of Organisations

A large majority of the organisations serving and administering the National Capital Territory of Delhi are not only permanent but have acquired a statutory status. Staff remain demotivated and financial management regimes remain ineffective. The burden of debt continues to increase. Since there is no threat to job security and career enhancement, personnel in these organisations fail themselves and the organisations by non-performance. The transfer of the Delhi Transport Corporation and Delhi Electric Supply Undertaking from central government to the Delhi Government had no impact on this state of affairs: the two .continued as before.

26.4.3 Political and Bureaucratic Bankruptcy

All major organisations such as the Delhi Development Authority, Delhi Transport Corporation, Delhi Vidyut Board etc. are managed by civil servants. Although they have inherited the well organised colonial bureaucracy, which worked extremely efficiently for the attainment of the objectives of the imperial government, present day bureaucrats lack professionalism. A nexus between the civil servants and their political masters leaves no room for accountability. Further, organisations do not have any coherent training policies for efficient administration and management. 'Training' is often an excuse for leisure trips abroad. For example, in 1996 five top officials of the DDA made a world trip at the expense of public exchequer to get training on the state of the art of urban planning, but three out of five left the organisation within three months of their return. Inside India a training programme is used by an employee to get time off from the job. There are no programme for integrated human resources development in a majority of the large organisations.

26.4.4 Lack of Policy Direction and Leadership

Top leadership whether elected or nominated or appointed lacks policy direction. Even if good policies are formulated, implementation is thwarted by the lack of commitment of the top leadership to serve the people. As far as appointed leadership is concerned, it is transferred on regular intervals resulting in the want of continuity of policy direction. Sometimes the civil servants are not even able to settle down before they are transferred to another place.

26.4.5 Breakdown in Communications

There is complete breakdown in communications between the Government of the National Capital Territory of Delhi and the adjoining states. For example, the Uttar Pradesh Government impounded Delhi Transport Corporation's buses plying on the Uttar Pradesh roads because the Delhi buses were allegedly exceeding the mileage allowed to them. In a retaliatory action the Delhi Government impounded Uttar Pradesh Transport Corporation's buses. There were no conflict resolution mechanisms that could be used to resolve such conflicts. Talks between the two transport secretaries failed. It was only after the Delhi Chief Minister and the Governor of Uttar Pradesh (the incoming elected government in Uttar Pradesh had not yet been sworn in) held meetings that the issue was sorted out. Communications between various organisations in Delhi also remains on the brink of a breakdown. For example, one of the major reasons why basic services in the slums and squatters of Delhi can not be provided is the lack of communications between major providers such as the Municipal Corporation of Delhi and New Delhi Municipal Council on one hand and the Department of Urban Development on the other.

26.5 Approaches to Develop and Provide Infrastructure in Mega Cities

26.5.1 Master Plan - Centred Approach

The basic premise of the Master Plan - Centred Approach to infrastructure development and provision is that coordination and cooperation among organisations works and integration between sectors of infrastructure is possible. The master plan remains at the centrestage by making proposals on almost all aspects of life including physical and social infrastructure. In India, as far as metropolitan cities are concerned, development authorities were created to turn this belief into reality. The Delhi Development Authority was the first to be established in 1957. The first Master Plan for Delhi became operative in 1962. The Authority modified the Master Plan in 1990. At present work is underway to further modify the Master Plan. But in both the 1962 and 1990 plans, no effort was made to propose mechanism which might facilitate coordination, cooperation and integration (see Kumar, 1996a). Traditional file based decision making regimes and occasional meetings are the only instruments of coordination, cooperation and integration.

Secondly, the Master Plan - Centred Approach to infrastructure relies heavily on the myth that the proposals entailed in the master plan will get automatically implemented either by the same agency, which prepared the master plan, or by a sister organisation. Therefore, the master plan continues to prescribe for its parent organisation and other organisations consequently creating "an amalgam of ... prescriptions" (Amos, 1993 : 149). This is true of development authorities in India in general, and this is true of Delhi. The Master Plan for Delhi 2001 is nothing but a bundle of prescriptions (for detailed analysis seeKumar, 1996a). Prescriptions can be found in various forms including the norms, standards, rulings for other organisations and recommendations.

Thirdly, in the case of the Delhi Development Authority problems of coordination occur because it has been working in "development areas". When these areas have been 'developed' (e.g. a housing scheme has been completed) they are denotified i.e. handed over to local authorities - the Municipal Corporation of Delhi or New Delhi Municipal Council. It is only then that local authorities begin to provide and maintain the utilities and services. This system worked well till early 1980s when local authorities had to provide utilities to a comparatively small number of people. But now things have come to a pass. Since local authorities are asked only after the areas have been physically developed, these authorities have refused to provide, maintain and sustain for a reasonable period of time. For example, there are many new developments in Delhi whereby the DDA has allowed flatted residential development (apartments), but when the cooperative societies approached the MCD for water connections, it has refused for it does not have adequate supplies for the existing areas under its jurisdiction. In this case residents have no other alternative but to draw water from ground which is rarely potable. On the other hand it often happens that electricity or telephone connections are provided by one organisation when planning permissions and building permissions have been consistently denied by another organisation. In Delhi there seems to be no coordination between the DDA, MCD, NDMC and the Delhi Vidyut Board (DESU).

Fourthly, unauthorised construction is rampant in Delhi. More than half the population live in slums and squatter colonies, a majority of which are unauthorise. According to Delhi Government's own survey of 1994, Delhi has 1,080 slum clusters including 480,000 hutments and a population of 2,100,000. Half of the population residing in 929 jhuggies had emerged before 1991, and for these the Delhi Government is seeking central government's permission for regularisation (Pandey, 1997 : 3) (retrospective legeal entitlement). In most of the unauthorised colonies, services are provided by the residents themselves. For instance, water supplies, sanitation and transportation services are first provided. The sewer lines are laid by half-trained masons and then illegally connected to the existing sewerage systems. Engineering considerations are rarely taken into account. Moreover, these zones are low lying areas where lack of drainage facilities leads to serious health hazards. Similarly pipes are burst to steal water municipal supplies. Punctured water pipes cause potable water to become non-potable. As far as transportation is concerned, Khranaga (road made up of bricks and mud) are laid.

This process of development continues for some years before local governments, under political pressure, regularize these colonies. This means that government must provide all basic services including water, sewerage, electricity, transportation and social services to these areas. But authorities can do very little other than let these settlements exist with insignificant improvements. For example, since layout plans are not prepared, right-of-ways of roads are extremely insufficient for comfortable movement of goods and people . Large pipes can not be laid because that requires major digging and disruption of inadequate thoroughfares. The existing pipes can neither bring in water supplies nor take out sewer through the planned sewerage system.

Then there are problems of inter-organisation and intra-organsation coordination. Although Delhi's water supplies are looked after by one organisation called the Delhi Water Supply and Sewage Disposal Undertaking, yet there is no coordination between its various departments. It is common experience that roads are dug for improvement or maintenance by water supply department's personnel for some days only to be reopened again after few days by the sewerage department's personnel.

Lastly, if there is anything contained in the Master Plan for Delhi 2001 which is violated with impunity, it is the misuse of open spaces. Temples are being constructed in the parks. Schools are set up in the parks in no time. Tents become permanently occupied in the parks. Even the current Chief Minister has constructed part of his official bungalow on the site earmarked as a park. There seems to be no space left unbuilt in the parks and playgrounds.

26.5.2 Programmes of Ad-hocism

In theory Delhi has a 90% water supply, and 85% sanitation coverage. But the aggregate figures provided are misleading: state governments calculate coverage on the basis of total water supplied and total population. These figures consequently are inadequate to reflect the accessibility based on income and location. According to a sample survey conducted by the National Institute of Urban Affairs less than 45 per cent of urban households have access to municipal water supply, which is considered to be a safe source of drinking water. Around 10% people use hand pumps and 20% use wells at considerable health risk (National Institute of Urban Affairs, 1989). Moreover, one out of every four persons depends on water source completely unsafe for human consumption.

In this section we discuss government led efforts to provide basic services through Environmental Improvement of Urban Slums (EIUS), Urban Community Development Programme (UCD) and the Sites and Services Scheme,. Other related programmes for urban poor are 'Slum Upgradation Programme, Slum Reconstruction Scheme, Jhuggi Jhompri Resettlement Scheme of Delhi and Night Shelter Scheme of Delhi.

The Jhuggi Jhompri Removal Scheme of Delhi

Until early 1960s, the problem of slums and squatters in Delhi was limited to the congested and dilapidated "Katras" of the walled city. There were only 61 hutment clusters as against 1,726 "Katras" in 1957. Although population wise, a substantial number of persons i.e. 47.5 per cent lived in hutments as compared to 52.5 per cent in the "Katras".

On the recommendations of the Advisory Committee constituted by central government to look at the problems associated with the growth of jhuggi-jhompris in Urban Delhi, a scheme named 'Jhuggi-Jhompri Removal Scheme' was initiated. Each squatter family was to be alloted a 99 year lease of an 80 sq.yd. plot containing a latrine, a water-tap and plinth on which the family could build a hut or house. The families with income higher than Rs.250 per month could get the plots by making full payment in lumpsum. For others, the cost of land was subsidezed to the extent of 50% and they had the option to make the payment in installments. This scheme was later modified due to two main difficulties. Firstly, it encouraged resale of plots by allotees and acted as an incentive to further squatting. Secondly, most of the families were unable to pay the monthly installment due to their poverty.

To deal with these difficulties, it was decided to completely eliminate the element of ownership in this scheme. Instead, the plots were given on rent which were to be subsidized to the extent of 50 per cent for families with income less than Rs.250 per month. Those families who had income more than Rs 250 p.m. were completely taken out of the scheme. Later on, it was realized that clearing areas without removing all squatting families was difficult. Therefore, in 1964, the government decided to allot 25 sq.yd camping sites even to the families with higher income on payment of full rent. Soon, this facility was also withdrawn as it led to fresh squatting on the cleared sites.

A study group set up by Home Ministry, in 1967, reviewed working of Jhuggi-Jhompri Removal Scheme and pointed out the following drawbacks :
- The progress for clearance and relocation was very slow. Only 16,000 families as against the target of 50,000 families were relocated.
- Government did not have any control on the resale of plots to the unauthorised persons.
- Provision of services and amenities in the Jhuggi-Jhompri colonies were inadequate and their gross neglect and inefficient management had caused failure of the scheme.

At present, the Jhuggi-Jhompri Removal Scheme now provides that only pre-July 1960 squatters are to be given 25 sq.yd plots in regular Jhuggi-Jhompri colonies in nearby localities at subsidezed rent of Rs 350 per month.

Environmental Improvement of Urban Slums

In the earlier national economic plans, slums and squatters were envisaged as something to be cleared off or removed. It took more than 25 years for the government to realise that clearing all the slums and squatters of urban India is not possible. In 1972 an alternative approach was signalled with Environmental Improvement of Urban Slums (EIUS) programme. Outlay on this scheme has increased from 50 crores in the Fifth Five Year Plan to Rs.269.55 crores in the Seventh Five Year Plan. It is expected that approximately 9 million slum dwellers will benefit from this programme (National Institute of Urban Affairs, 1989 : 32 - 34).

The EIUS programme is aimed at provision of basic infrastructure services to the squatter settlements in a phased manner. The first phase included the following components :
- Pay and use community toilet complexes at the settlement level (1 WC and 1 Bath for 30 families).
- Water supply through hydrants/ deep hand pumps (One for 100 families).
- Street lights in the settlement (minimum distance between two light poles to be 30m).

• Dhalaos and dustbins for the collection of rubbish.

In the second phase of the programme, storm water drains and paved pathways are to be provided. As a matter of policy, only squatter settlements in the public land were covered.

A Task Force in Housing and Urban Development, constituted by the National Planning Commission in 1983, reviewed the state of affairs of this programme. A detailed analysis revealed various lacunae of the EIUS programme. Firstly, the nature of improvement works and conditions under which works were executed indicated lack of attention to detail and absence of awareness of the field realities on the part of authorities. Secondly, routinisation of the execution, greater concern to meet financial targets at any cost, and symbolic execution of projects emerged prominently in this programme. Eventually it became evident that this was not a programme which could substantially ameliorate the quality of life of the urban poor. Their huts have developed at locations earmarked in the Master Plan for other land uses. This programme did not even attempt to establish a linkage between its objectives and proposals of the Master Plan. This subject always remained outside the purview of town planning framework as aberrations in the Master Plan. The critical issue of security of tenure of the residents of squatter remained unanswered. In the absence of security of tenure, residents were reluctant to invest in the shelter improvement. Thus, matching the perfunctory and half hearted manner of the programme implementation, the community response to this programme was equally dismal. Decisions about the nature, location and management of facilities were always unilateral with the involvement of the residents at later stages of programme implementation or none at all. A survey conducted by TCPO in 1990 revealed that out of the total settlements covered, only 38 per cent households displayed any sort of active participation (Town and Country Planning Organisation, 1994).

Environmental Improvement of Urban Slums of Delhi

Environmental Improvement of Urban Slums started in Delhi in 1972 under the minimum needs programme (MNP) of government of India. The main aim of this scheme was to upgrade the quality of environment without relocating the squatters. This was not a complete change in approach of the government as relocation was also being implemented parallel to it.

About 20,000 households were covered in about 140 squatter settlements within a decade. The scheme basically concentrated on physical upgradation without incorporating other related activities or aspects such as employment generation, education, health and nutrition.

This scheme had a larger coverage than Jhuggi-Jhompri Removal

Scheme. The major reason for its relative success was its lower per capita cost i.e. one third of the cost of relocation carried out Jhuggi-Jhompri Removal Scheme. But it lost its impact because of the ad hoc nature in which sites for upgradation were selected. Another paradoxical feature of this scheme was the way in which many improved squatter settlements were later on removed and people relocated after a few years. Another cause of its weak impact on Delhi's Jhuggi-Jhompri clusters was the fact that the inner city slums (Katra's etc) also formed a major part of the target area. The share allocated for squatters was consequently reduced, although it proved very cost effective in the provision of physical amenities and facilities (Misra and Gupta, 1981). Later efforts to improve the quality of life of urban poor in Delhi included the Urban Basic Services Programme (UBS) and its revised version i.e. the Urban Basic Services Programme for Poor (UBSP).

Delhi Urban Community Development (UCD) Project

In 1958 the the pilot project Delhi Urban Community Development was launched with Ford Foundation support. Encouraged by the success of this project, the Ford Foundation sponsored three other similar projects in other cities. Out of the four, only the Baroda Project survived. In other cities, the project was either abandoned or became marginalised as newer programmes were introduced. The Baroda Project still continues now with the help of Local Citizen's Council.

In the late 1960s the government of India established its own version of the UCD across the country. Fourteen projects were started in 1966 and another six in 1967. Each project had one project officer, eight community organisers and some voluntary workers from the community. The majority of the staff was taken from rural community development projects. The expected coverage at the time was 50,000 population with a small budget of Rs.65,000/- per year. Initially 50% of the expenditure was to be borne by the Union Government and 50% by state governments and local bodies. By 1970 the Union Government withdrew its share of financing. Although most of these projects survived, there was hardly any expansion or replication of these projects in the states. Gujrat state was an exception which started 11 new projects on its own. In 1976, the UNICEF associated itself with this programme, starting with the Hyderabad UCD.

Sites and Services Scheme

The Sites and Services Scheme is basically a housing scheme intended to provide land and other technical and financial assistance to the urban poor.

The Scheme's components include provision of residential plots, toilet and bath units, construction of low cost housing units, provision of commercial and industrial sites with buildings, off- site infrastructure facilities such as access roads, trunk water and sewer lines, provision of community facilities, supply of self help building materials, provision of small scale business through small industries and cottage industry sheds, and provision of maternity and child health services. By 1989, 91,664 plots have been sanctioned by the Housing and Urban Development Corporation (HUDCO) under the Sites and Services Scheme. The eligibility criterion for the EWS (economically weaker sections) to get HUDCO loans under this scheme is that a household's income must not exceed Rs.700 per month.

26.6 The Convergence Approach

The major objectives of the Convergence Approach are to improve the reach and coverage of the current programmes and services for the urban poor, to enhance the effectiveness of current programmes and services through community control, and to monitor and evolve community based innovative action plans. Difficulties in assessing the actual felt needs of the urban poor coupled with drawbacks in the delivery of services made it obvious that public participation is necessary to make a success of such programmes. It was realized that government should act as a facilitator and not provider of services. The thrust of this programme is make communities realise that they have a stake in the effort. This stake can either be created simply through financial sharing or physical self help or coordination.

26.6.1 Urban Basic Services Programme

In July 1986 as a part of the national Seventh Five Year Plan, a new programme called the Urban Basic Services Programme (UBS), was launched. It began as a centrally sponsored scheme, with the UNICEF's assistance. It came into being as a combined form of 'UCD' (Urban Community Development), Low Cost Sanitation, and the Integrated Development of the Small and Medium Towns, all supported by the UNICEF. The programme started with the organisation of small homogeneous groups as they were, and organised them into Block Vikas Mandals (BVM's) for the ultimate union of all such groups into an incorporation of a so-called slum community. These groups were pursued for their involvement in the programme for the delivery and management of services. It was found after three years of the programme that in almost all the slums, communities were ready to participate in the activities such as environmental improvement, health and water supply. Provision of services was fully or partially financed by the agency concerned, and the organisational support for the community was provided by the UBSP (Kumar, 1991 : 57).

Later on, in 1991, a revised and expanded form of UBS programme was started named as the UBSP (Urban Basic services Programme for poor). This decision by government did not result in the closure of previous UBS programme. Instead, it created two parallel programmes with almost same objectives i.e., UBS and UBSP. This has resulted in the marginalisation of the older programme (UBS) which is now limited to very few activities such as distribution of ORS (Oral Rehydration Salts) packets and water purification tablets. Since the Eighth Five Year Plan, this programme has been extended to 500 cities and towns (excluding Calcutta) with a budget of 100 crores. At present, the UBSP covers approximately 290 towns and cities. It is claimed that 31.31 lakh persons in the selected 2,466 slum pockets are being covered under these towns and cities (National Institute of Urban Affairs, 1994b : 3).

In the UBSP, modus operandi for achieving the goal of community participation follows a very systematic methodology in which community based structures at various hierarchies of settlement are formed (Kumar,1991 : 58). To start with, a Community Organiser who is responsible for an average of 2,000 slum households is appointed. He facilitates formation of small Neighbourhood Groups (NHGs) of 20-25 households each. These NHGs then select their representatives who are known as the Resident Community Volunteer's (RCVs). For every 200-250 households, a Neighbourhood Committee (NHC) comprising of respective RCVs is formed. The NHC is the first level of community Organisation which is followed by the formation and registration of Community Development Society (CDS) at the Slum Unit Level. Community structures thus formed are financially empowered with the funds from the UBS who name the beneficiary families by assessing their needs and also play a crucial role in providing feedback and implementing need based activities.

Howwever, although there have been numerous attempts to improve the quality of life of the urban poor of India, they lack a coherent effort as far as basic approach to the problem is concerned. Moreover, due to limited institutional capacity and stunted managerial effort, the programmes have tended to degenerate , specially the ones which involve community organisation as a necessary component.

26.6.2 *Networking Approach*

Economic activity flourishes in the cities with adequate levels of infrastructure. The World Bank believes that infrastructure development and economic productivity reinforce each other. On the basis of the Networking Approach many programmes have been implemented in Indonesia. An integrated approach to urban infrastructure developments in Indonesia calls for interagency coordination for the provision of water supply, drainage, sanitation, solid waste, roads, low cost community housing improvement and urban market infrastructure. The basic approach heavily relies on the fact that the seven sub-sectors of urban infrastructure should coordinate among themselves in respect of "systematic investment programming and cost effective implementation procedures (Dimitriou, 1991 : 195). But such coordinative efforts have begun to flounder because intersectoral coordination is only restricted to low cost housing, water supply, roads, and waste disposal.

Sections two and three has revealed that organisations and approaches to provide basic services have been developing independent of each other (see Tables 26.4 and 26.5).

26.7 Principles for Building Organisations to Develop and Provide Infrastructure

26.7.1 *Mechanisms of Coordination*

As long as more than one organisation is involved in the development and provision of infrastructure, functional and jurisdictional fragmentation will take place. "Fragmentation occurs where there is differentiation without appropriate means of integration ..." (Stewart and Clark, 1996 : 12). Therefore there will be a need for coordination between various organisations. Communicating and decision making through file movement is simply not effective. White (1989 : 529) proposes that organisation "needs to promote consultations and negotiations about the rules rather than simply holding the several units accountable to predefined rules". Efforts should be made to create an environment whereby information is freely communicated and hindrances to communications are removed even by taking action against the erring employees. The providers should have "a thorough knowledge of how different bodies obtain and deploy resources [and] how other bodies manage resources. [The existing elitist role of planning] implies the prescription of resource allocations in accordance with the plans" (Amos, 1993 : 149).

Table 26.4   Organisations and approaches for the development and provision of infrastructure in Delhi

Organisations	Date of Establishment	Approaches	Year of Launch
Delhi Vidyut Board	1903	Master Plan Centred Approach	1955
Delhi Water Supply & Sewage Disposal Undertaking	1926	Programmes of Ad-hocism	1960s to 1980s
Delhi Transport Corporation	1935	Convergence Approach	1986 and 1991
Delhi Development Authority	1957	Networking Approach	Not Yet Started

Table 26.5   Non-conformist efforts for the development and provision of infrastructure in Delhi

Approach	Organisations	Old/New Organisations
Master Plan Centred Approach	Delhi Development Authority	New Organisation
Programmes of Ad-hocism	Various Organisations	Existing Bureaucracies
Covergence Approach	Local Self Government Department	Existing Bureaucracies
Networking Approach	To be identified	Not yet identified

26.7.2 Flexible Structures for Communications and Decision Making

Whenever there has arisen a need to provide new service or implement new programme, government has either created new organisations such as the Delhi Development Authority or tried to implement the new approaches through the existing bureaucracies. Both methods have not been very effective (Kumar, 1996a). New approaches or distinct programmes will be introduced in the future also but a better way to deal with such situations is to keep the existing organisational structures flexible. This will allow organisational change when required. Introduction of brand new organisations should be allowed only under extraordinary circumstances.

26.7.3 *Organisational Changes and Policy Choices*

All four approaches to develop and provide infrastructure have been primarily initiated by the international agencies such as the Ford Foundation, the UNICEF etc. For instance the Master Plan Centred Approach was led by the Ford Foundation team. The ad-hoc programmes have been initiated by the World Bank. More recently the Convergence Approach was launched by the UNICEF and the Networking Approach by the World Bank. Introduction of new programmes by international agencies in itself is not bad. What is bad is the introduction of such programmes through the existing bureaucracies without any organisational change. Therefore it is argued here that state governments should make appropriate organisational changes before accepting new initiatives. Take the example of the Convergence Approach whereby the UBS was launched in 1986. This Approach necessarily required close working relationships between the lead organisation implementing the UBS and the existing providers of services. That relationship did not exist prior to the launch of the programme. Not only did the authorities not address this issue, they simply never got around to recognising it. At least in Delhi the result was for all to see. Providers simply did not respond to requests through letters and meetings attended by junior officers of the lead organisation (see Kumar, 1991). Further, a community participation approach was implemented through a tall hierarchical organisation that extended upto Lt. Governor. Similarly the Networking Approach will not work if necessary organisational changes are not brought about including getting rid of the budget-based decision making processes within different organisations.

26.7.4 *Regulatory Nature of Providers*

Most of the providers notably the Delhi Vidyut Board, the Delhi Transport Corporation and the Delhi Water Supply and Sewage Undertaking have failed more on enforcement than on provision. System losses are as high as 40 per cent of the total supplies. One reason is that the enforcement officers can not be spared for this work. A large majority of employees are engaged in production and production related activities. It is contended that major providers should concentrate on provision and enforcement rather than production. Production may be carried out by public or private organisations through the process of tendering, system of franches or other means. Furthermore, organisations should focus on creating people friendly policies, acts and regulations.

26.7.6 Sustainability for Self Financing

First, lack of funds is not a problem that organisations in Delhi face. For example, every year money allocated for the MRTS has lapsed, because it was not spent. Last year the entire amount allocated for the construction of the MRTS was not utilized. Nonetheless the major providers have to be self financing. For this, market pricing mechanisms (Foster and Plowden, 1996) could be adopted to better serve the people of Delhi. But, like Mitlin, Satterthwaite and Stephens (1996 : 3), this author also does not believe that "all can join the market and those who cannot join the market cannot be helped effectively by intervention from either the state or from development assistance agencies". (Mitlin, Satterthwaite and Stephens, 1996 : 3). But wherever subsidies are necessary these should be provided directly. Furthermore, user charges for financing capital costs, maintenance and operating expenditure, local taxes, tax sharing between central and local governments, allocation of central grants, subsidies, loans and investment equity capital, private sector, NGOs all can be major sources of infrastructure financing (Cheema, 1994).

26.7.7 Interdependence of Services

Most of the services are interdependent. "For instance the use of pit latrines may depend upon an adequate supply of piped potable water so that residents do not have to depend upon shallow wells likely to be polluted by the latrines" (Amos, 1993 : 137). In Delhi's many resettlements colonies people are provided with dry pit latrines but due to non-availability of regular potable water supplies, residents have to depend upon the shallow handpumps.

26.8 Concluding Remarks

Policy concerns has resulted in four main approaches. The first approach used a master plan as the coordinating instrument for the development of infrastructure. Distinct development programmes resulted in the second approach to provide infrastructure to the poor. On the initiative of the UNICEF the third approach was developed out of the belief that development of community structures would facilitate sustained development and provision of basic services. This is known as the Convergence Approach. The most recent approach of 1990s which has World Bank blessing is called the Networking Approach, whereby integrated provision of infrastructure is advocated. One common theme of these approaches is that they all addressed the issue of coordination, cooperation, integration and balanced development of infrastructure. Needs and preferences of people were given in the form of standards either by centre, state or regional organisations. Not a single approach involved people in the development and provision of infrastructure including the UBS and its clone UBSP.

A major argument of this paper is that the design of organisations meant for the provision of infrastructure should be based on peoples' preferences leading to the formation of new approaches. Obviously each approach would require a new internal organisational framework. Thus organisations have to be flexible and their internal forms must be able to change along with the changes in policies or approaches. Tall hierarchical organisations are surely unsuitable for programmes whose main purpose is to provide services through the mechanisms of community participation.

References

Advisory Commission on Intergovernmental Relations (1988) Metropolitan Organisation : The St. Louis Case, a commission report, *AICR, Washington, D.C.*

Amos, F.J.C. *(1989) Strengthening Municipal Government*, Cities, *Vol.8, No.3, pp. 202 - 208.*

Amos, F.J.C. *(1993) 'Planning and Managing Urban Services'*, in Devas, N. and Rakodi, C. *(eds.)* Managing Fast Growing Cities, New Approaches to Urban Planning and Management in the Developing World, *Longman Scientific and Technical, New York.*

Brennan, E. *(1994) 'Mega-City Management and Innovation Strategies : Regional Views'*, in Fuchs, R.J., Brennan, E., Chamie, J., Lo, F.C. and Uitto, J.I. *(eds.)* Mega-City Growth and the Future, *United Nations University Press, Tokyo.*

Cheema, G.S. *(1994) 'Priorities of Urban Management in Developing Countries : the research agenda for the 1990s'* in Fuchs, R.J., Brennan, E., Chamie, J., Lo, F.C. and Uitto, J.I. *(eds.)* Mega-City Growth and the Future, *United Nations Univ. Press. Tokyo.*

Chitnis, S. and Suvannathat, C. *(1984) 'Schooling for the Children of Urban Poor'*, in Richards, P.J. and Thomson, A.M. *(eds.)*Basic Needs and the Urban poor, An ILO-WEP Study, *Croom Helm Limited, Kent.*

Clark, G. *(1991) Urban Management in Developing Countries, a critical role*, Cities, *Vol. 8, No.2, pp. 93 - 107.*

Cotton, A. and Franceys, R. *(1991)* Services for Shelter, *Liverpool Planning Manual 3, The Liverpool University Press, Liverpool.*

Cotton, A. and Franceys, R. (1994) Infrastructure for the Urban Poor, Policy and Planning Issues, *Cities*, Vol. 11, No.1, pp. 15 - 24.

Davey, K. *(1993) 'The Institutional Framework for Planning, and Role of Local Government'*, in Devas, N. and Rakodi, C. *(eds.)* Managing Fast Growing Cities, New Approaches to Urban Planning and Management in the Developing World, *Longman, New York.*

Davey, K. *(1996) 'Conclusions : Effectiveness and Governance'*, in Davey, K. et al *(eds.)* Urban Management, the Challenge of Growth, *Avebury, Aldershot, Hants.*

Dimitriou, H.T. *(1991) An Integrated Approach to Urban Infrastructure Development, a review of the Indonesian experience*, Cities, *Vol. 8, No.3, pp. 193 - 208.*

Fogelson, R. *(1967)* Fragmented Metropolis, Los Angeles, *Harvard University Press, Cam-*

    bridge.

Foster, C.D. and Plowden, F.J. (1996) The State Under Stress, Can the Hollow State be Good Government?, *Open University Press, Buckingham, Philadelphia.*

Government of the National Capital Territory of Delhi (1997) Government of Delhi at your Doorsteps, The Times of India, *Vol. CLIX, No.38, p.8.*

Government of India (1991) The Constitution (Sixty Ninth Amendment) Act 1991, *Government of India, New Delhi.*

Goyal, P. (1993) Delhi's March Towards Statehood, *UBSPD, New Delhi.*

Graham, S. and Marvin, S. (1995) 'More Than Ducts and Wires, Post-Fordism Cities and Utility Networks', in Healey, P. Cameron, S. Davoudi, S. Graham, S. and Madani-Pour, A. (eds.) Managing Cities, A New Urban Context, *John Wiley and Sons, Chichester.*

HUDCO (1996) Adequate Shelter and Services for All, *HUDCO, New Delhi.*

King A.D. (1976) Colonial Urban Development, Culture, social power and environment, *Routledge and Kegan Paul, London.*

Kumar, A. (1991) Delivery and Management of Basic Services to the Urban Poor : The Role of the Urban Basic Services, Delhi, Community Development Journal, *Vol. 26 No.1, 51 - 60.*

Kumar, A. (1996a) Does the Master Plan for Delhi has a Coherent Policy Framework?, Urban India, *Vol. 26, No. 1, pp. 11 - 45.*

Kumar, A. (1996b) Unified Metropolitan Transport Authority for Delhi, *in Vajpeyi, S.C. and Verma, S.P. (eds.) 'National Capital Territory of Delhi, a study on landmarks in state public administration since independence', IIPA, New Delhi.*

Linn, J.F. and Wetzel, D.L. (1994) 'Financing Infrastructure in Developing Country Mega - Cities', in Fuchs, R.J. Brennan, E,. Chamie, J., Lo, F.C. and Uitto, J.I. (eds.) Mega City Growth and the Future, *United Nations University Press, Tokyo.*

Lyons, W.E. and Lowery, D. (1989) Governmental Fragmentation versus Consolidation : Five Public Choice Myths about how to Create Informed, Involved and Happy Citizens, Public Administration Review, *Vol.49, No.6, pp.533-544.*

Misra, G.K. and Gupta, R. (1981) Resettlement Policies of Delhi, *SPA New Delhi.*

Mitlin, D., Sallerthwaite, D, and Stephens, C. (1996) Introduction, Environment and Urbanisation, *Vol.8, No. 2, pp.3-7.*

Muddassir, S.M. and Risbud, N. (1993) Health -Infrastructure Relationships in Squatter Settlements of Delhi, ITPI Journal, *Vol. 12, No. 2 (156), pp. 39 - 52.*

National Institute of Urban Affairs (1989) Urban Poverty, A Status, *New Delhi.*

National Institute of Urban Affairs (1994a) Convergence : Reaching the Urban Poor, Urban Poverty, *March - June 1994, pp. 1 - 16.*

National Institute of Urban Affairs (1994b) Composite Credit Mechanism for the Urban Poor, Urban Poverty, *December 1993 to February 1994, pp. 1 - 12.*

Ramasubban, K.S. (1993) 'Water Supply and Sanitation in Urban Environment in India', in Sivaramakrishnan, K.C. (ed.) Managing Urban Environment in India, Towards, An agenda for Action, *Vol.2 : Land, Water Supply and Sanitation, Times Research Foundation, Calcutta.*

Pandey, R. (1997) Fate of 10 Lakh Jhuggies - dwellers Hangs in Balance, The Pioneer, *Vol.7, No.14, p.3.*

Prasad, M. (1996) Infrastructure is Vital for the Growth of the Country, Cities, *Vol. 21, No. 21, pp. 4 - 5.*

Seabrook, J. (1996) In the Cities of the South, Scenes from a Developing World, Verso, *London.*

Sharma, B.K. (1996) Impact of Economic Development on Environment in Delhi, Yojana, *Vol. 40, No. 6, pp. 34 - 36.*

Stephens, C. (1996) Healthy Cities or Unhealthy Islands : the health and social implications of urban inequality, Environment and Urbanisation, *Vol.8, No.2, pp. 9 - 30.*

Stoker, G. and Young, S. (1993) Local Choice for a Balanced Strategy,

Stewart, J. and Clark, M. (1996) Elected special - purpose Authorities : The case considered, Local Government Studies, *Vol.22, No.1, pp.1 - 18.*

Terry, G.R. and Franklin, S.G. (1987) Principles of Management, *All India Traveller Bookseller, New Delhi.*

Town and Country Planning Organisation (1994) Environmental Improvement of Urban Slums, annual report 1992-93, *TCPO, New Delhi.*

Wegelin, E.A. (1990) New Approaches in Urban Services Delivery : A comparison of Emerging Experiences in Slected Asian Countries, Cities. *7 (3) : 244 - 258. also see cities 8(2) 142-150.*

World Bank (1994) World Development Report : Infrastructure for Development, *Oxford University Press, New York.*

White, L.G. (1989) Public Management in a Pluralistic Arena, Public Administration Review, *Vol.49, No.6, pp.522-532.*

Yates, D. (1977) The Ungovernable City, the politics of Urban problems and policy making, *the MIT Press, Cambridge.*

Printed and bound by CPI Group (UK) Ltd, Croydon, CR0 4YY

21/10/2024

01777086-0009